THE Sustainment & Multifunctional Logistician's

SMARTbook

Warfighter's Guide to Logistics, Personnel Services, & Health Services Support

2nd Revised Edition

The Lightning Press

Norman M. Wade

The Lightning Press

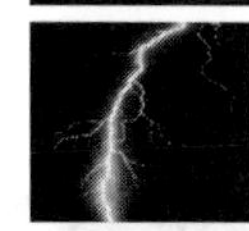

2227 Arrowhead Blvd.
Lakeland, FL 33813
24-hour Voicemail/Fax/Order: 1-800-997-8827
E-mail: SMARTbooks@TheLightningPress.com
www.TheLightningPress.com

2nd Revised Edition

The Sustainment & Multifunctional Logistician's SMARTbook

Warfighter's Guide to Logistics, Personnel Services & Health Services Support

(Updated and retitled second revised edition of The Combat Service Support & Deployment SMARTbook.)

ISBN: 978-0-9824859-2-7

About our cover photo: 1st Lt. Kurt Shingledecker and his platoon participate in an air-assault mission conducted by the 101st Airborne Division, July 17, 2008, in central Iraq. (U.S. Air Force photo by Photo by Spc. Richard Del Vecchio) www.army.mil. Other images courtesy Department of Defense.

Printed and bound in the United States of America.

Preface

Sustainment is the provision of **logistics, personnel services, and health services support** necessary to support full spectrum operations until mission accomplishment. The provision of sustainment is an integrated process (people, systems, materiel, health services, and other support) inextricably linked to operations. When optimized, sustainment operations ensure **strategic and operational reach and endurance** for forces in any operational environment.

From a **strategic** perspective sustainment builds combat readiness, delivers a combat ready force to the combatant commander as part of the joint force, and maintains combat power and endurance across the depth of the operational area. At the **operational and tactical levels**, sustainment is provided by highly trained modular sustainment forces, integrated and synchronized with the operational plan. Logisticians today must be prepared to conduct a wide-ranging array of concurrent operations to support deployment, employment, sustainment, redeployment, and reconstitution.

Joint logistics delivers sustained logistic readiness for the combatant commander and subordinate joint force commanders through the integration of national, multinational, Service, and combat support agency capabilities. The integration of these capabilities ensures forces are physically available and properly equipped, at the right place and time, to support the joint force.

Force projection is the military element of national power that systemically and rapidly moves military forces in response to requirements across the spectrum of conflict. It is a demonstrated ability to **alert, mobilize, rapidly deploy, and operate** effectively anywhere in the world. Today's operational environment is characterized by a wider spectrum of potential contingencies, increased uncertainty, and a more complex range of operational conditions. The situation demands swift action; through rapid strategic response, the combatant commander immediately begins to neutralize the early advantages of the adversary.

SMARTbooks - The Essentials of Warfighting!

Recognized as a doctrinal reference standard by military professionals around the world, SMARTbooks are designed with all levels of Soldiers, Sailors, Airmen, Marines and Civilians in mind.

SMARTbooks can be used as quick reference guides during actual tactical combat operations, as study guides at military education and professional development courses, and as lesson plans and checklists in support of training. Serving a generation of warfighters, military reference SMARTbooks have become "mission-essential" around the world. Visit **www.TheLightningPress.com** for complete details!

SMARTregister for Updates

Keep your SMARTbooks up-to-date! The Lightning Press will provide e-mail notification of updates, revisions and changes to our SMARTbooks. Users can register their SMARTbooks online at **www.TheLightningPress.com**. Updates and their prices will be announced by e-mail as significant changes or revised editions are published.

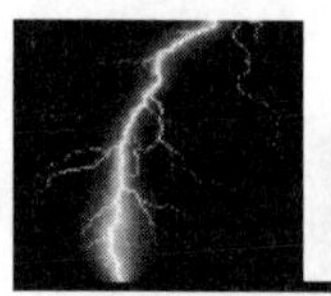

References

The following references were used to compile The Sustainment & Multifunctional Logistician's SMARTbook. All references are considered public domain, available to the general public, and designated as "approved for public release; distribution is unlimited." The Sustainment & Multifunctional Logistician's SMARTbook does not contain classified or sensitive material restricted from public release.

Field Manuals (FMs)

FM 1-0	21 Feb 2007	Human Resources Support
FM 1-04	15 Apr 2009	Legal Support to the Operational Army
FM 1-05	18 Apr 2003	Religious Support
FM 1-06	21 Sep 2006	Financial Management Operations
FM 3-0	27 Feb 2008	Operations
FM 3-04.500	23 Aug 2006	Army Aviation Maintenance
FM 3-35	15 Jun 2007	Army Deployment and Redeployment
FM 3-35.1	1 Jul 2008	Army Pre-positioned Operations
FM 4-01.011	31 Oct 2002	Unit Movement Operations
FM 4-01.30	1 Sep 2003	Movement Control
FM 4-02	13 Feb 2003	Force Health Protection in a Global Environment
FM 4-02.1	28 Sep 2001	Combat Health Logistics
FM 4-30.13	1 Mar 2001	Ammunition Handbook: Tactics, Techniques and Procedures for Munitions Handlers
FMI 4-30.3	28 Jul 2004	Maintenance Operations and Procedures
FM 4-30.31	19 Sep 2006	Recovery and Battle Damage Assessment and Repair
FM 4-90.1	TBP	Brigade Support Battalion
FM 4-90.7	10 Sep 2007	Stryker Brigade Team Logistics
FM 4-93.2	4 Feb 2009	The Sustainment Brigade
FMI 4-93.4	15 Apr 2003	Theater Support Command
FM 4-93.41	22 Feb 2007	Army Field Support Brigade Tactics, Techniques and Procedures
FM 5-0	20 Jan 2005	Army Planning and Orders Production
FM 10-27	20 Apr 1993	General Supply in Theaters of Operation
FM 55-1	3 Oct 1995	Transportation Operations

Joint Publications (JPs)

JP 4-0	18 Jul 2008	Joint Logistics
JP 4-02	31 Oct 2006	Health Service Support

Table of Contents

Chap 1 The Sustainment Warfighting Function

Chap 2

Sustainment Brigade Operations

Chap 3

Brigade Combat Team (BCT) Sustainment

Chap 4

Sustainment Planning

Chap 5 Joint Logistics

Chap 6

Deployment & Redeployment Operations

Index

Chap 1

I. The Sustainment Warfighting Function

Ref: FM 4-0, Sustainment (Aug '09), introduction and chap. I.

Sustainment is the provision of logistics, personnel services, and HSS necessary to maintain operations until mission accomplishment (FM 3-0). The Sustainment operational concept supports the Army's operational concept of full spectrum operations as described in FM 1 and FM 3-0. The provision of sustainment is an integrated process (people, systems, materiel, health services, and other support) inextricably linked to operations. From a strategic perspective sustainment builds Army combat readiness, delivers a combat ready Army to the CCDR as part of the joint force, and maintains combat power and endurance across the depth of the operational area.

The Sustainment Warfighting Function

The Sustainment WFF is related tasks and systems that provide support and services to ensure freedom of action, extend operational reach, and prolong endurance (FM 3-0):

Logistics

Personnel Services

Health Services Support (HSS)

Ref: FM 4-0, Sustainment, chap. 1.

The Sustainment WFF consists of three major sub-functions—logistics, personnel services, and health services. It should be noted that there is a realignment of some of the former CSS BOS tasks across several Army WFFs and vice versa. Some of these realignments include explosive ordnance disposal and force health protection. *See p. 1-3 for further discussion.*

The Sustainment Operational Concept

The Army sustainment operational concept is the provision of logistics, personnel services, and health service support necessary to maintain and prolong operations until successful mission completion. This is accomplished through the integration of national and global resources and ensures Army forces are physically available and properly equipped, at the right place and time, to support the combatant commander (CCDR). The concept leverages host nation (HN) and multinational support, contracting, and other available capabilities to reduce over burdening military resources and at the same time maintaining a campaign quality Army.

I. The Sustainment Warfighting Function (WFF)

Ref: FM 4-0, Sustainment (Aug '09), pp. 1-4 to 1-7.

The sustainment WFF is related tasks and systems that provide support and services to ensure freedom of action, extend operational reach, and prolong endurance (FM 3-0). The endurance of Army forces is primarily a function of their sustainment. Sustainment determines the depth and duration of Army operations. Successful sustainment enables freedom of action by increasing the number and quality of options available to the commander. It is essential to retaining and exploiting the initiative.

The sustainment WFF consists of three major sub-functions: logistics, personnel services, and health services support. A summary of the categories and functions of sustainment is outlined below.

I. Logistics

Logistics is the planning and executing the movement and support of forces. It includes those aspects of military operations that deal with: design and development, acquisition, storage, movement, distribution, maintenance, evacuation, and disposition of materiel; movement, evacuation, and hospitalization of personnel; acquisition or construction, maintenance, operation, and disposition of facilities; and acquisition or furnishing of services (JP 4-0).

See pp. 1-45 to 1-58 for full discussion of the logistics sub-function.

A. Supply. Supply is the procurement, distribution, maintenance while in storage, and salvage of supplies, including the determination of kind and quantity of supplies. There are ten classes in the U.S. supply system. While munitions is a class of supply, it is unique due to the complexities of activities associated with its handling. Munitions require special shipping and handling, storage, accountability, surveillance, and security.

B. Field Services. Field services maintain combat strength of the force by providing for its basic needs and promoting its health, welfare, morale, and endurance. They include clothing repair and exchange, laundry and shower support, mortuary affairs (MA), aerial delivery, food services, billeting, and sanitation.

C. Transportation. Transportation is the moving and transferring of personnel, equipment, and supplies to support the concept of operations, including the associated planning, requesting, and monitoring.

D. Maintenance. Maintenance is all actions taken to retain materiel in a serviceable condition or to restore it to serviceability. It consists of two levels, field and sustainment maintenance. It includes inspection, testing, servicing, and classification as to serviceability, repair, rebuilding, recapitalization, reset, and reclamation.

E. Distribution. Distribution is defined as the operational process of synchronizing all elements of the logistics system to deliver the right things to the right place and right time to support the CCDR. It is a diverse process incorporating distribution management and asset visibility.

F. Operational Contract Support. Operational contract support is the process of planning for and obtaining supplies, services, and construction from commercial sources in support of operations along with the associated contractor management functions.

G. General Engineering Support. General engineering includes those engineering capabilities and activities, other than combat engineering, that modify, maintain, or protect the physical environment.

II. Personnel Services

Personnel services include HR support, religious support, FM, legal support, and band support. Personnel services are those sustainment functions maintaining Soldier and Family readiness and fighting qualities of the Army force. Personnel services complement logistics by planning for and coordinating efforts that provide and sustain personnel (FM 3-0).

See pp. 1-59 to 1-68 for full discussion of the personnel services sub-function.

A. Human Resource (HR) Support. HR support includes the human resources functions of manning the force, HR services, personnel support, and HR planning and operations. HR support maximizes operational effectiveness and facilitates support to Soldiers, their Families, Department of Defense (DOD) civilians, and contractors who deploy with the force.

B. Financial Management (FM) Operations. FM is comprised of two mutually supporting core functions: Finance and Resource Management operations. Finance operations include developing policy, providing guidance and financial advice to commanders; disbursing support to the procurement process; banking and currency; accounting; and limited pay support. RM operations include providing advice to commanders; maintaining accounting records; establishing a management internal control process; developing resource requirements; identifying, acquiring, distributing, and controlling funds; and tracking, analyzing, and reporting budget execution.

C. Legal Support. Legal support is the provision of professional legal services at all echelons. Legal support encompasses all legal services provided by judge advocates and other legal personnel in support of units, commanders, and Soldiers in an area of operation (AO) and throughout full spectrum operations.

D. Religious Support. Religious support facilitates the free exercise of religion, provides religious activities, and advises commands on matters of morals and morale. The First Amendment of the U.S. Constitution and Army Regulation (AR) 165-1 guarantee every American the right to the free exercise of religion. Commanders are responsible for those religious freedoms within their command.

E. Band Support. Army bands provide critical support to the force by tailoring music support throughout military operations. Music instills in Soldiers the will to fight and win, foster the support of our citizens, and promote our national interests at home and abroad.

III. Health Services Support

Health services support (HSS) is all support and services performed, provided, and arranged by the AMEDD to promote, improve, conserve, or restore the mental and physical well being of personnel in the Army and, as directed in other Services, agencies and organizations. Army Health System (AHS) support includes both HSS and force health protection (FHP). The HSS mission is a part of the sustainment WFF. The FHP mission falls under the protection WFF.

See pp. 1-69 to 1-72 and FM 4-02.12 for a full description of the HSS sub-function.

*Related WFF Functions

As a result of the movement from battlefield operating systems to the WFF construct, some tasks are realigned. Two of those tasks include:

A. Explosive Ordnance Disposal (EOD). Explosive Ordnance Disposal is the detection, identification, on-site evaluation, rendering safe, recovery and disposal of explosive ordnance (EO) / improvised explosive devices (IEDs), weapons of mass destruction (WMD) which threaten forces, citizens, facilities, critical infrastructure, or operations. From a WFF perspective, EOD falls under the Protection WFF.

See FM 4-30.50 for details.

B. Internment and Resettlement (I/R) Operations. I/R operations are included under the Sustainment WFF (FM 3.0). While not a major sub-function of the sustainment WFF; I/R are supported by logistics, personnel services, and HSS. The Army is the DOD executive agent (EA) for all detainee operations. Within the Army, and through the CCDR, the Military Police (MP) are tasked with coordinating shelter, protection, accountability, and sustainment for detainees. The I/R function addresses MP roles when dealing with detainees, dislocated civilians, and US military prisoners. The MPs perform the internment and resettlement functions of collecting, evacuating, and securing detainees.

See p. 1-44, FM 3-19.1, MP Opns and FM 3-19.40, Internment/Resettlement.

II. Principles of Sustainment

Ref: FM 4-0, Sustainment (Aug '09), pp. 1-1 to 1-4. See also p. 5-9 for the principles of joint logistics.

The principles of sustainment are essential to maintaining combat power, enabling strategic and operational reach, and providing Army forces with endurance.

While these principles are independent, they are also interrelated. For example, in order for commanders to provide responsive sustainment, they must be able to anticipate requirements based on their knowledge and understanding of future operations. Simplicity in planning and executing sustainment increases survivability, improves efficiencies through economy, and facilitates a continuity of resources thus reducing complexity and confusion. When the execution of plans does not proceed as expected, commanders may improvise to meet mission requirements. The most essential principle is integration. Without deliberate integration of Army sustainment with Joint and MNFs and OGA the achievement of these principles becomes impossible.

1. Integration

Integration is the most critical principle. Integration is joining all of the elements of sustainment (tasks, functions, systems, processes, and organizations) to operations assuring unity of purpose and effort. It requires deliberate coordination and synchronization of sustainment with operations across all levels of war. Army forces integrate sustainment with joint forces and multinational operations to maximize the complementary and reinforcing effects from each service component's or nation's competencies and resources. Integration of sustainment occurs throughout the operations process—plan, prepare, execute, and assess. One of the primary functions of the sustainment staff is to ensure the integration of sustainment with operations plans. Not properly integrating sustainment and operations could result in mission failure.

2. Anticipation

Anticipation is the ability to foresee events and requirements and initiate necessary actions that most appropriately satisfy a response. Anticipation of sustainment facilitates responsive support. It is based on professional judgment resulting from experience, knowledge, education, intelligence, and intuition. Sustainment commanders and staffs visualize future operations and identify appropriate required support. They must then start the process of acquiring the materiel or placement of support that bests sustains the operation. Anticipation is facilitated by automation systems that provide the common operational picture upon which judgments and decisions are based. Anticipating sustainment also means staying abreast of operational plans (OPLANs), continuously assessing requirements, and tailoring support to meet current operations and the changing operational environment.

3. Responsiveness

Responsiveness is the ability to meet changing requirements on short notice and to rapidly sustain efforts to meet changing circumstances over time. It is providing the right support in the right place at the right time. It includes the ability to see and forecast operational requirements. Employing appropriate information systems enables the commander to make rapid decisions. Responsiveness involves identifying, accumulating, and maintaining sufficient resources, capabilities, and information necessary to meet rapidly changing requirements. A responsive sustainment system is crucial to maintaining endurance; it provides the commander with flexibility and freedom of action. It also maintains the tempo of operations and the ability to retain and exploit the initiative. Through responsive sustainment, commanders maintain operational focus and pressure, set the tempo of friendly operations to prevent exhaustion, replace ineffective units, and extend operational reach.

4. Simplicity

Simplicity strives to minimize the complexity of sustainment. Simplicity relates to processes and procedures. Unnecessary complexity of processes and procedures compounds the confusion. Simplicity fosters efficiency throughout the operations process and allows for more effective control of sustainment. Clarity of tasks, standardized and interoperable procedures, and clearly defined command relationships contribute to simplicity. Simplicity enables economy and efficiency in the use of resources, while ensuring effective support of forces.

5. Economy

Economy means providing sustainment resources in an efficient manner to enable a commander to employ all assets to generate the greatest effect possible. The commander achieves economy through efficient management and discipline by prioritizing and allocating resources. Staffs look for ways to eliminate redundancies and capitalize on joint interdependencies. They also apply discipline in managing resources minimizing waste and unnecessary stockpiling. Disciplined sustainment assures the greatest possible tactical endurance of the force and constitutes an advantage to commanders who achieve economy of force in sustainment. Staffs also achieve economy by contracting for support or using HN resources that reduce or eliminate the use of limited military resources. Economy reflects the reality of resource shortfalls, while recognizing the inevitable friction and uncertainty of military operations. Economy enables strategic and operational reach by reducing unnecessary use of transportation requirements. Additionally, it reduces unnecessary storage and warehouse support.

6. Survivability

Survivability is the ability to protect personnel, information, infrastructure, and assets from destruction or degradation. It includes all aspects of protecting personnel (includes FHP), materiel, and organizations while deceiving the enemy. The ability of adversaries to disrupt the flow of sustainment could significantly degrade forces' ability to conduct operations as well as sustain them. Planners integrate survivability with operational planning to maximize survivability. Dispersion and decentralization of sustainment functions enhances survivability. The commander may have to balance risk with survivability in considering redundant capabilities and alternative support plans. The ability to protect lines of communications promotes survivability, helping to ensure operational reach and endurance.

7. Continuity

Continuity is the uninterrupted provision of sustainment across all levels of war. Continuity is achieved through a system of integrated and focused networks linking sustainment to operations. Continuity is enabled through joint interdependence, linked organizations, distribution systems, and information systems. Continuity assures confidence in sustainment allowing commanders freedom of action, operational reach, and endurance. It requires commanders to track resources and make critical decisions eliminating backlogs or bottlenecks. Sustainment staffs at all levels work hand in hand with operational staffs ensuring synchronization of requirements over the entire course of the operation.

8. Improvisation

Improvisation is the ability to adapt sustainment operations to unexpected situations or circumstances affecting a mission. It includes creating, inventing, arranging, or fabricating what is needed from what is available. It may also involve changing or creating methods that adapt to an enemy that quickly evolves. This requires commanders, their staffs, and Soldiers to improvise other possible means to accomplish an operation. The sustainment commander must apply operational art to visualize complex operations and understand what is possible at the tactical level. These skills enable commanders to improvise operational and tactical actions when enemy actions or unexpected events disrupt sustainment operations.

III. Changes & New Concepts: FM 4-0

Ref: FM 4-0, Sustainment (Aug '09), pp. vi to x.

FM 4-0, Combat Service Support, was published under the new Army doctrine numbering system on August 2003. The manual served two purposes. First, it linked CSS and operational doctrine. Second, it served as the bridge between joint doctrine and Army CSS doctrine. Since its publication in 2003, the world and the Army have changed.

The new FM 4-0, Sustainment, implements the changes to our doctrine as a result of the conversion to a modular Army. It discusses from a broad perspective how sustainment is provided to the modular Army while conducting full spectrum operations. It also serves as the bridge between joint and Army sustainment doctrine.

Combat Service Support (CSS) to Sustainment (WFF)

One of the most notable changes is the title of the FM from Combat Service Support to Sustainment. The Army made a conscious decision to rescind the terms combat arms, combat support, and combat service support. It now uses the appropriate Warfighting Function (WFF) to describe unit types and functions. The six WFFs replace the battlefield operating systems and consist of movement and maneuver, intelligence, fires, sustainment, command and control (C2), and protection. The six WFFs also make up the elements of combat power tied together by leadership and information. The sustainment WFF describes both unit types and functions.

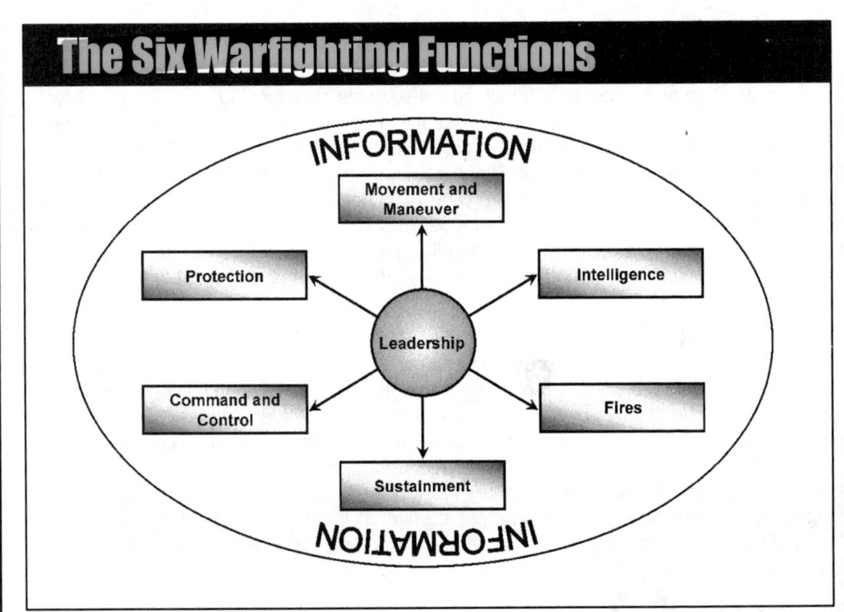

Ref: FM 4-0, Sustainment, fig. 1, p. VII.

The Operational Environment

The operational environment is defined as the composite of the conditions, circumstances, and influences that affect the employment of capabilities and bear on the decisions of the commander (FM 3-0). The operational environment directly impacts the means by which the Army sustains operations as a part of the joint and multinational force (MNF). It is interconnected and increasingly global. It is also extremely fluid, with continually changing coalitions, alliances, partnerships, and new threats constantly appearing and disappearing. The constantly changing operational environment presents many challenges to sustainment of forces. These challenges include providing support in varied physical environments (terrain, climate, and urban areas), working among multicultural populations and operating in areas where it may be difficult to discern the enemy from non combatants. In today's global situations, the physical environment alone in one area of responsibility (AOR) may differ vastly from another. The biggest challenge may be providing responsive sustainment to a rapidly deployable Army force to meet threats worldwide. Overcoming this challenge requires more than ever, joint sustainment interdependence. All Services require logistics, personnel services, and HSS to maintain operational readiness and combat effectiveness. Working together as a joint team enables the U.S. military to reduce redundancies and increase efficiencies in sustainment operations.

Therefore, the operational environment requires forging strong sustainment alliances and coalitions. It will be rare that U.S. forces operate alone. As such, sustainment commanders must consider and plan for interoperability with our allies and coalition forces. Commanders may be required to share technology, processes, and procedures to ensure that our partners can deliver the same decisive operations. Commanders should also consider contracting and host nation support (HNS) options as possible sources of support. Contracting provides commercial supplies, services, and minor construction to supplement military capabilities, giving the mission commander operational flexibility. HNS provides trained, skilled labor to augment sustainment operations.

The operational environment includes the spectrum of conflict which ranges from peace to general war. We will continue to see natural or man-made disasters. As a result, the U.S. military will find itself providing stability to weakened, failed, or defeated governments. Sustainment provides the necessities of life. During Stability operations, sustainment may be critical to influencing military strategy and gaining support of affected populations. While sustainment may not be the decisive operation, it can serve to shape the environment. Sustainment commanders must understand how to maximize benefits afforded by working with nongovernmental organizations (NGO), other governmental agencies (OGA), and intergovernmental organizations (IGO).

We face a variety of threats to our homeland ranging from natural disasters to direct attacks. The employment of our military to assist in responding to these threats is a reality. As a result, the Army reserves and active components must plan for and be prepared to provide the required sustainment resources.

The Modular Force

Numerous operations conducted over the past two decades have demonstrated that Army of Excellence organizations were not as flexible and responsive as Joint Force Commanders (JFCs) required. They met JFC needs, but at high costs in organizational turbulence, inefficiency, and slower response times than desired.

The Army modular organizations provide a mix of land combat power capabilities that can be organized for any combination of offensive, defensive, stability, or civil support operations as part of a joint campaign. The modular force has brought about many changes to the Army's capability to provide sustainment. These changes cut across all of the sustainment functions and represent the Army's imperatives.

Army Force Generation (ARFORGEN)

ARFORGEN is the Army's system for generating land power capabilities that respond to the operational needs of JFCs and sustaining those capabilities as long as required. It is a shift from tiered readiness to cyclic readiness and represents a change of the way the Army generating force performs its Title 10 functions. The generating force resets operational forces upon redeployment.

See p. 6-7 for further discussion of the Army Force Generation (ARFORGEN) process.

IV. Joint Sustainment and the Army Sustainment Warfighting Function

FM 4-0 serves as a bridge between joint doctrine and Army doctrine. As such, it is important to understand the linkage between sustainment as a joint function and sustainment as an Army WFF.

The joint function sustainment is one of six joint functions; C2, intelligence, fires, movement and maneuver, protection, and sustainment. Joint functions are related capabilities and activities grouped together to help Joint Force Commanders (JFC) integrate, synchronize, and direct joint operations. The joint functions are mirrored by the Army WFFs. Joint Pub 3-0 describes joint sustainment as "the provision of logistics and personnel services necessary to maintain and prolong operations until mission accomplishment". Sustainment is primarily the responsibility of the supported CCDR and Service component commanders in close cooperation with the Services, Chief of Staff of the Army (CSA), and supporting commands (JP 4-0).

The Army sustainment WFF is one of six WFFs. A WFF is a group of tasks and systems (people, organizations, information, and processes) united by a common purpose that commanders use to accomplish missions and training objectives.

First it should be noted that the joint functions are capability based. The JFC must rely on the Services to provide the capabilities upon which he/she conducts joint operations. On the other hand, Army forces own their functional capabilities and generally describe those capabilities as functions or tasks. Secondly, the joint function sustainment is split between two staff elements, the Joint Staff (J-1), Personnel and Joint Staff (J-4), Logistics.

See chap. 5, Joint Logistics (pp. 5-1 to 5-24), for further discussion.

Sustainment Crosswalk

SUSTAINMENT

JOINT SUSTAINMENT FUNCTION | ARMY SUSTAINMENT WARFIGHTING FUNCTION

Joint	Joint Sustainment	Army Sustainment	Army
JOINT LOGISTICS CAPABILITIES	SUPPLY DEPLOYMENT / DISTRIBUTION	SUPPLY TRANSPORTATION DISTRIBUTION	LOGISTICS
	MAINTENANCE	MAINTENANCE	
	LOGISTICS SERVICES	FIELD SERVICES	
	OPERATIONAL CONTRACTING	OPERATIONAL CONTRACTING	
	ENGINEERING	GENERAL ENGINEERING	
	HEALTH SERVICES	ARMY HEALTH SYSTEMS SUPPORT HOSPITALIZATION DENTAL TREATMENT BEHAVIORAL HEALTH LABORATORY SERVICES CBRNE TREATMENT MEDICAL EVACUATION MEDICAL LOGISTICS	ARMY HEALTH SERVICES
PERSONNEL SERVICES	PERSONNEL	HUMAN RESOURCES SUPPORT	PERSONNEL SERVICES
	LEGAL	LEGAL SUPPORT	
	CHAPLAIN	RELIGIOUS SUPPORT	
	FINANCE	FINANCIAL MANAGEMENT	
		BAND SUPPORT	

Ref: FM 4-0, Sustainment, fig. 2, p. VIII.

II. Sustainment Roles & Responsibilities

Ref: FM 4-0, Sustainment (Aug '09), chap. 2.

Unified action is the synchronization, coordination, and/or integration of the activities of governmental and nongovernmental entities with military operations to achieve unity of effort (FM 3-0). Unified action involves the application of all instruments of national power, including actions of OGA and multinational military and non military organizations. It requires joint integration. All Services must effectively operate together. Sustainment of unified action is pivotal to achieving campaign end state. The combination of diverse sustainment capabilities generate and sustain combat power that is more potent than the sum of its parts. The integration of sustainment capabilities maximizes efficiencies in delivering support to campaigns. This chapter will discuss the roles and functions of the Army, joint, interagency, intergovernmental, and multinational organization's contribution to the sustainment of unified actions.

I. Joint Interdependence

Joint interdependence is the purposeful reliance by one Service's forces on another Service's capabilities to maximize the complementary and reinforcing effects of both. Army forces operate as part of an interdependent joint force. One area of joint interdependence is joint sustainment. Sustainment is inherently joint. All services use logistics, personnel services, and health services to support their forces. The mutual reliance on joint sustainment capabilities makes for a more effective utilization of sustainment resources. Combinations of joint capabilities defeat enemy forces by shattering their ability to operate as a coherent, effective force.

Sustainment of Joint Forces

Sustainment of joint forces is the deliberate, mutual reliance by each Service component on the sustainment capabilities of two or more Service components (FM 3-0). Effective sustainment determines the depth to which the joint force can conduct decisive operations; allowing the JFC to seize, retain, and exploit the initiative (JP 3-0). It provides JFCs with flexibility to develop any required branches and sequels and to refocus joint force efforts as required. Sustainment is crucial to supporting operations. CCDRs and their staffs must consider a variety of sustainment factors including defining priorities for sustainment and common user logistics (CUL) functions and responsibilities.

Common User Logistics (CUL)

CUL is materiel or service support shared with or provided by two or more Services, DOD agencies, or multinational partners to another Service, DOD agency, non-DOD agency, and/or multinational partner in an operation (JP 4-07). It can be restricted by type of supply and/or service and to specific unit(s) times, missions, and/or geographic areas. Service component commands, DOD Agencies (such as Defense Logistics Agency (DLA)), and Army commands (such as USAMC), provide CUL to other service components, multinational partners, and other organizations (such as NGOs).

The Army Service Component Command (ASCC) is responsible for providing support to Army forces and common-user logistics to other Services as directed by the CCDR and other authoritative instructions. The TSC is the logistic C2 element assigned to the ASCC and is the Army logistic headquarters (HQ) within a theater of operations. When directed, the TSC provides lead Service and executive agency support for designated CUL to OGA, MNFs, and NGOs. Additionally, the MEDCOM(DS) provides AHS support to other services when directed.

II. Title 10 Responsibility

Ref: FM 4-0, Sustainment (Aug '09), pp. 2-2 to 2-4.

Each Service retains responsibility for the sustainment of forces it allocates to a joint force. The Secretary of the Army exercises this responsibility through the Chief of Staff the Army (CSA) and the ASCC assigned to each combatant command. The ASCC is responsible for the preparation and administrative support of Army forces assigned or attached to the combatant command.

The ASCC is responsible for all Army Title 10 functions within the CCDR's AOR. When an ASCC is in support of a GCC, it is designated as a Theater Army. The Theater Army commander exercises administrative control (ADCON) over all Army forces within the CCDR's AOR. He/She is responsible for preparing, training, equipping, administering, and sustaining Army forces assigned to combatant commands.

Title 10, United States Code (USC) specifies that individual services retain responsibility for sustainment, but the purposeful combination of complimentary service capabilities to create joint interdependent forces is often the most effective and efficient means by which to sustain a joint force. Additionally, common user support may be controlled and provided by other means. Options for executing sustainment of a joint force include:

- **Executive agent (EA).** The Secretary of Defense (SECDEF) or the Deputy Secretary of Defense, may designate the head of a DOD component (such as Chief of a Service, CCDR, or director of a Combat Support Agency) as an EA for specific responsibilities, functions, and authorities to provide defined levels of support for operational missions, administrative, or other designated activities that involve two or more DOD components. By definition, the designation as an EA makes that organization responsible for a joint capability within the boundaries of the EA designation.
- **Lead Service.** The CCDR may choose to assign specific CUL functions, to include both planning and execution to a Lead Service. These assignments can be for single or multiple common user functions and may also be based on phases and/or locations within the AOR. The CCDR may augment the Lead Service logistics organization with capabilities from another component's logistics organizations as appropriate. The Lead Service must plan issue procedures and sustainment funding for all items issued to other Services as well as a method for collecting items from other Services.
- **Subordinate logistics command.** The CCDR may assign joint logistics responsibilities to a subordinate Service component and establish a joint command for logistics (see JP 4-0). In order for the subordinate logistics command to be successful, the CCDR must augment it with the capabilities needed to integrate and control the delivery of theater support to meet the joint force requirements.

The Theater Army's ADCON of Army forces entails providing administrative (legal, human resources, and finance) and logistics support to Army forces. When designated as an EA, the Army also enters into inter-Service, interagency, and intergovernmental agreements for certain responsibilities. These may include:

- General engineering support
- Common-user land transportation
- Disaster assistance
- Force protection
- Mortuary services
- Detainee operations
- Bulk fuel management
- Postal operations

The Theater Army commander's principal focus is on operational-level theater support involving force generation and force sustainment during campaigns and other joint operations. The Theater Army commander matches sustainment requirements for a campaign to the capabilities of the Army forces. In all joint operations, sustainment is a service responsibility unless otherwise directed by EA directives, CCDR lead service designations, or interservice support agreements (ISSAs).

Title 10 also includes two other CCDR's responsibilities and authorities that overlap the military departments' Title 10 functions. These are joint training and directive authority for logistics (DAFL). We will only discuss DAFL.

Directive Authority for Logistics (DAFL)

DAFL is the CCDR authority to issue directives to subordinate commanders. It includes peacetime measures to ensure the effective execution of approved OPLANs, effectiveness and economy of operation, prevention or elimination of unnecessary duplication of facilities, and overlapping of functions among the Service component commands.

CCDRs exercise DAFL and may delegate directive authority for a common support capability. The CCDR may delegate directive authority for as many common support capabilities to a subordinate JFC as required to accomplish the subordinate JFC's assigned mission. When the CCDR gives a service component CUL responsibility, he/she must specifically define the responsibilities. For some commodities or support services common to two or more Services, one Service may be given responsibility for management based on DOD designations or ISSAs. However, the CCDR must formally delineate this delegated directive authority by function and scope to the subordinate JFC or Service component commander.

The CCDR may elect to assign responsibility to establish a joint command for logistics to a subordinate Service component. The senior logistics HQ of the designated Service component will normally serve as the basis for this command. This may be an organization joint by mission (such as campaigns, major operations, and humanitarian missions), but not by design. When exercising this option, the CCDR retains DAFL and must specify the control and tasking authorities being bestowed upon the subordinate joint command for logistics, as well as the command relationships it will have with the Service components. *See also p. 5-17.*

Other Combatant Command Authority

Besides logistics, the CCDR has the authority to direct certain other functional activities. These activities include personnel services and health services support.

Personnel Services

The CCDR's command authority allows him/her to direct and approve those personnel services necessary to carry out assigned missions. It also allows for the standardization of personnel policies within the command.

Health Services Support (HSS)

The CCDR requires medical capabilities that are scalable to the requirement, interoperable with other medical forces, and capable of rapid deployment into the joint operations area (JOA). Each Service has organic medical units with capabilities that are tailored for their traditional roles and missions and are normally capable of meeting other Services' requirements. The joint force surgeon provides recommendations to the JFC on the effective employment of all HSS capabilities. The AHS provides support to other Services when directed by the CCDR.

The theater surgeon recommends a theater evacuation policy through the CCDR and the Joint Chiefs of Staff (JCS) for approval by the SECDEF. The evacuation policy establishes the number of days an injured or ill Soldier may remain in the theater before returning to duty.

III. Strategic Level Providers

Strategic providers enable U.S. forces to maintain combat power and enable strategic reach and ensure endurance. Strategic providers include DOD agencies and commands. They link the national base to the theater. These agencies and commands provide sustainment to joint and Army forces.

1. Industrial Base

The industrial base consists of privately owned and government-owned industrial capability and capacity for manufacture, maintenance, modification, overhaul, and/or repair of items required by the U.S. and selected allies. It includes the production base and maintenance base. Active plants and production lines have some capability to surge. Repair parts manufacturers may be able to surge production for items that sustain deployed weapon systems. National policy requires the use of commercial materiel as much as possible.

2. Defense Logistics Agency (DLA)

DLA provides support for military departments and the GCC during peace and war. DLA is the focal point for the industrial base and is the EA for consumable supply items. DLA procures, stores, and distributes items to support the military Services and other customers. It also buys and distributes hardware and electronic items used in the maintenance and repair of military equipment. Excluded supply items are munitions, missiles, and military Service unique items. *See also p. 2-7.*

DLA provides contract administration services to all DOD components and administers and supervises:

- The Federal Catalog System
- The Defense Personal Property Reutilization Program
- The DOD Industrial Plant Equipment Reserve
- The Defense National Stockpile

DLA provides reutilization and marketing services at the strategic through operational levels. Initially, salvage and excess materiel destined for the Defense Reutilization and Marketing Office is collected in theater areas. As the theater matures, DLA-directed activities may use HN or contractor support to assist in retrograding this materiel for inspection, classification, and disposal.

3. U.S. Transportation Command (USTRANSCOM)

USTRANSCOM is responsible for providing common-user and commercial air, land, and sea transportation (including patient movement), terminal management, and aerial refueling to support global deployment, employment, sustainment, and redeployment of U.S. forces. This ability makes possible projecting and maintaining national power where it is needed with the required speed, agility, high efficiency, and accuracy. The USTRANSCOM commander has the authority to procure commercial transportation services (such as Logistics Civil Augmentation Program (LOGCAP)), through its transportation component commands and to activate, with approval of the SECDEF, the Civil Reserve Air Fleet and Ready Reserve Fleet.

USTRANSCOM is DOD's Distribution Process Owner (DPO). As the DPO, USTRANSCOM is responsible for monitoring and managing the global distribution network. The DPO ensures the flow of force movement and sustainment for the supported GCC.

USTRANSCOM is composed of three component commands which remain under the combatant command of USTRANSCOM in contingency operations; Air Mobility Command (AMC), Military Sealift Command (MSC), and the Military Surface Deployment and Distribution Command (SDDC). These component commands provide inter-modal transportation to meet our national security objectives. While these commands normally remain under the C2 of USTRANSCOM in contingency operations, operational command (OPCON) or tactical control (TACON) could be delegated to a theater upon request of the GCC and approval of Commander, USTRANSCOM.

4. Defense Finance and Accounting Service (DFAS)

The DFAS is responsible for the delivery of responsive accounting and finance services. DFAS is an agency supporting the Office of the Under Secretary of Defense-Comptroller, the principal advisor to the SECDEF for budgetary and fiscal matters. It is DFAS's responsibility to coordinate and collaborate with all civilian defense agencies, military services, and the combatant commands that provide warfighting capabilities.

5. U.S. Army Human Resources Command (USAHRC)

The USAHRC integrates, manages, monitors, and coordinates HRS to develop and optimize Army human resources across the spectrum of conflict. The commander of USAHRC is the Army functional proponent for the military personnel management system and operates within the objectives set by the Army G-1. USAHRC major functions include the following:

- Execute the nine major functional categories of the Army personnel life cycle: force structure, acquisition, individual training and development, distribution, deployment, sustainment, professional development, compensation, and transition
- Man the force and provide personnel support and human resources services to Soldiers, their Families, and organizations
- Synchronize all military personnel activities to achieve efficient and cost effective execution of all human resources processes on an Army-wide basis to ensure current and future personnel requirements are defined
- Interact with human resource organizations, including U.S. Army training centers, U.S. Army garrisons, divisions and corps, installations, and forward deployed bases to ensure policy, procedures, and service delivery systems support operational requirements at all levels

U.S. Army Finance Command (USAFINCOM)

The USAFINCOM is an operating agency of the Assistant Secretary of the Army (Financial Management & Comptroller) (ASA(FM&C)). USAFINCOM provides advice and management information to the ASA(FM&C) and interacts between the Army staff, Army commands, units, and DFAS on matters concerning finance and accounting policy, systems, procedures, and reporting.

IV. The Generating Force

The generating force consists of those Army organizations whose primary mission is to generate and sustain the operational Army's capabilities for employment by JFCs. The generating force activities include support of readiness, ARFORGEN, and the routine performance of functions specified and implied in Title 10. As a consequence of its performance of functions specified and implied by law, the generating force also possesses operationally useful capabilities for employment by or in direct support of JFCs. Generating force capabilities include analyzing, understanding and adapting, and generating operational forces tailored to the specific context in which they will be employed.

The generating force's ability to develop and sustain potent land power capabilities is useful in developing partner security forces and governmental institutions, with its capability to develop, maintain, and manage infrastructure. Moreover, an increasingly pervasive information environment, combined with improved transportation capabilities, allow the effective application of capabilities from outside the area of operations.

See following pages (p. 1-14 to 1-15) for further discussion of generating support to the operating force.

The Generating Force

Ref: FM 4-0, Sustainment (Aug '09), pp. 2-5 to 2-10.

The generating force consists of those Army organizations whose primary mission is to generate and sustain the operational Army's capabilities for employment by JFCs. The generating force activities include support of readiness, ARFORGEN, and the routine performance of functions specified and implied in Title 10. As a consequence of its performance of functions specified and implied by law, the generating force also possesses operationally useful capabilities for employment by or in direct support of JFCs. Generating force capabilities include analyzing, understanding and adapting, and generating operational forces tailored to the specific context in which they will be employed.

Generating Support to the Operating Force

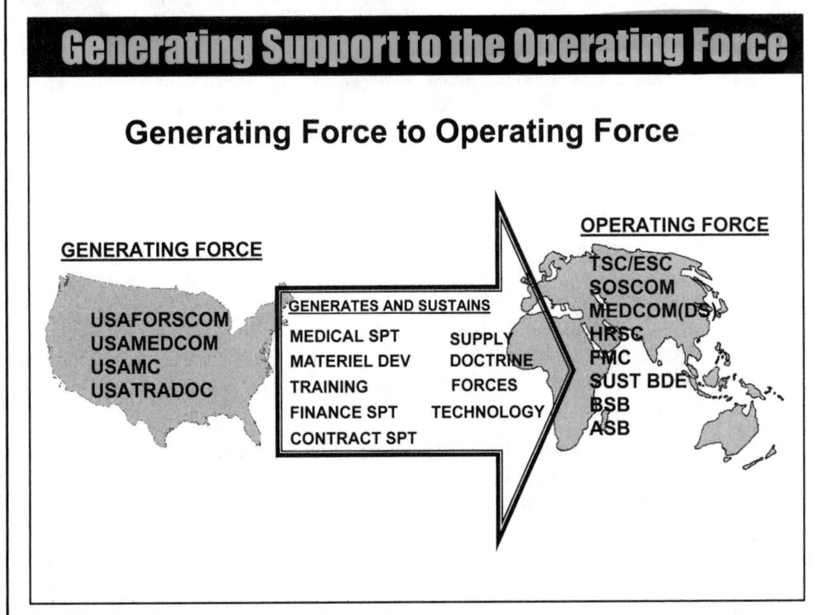

Ref: FM 4-0, Sustainment, fig. 2-1, p. 2-6.

The generating force's ability to develop and sustain potent land power capabilities is useful in developing partner security forces and governmental institutions, with its capability to develop, maintain, and manage infrastructure. Moreover, an increasingly pervasive information environment, combined with improved transportation capabilities, allow the effective application of capabilities from outside the area of operations.

The generating force is responsible for moving Army forces to and from ports of embarkation. They also provide capabilities to assist in the management and operation of ports of embarkation and debarkation and provide capabilities to GCC to conduct reception, staging, onward movement, and integration (RSOI).

Generating force organizations provide a continuum of support that integrates the sustainment base with operating forces. For example, USAMC is part of the generating force. It deploys certain organizations forward with an operational focus and role, such as the Army Field Support Brigade (AFSB), Logistic Support Element, Army Field Support Battalion, Brigade Logistics Support Team, and Contracting Support Brigade (CSB).

1. U.S. Army Training and Doctrine Command (USATRADOC)

USATRADOC is designated by the Secretary of the Army (SECARMY) as an Army command (ACOM) under the direction of Headquarters, Department of the Army (HQDA) and is assigned to carry out certain roles and functions of the SA pursuant to 10 USC 3013(b) with regard to Army forces worldwide. It recruits and trains uniformed personnel, manages the Army's education system, and operates the Army's training centers. TRADOC leads Army requirements determination; integrates doctrine, organization, training, materiel, leadership, and education' personnel and facilities (DOTLM-PF) developments to support required capabilities; and coordinates, synchronizes, and integrates Army capabilities developments with other ACOMs, the Combatant commands, the Joint Staff, and other military departments (AR 10-87).

2. U.S Army Materiel Command (USAMC)

USAMC equips and sustains the Army, providing strategic impact at operational speed. It provides logistics technology, acquisition support, contracting and contractor management, and selected logistics support to Army forces. It also provides related common support to other Services, multinational, and interagency partners. The capabilities of USAMC are diverse and are accomplished through its various major subordinate commands and other subordinate organizations.

See also p. 2-7. USAMC has several subordinate commands, see p.1-16.

3. U.S. Army Research Development and Engineering Command (RDECOM)

The RDECOM rapidly transitions state of the art technology to the force. RDECOM develops supplies and equipment from combat rations, clothing, battledress, to weapons, vehicles, and future combat systems for the force.

4. U.S. Army Security Assistance Command (USASAC)

USASAC manages Army security assistance that provides program management, including planning, delivery, and life cycle support of equipment, services, and training to and co-production with U.S. allies and multinational partners.

5. U.S. Army Chemical Materials Agency (USACMA)

The USACMA provides safe and secure storage of the chemical stockpiles and recovered chemical warfare material; and destroys all chemical warfare materials.

6. U.S. Army Medical Command (USAMEDCOM)

The USAMEDCOM provides Army health support for mobilization, deployment, sustainment, redeployment, and demobilization. USAMEDCOM is a direct reporting unit. The USAMEDCOM integrates the capabilities of its subordinate operational Army medical units with generating force assets such as medical treatment facilities (MTFs) and research, development, and acquisition capabilities. The USAMEDCOM also has regional medical commands responsible for oversight of day-to-day operations in military treatment facilities, exercising C2 over the MTFs in the supported region.

7. U.S. Army Forces Command (USAFORSCOM)

The USAFORSCOM is the Army component of U.S. Joint Forces Command. Joint Forces Command's mission is to provide U.S. military forces where needed throughout the world and to ensure they are integrated and trained as unified forces ready for any assigned task. The FORSCOM commander functions as commander of the Army forces of this unified command and provides military support to civil authorities, including response to natural disasters and civil emergencies. FORSCOM trains, mobilizes, deploys, and sustains combat ready forces capable of responding rapidly to crises worldwide.

USAMC Subordinate Commands

Ref: FM 4-0, Sustainment (Aug '09), pp. 2-7 to 2-9. See also p. 2-7.

USAMC has several subordinate commands. These commands are discussed below:

1. Surface Deployment and Distribution Command (SDDC)

SDDC is a subordinate command of USAMC and the ASCC of U.S. TRANSCOM. It provides inter-modal transportation to meet national security objectives.

2. Aviation and Missile Life Cycle Management Command (AM LCMC)

The AM LCMC integrates functions across their commodity and sustains aviation, missile, and unmanned aerial vehicle systems, ensuring weapons systems readiness with seamless transition to combat operations. It assists materiel developers (PEO/PM) with the development, acquisition, and fielding of aviation, missile systems, and related equipment.

3. CECOM Life Cycle Management Command (CECOM LCMC)

The CECOM LCMC integrates functions across their commodity and sustains command, control, communications, computers, intelligence, surveillance, and reconnaissance (C4ISR) information systems for joint interoperability. It assists materiel developers (PEO/PM) with the development, acquisition, and fielding of C4ISR systems.

4. Tank-Automotive & Armaments Life Cycle Management Command (TA LCMC)

The TA LCMC integrates functions across their commodity and sustains Soldier and ground systems for the operating force. This is accomplished through the integration of effective and timely acquisition, logistics, and technology. The TA LCMC is responsible for integration of initial fielding requirements with current item sustainment.

5. Joint Munitions & Lethality Life Cycle Management Command (JM&L LCMC)

The JM&L LCMC manages research, development, production, storage, distribution, and demilitarization of conventional ammunition. The JM&L LCMC consists of the Program Executive Office for Ammunition, the Joint Munitions Command (JMC), and the Armament, Research, Development and Engineering Center (ARDEC). The JM&L LCMC serves as the Single Manager for Conventional Ammunition (SMCA) and serves as the SMCA Field Operating Activity. PEO Ammunition develops and procures conventional and leap ahead munitions to increase combat firepower to execute total ammunition requirements. ARDEC is the principal researcher, technology developer, and sustainer of current and future munitions.

6. U.S. Army Sustainment Command (ASC)

The ASC is responsible for coordinating generating force support to the operating force. The ASC works in close coordination with other USAMC and national level sustainment and distribution organizations such as DLA and USTRANSCOM and the respective deployed TSCs. The ASC executes its operating force mission through its deployable AFSBs. The AFSB integrates system support contracting into the overall theater support plan. Theater support contracting, to include LOGCAP, is planned, coordinated, and executed by the ASCC supporting CSB under the C2 of U.S. Army Contracting Command's (USACC) Expeditionary Contracting Command (ECC).
The ASC provides continuous equipment and materiel readiness to CONUS forces through planning, resourcing, and materiel management in accordance with the ARFORGEN process. This is achieved by synchronizing strategic materiel management and by integrating acquisition, logistics, and technology.

7. U.S. Army Contracting Command (USACC)

The SECARMY directed the establishment of the USACC to consolidate most Army contracting under a single command. This includes most of the active component contingency contracting force structure.

V. The Operating Force

The operating forces are those forces whose primary missions are to participate in combat and the integral supporting elements thereof (see FM1-01). By law, operational Army units are typically assigned to CCDRs. The Army normally executes its responsibilities to organize, train, and equip operational Army units through ASCCs.

A. Army Service Component Command (ASCC)/ Theater Army (TA)

When an ASCC is in support of a GCC, it is designated as a Theater Army. The TA is the primary vehicle for Army support to joint, interagency, intergovernmental, and MNFs. The TA HQ performs functions that include RSOI; logistics over-the-shore operations; and security coordination.

The TA exercises ADCON over all Army forces in the area of responsibility unless modified by DA. This includes Army forces assigned, attached, or OPCON to the combatant command. As such, the TSC is assigned to the TA. The TA coordinates with the TSC for operational sustainment planning and management. The TA defines theater policies and coordinates with the TSC for technical guidance and execution of force projection and sustainment.

B. Corps

The corps provides a HQ that specializes in operations as a land component command HQ and a joint task force for contingencies. When required, a corps may become an intermediate tactical HQ under the land component command, with OPCON of multiple divisions (including multinational or Marine Corps formations) or other large tactical formations.

The corps HQ is designed to, in priority, C2 Army forces, leverage joint capabilities, and C2 joint forces for small-scale contingencies. Its primary mission is to C2 land forces in land combat operations. The corps HQ has the capability to provide the nucleus of a joint HQ. However, the ability of the corps to transition to a joint task force (JTF) HQ or joint force land component command (JFLCC) HQ is heavily dependent on other Service augmentation. The transition of a modular corps HQ to a joint HQ relies on a timely fill of joint positions, receipt of joint enabling capabilities, and comprehensive pre-activation training as a joint HQ.

C. Division

Divisions are the Army's primary tactical warfighting HQ. Their principal task is directing subordinate brigade operations. Divisions are not fixed formations. Therefore, they may not have all types of BCTs in an operation or they may control more than one of a particular type of BCT. A division can control up to six BCTs with additional appropriate supporting brigades during major combat operations. The types of support brigades are combat aviation, fires, maneuver enhancement, battlefield surveillance, and sustainment. The Sust Bde normally remains attached to the TSC but supports the division. The division may have OPCON of a Sust Bde while conducting large-scale exploitation and pursuit operations.

D. Brigade Combat Team (BCT)

As combined arms organizations, BCTs form the basic building block of the Army's tactical formations. They are the principal means of executing engagements. Three standardized BCT designs exist: heavy, infantry, and Stryker. Battalion-sized maneuver, fires, reconnaissance, and Brigade Support Battalion (BSB) are organic to BCTs.

See following pages (pp. 1-18 to 1-19) for a listing and discussion of operating force sustainment organizations.

E. Operating Force Sustainment Organizations

Ref: FM 4-0, Sustainment (Aug '09), pp. 2-11 to 2-14.

The operating forces are those forces whose primary missions are to participate in combat and the integral supporting elements thereof (see FM1-01). By law, operational Army units are typically assigned to CCDRs. The Army normally executes its responsibilities to organize, train, and equip operational Army units through ASCCs.

See previous page (p. 1-17) for discussion and listing of "warfighting" organizations.

1. Theater Sustainment Command (TSC)

The TSC serves as the senior Army sustainment HQ for the Theater Army. The TSC provides C2 of units assigned, attached, or under its OPCON. The mission of the TSC is to provide theater sustainment (less medical).

The TSC is capable of planning, preparing, executing, and assessing logistics and human resource support for Army forces in theater or JFC. It provides support to full spectrum operations. As the distribution coordinator in theater, the TSC leverages strategic partnerships and joint capabilities to establish an integrated theater-level distribution system that is responsive to Theater Army requirements. It employs Sust Bdes to execute theater opening (TO), theater sustainment, and theater distribution operations.

See pp. 2-71 to 2-79 for further discussion.

2. Expeditionary Sustainment Command (ESC)

Expeditionary Sustainment Commands (ESC) are force pooled assets that, under a command and control relationship with the TSC, command and control sustainment operations in designated areas of a theater. The ESC plans, prepares, executes, and assesses sustainment, distribution, theater opening, and reception, staging, and onward movement operations for Army forces in theater.

See pp. 2-71 to 2-79 for further discussion.

3. Human Resource Sustainment Center (HRSC)

The HRSC is a multifunctional, modular organization (staff element) assigned to a TSC that provides HR support to the theater. The HRSC integrates and ensures execution of HR support for postal, casualty, and personnel accountability operations throughout the theater as defined by the policies and priorities established by the ASCC G-1. The HRSC has a defined role to ensure that the theater HR support plan is developed and then supported with available resources (see FMI 1-0.02).

4. Financial Management Center (FMC)

The FMC Director, in coordination with the Theater Army G8, is the principal advisor to the commander on all aspects of finance operations. The FMC is assigned to the TSC and provides technical oversight and coordination of all theater finance operations to include companies and detachments in theater.

5 Army Field Support Brigade (AFSB)

AFSBs are assigned to the ASC. An AFSB provide integrated and synchronized acquisition logistics and technology (ALT) support, less medical, to Army operational forces. AFSBs are regionally aligned to a Theater Army and will normally be in direct support to the TSC or the lead theater logistics commander. AFSBs serve as ASC's link between the generating force and the operational force. They are responsible for the integration of ALT capabilities in support of operational and tactical commanders across full spectrum operations. The AFSB integrates theater support contracting into the overall ALT support plan, in coordination with the USA ECC CSB Commander/Principal Assistant Responsible for Contracting (PARC), supporting the ASCC

See FM 4-93.41 for further discussion.

6. Sustainment Brigade (SUST BDE)

When deployed, the SUST BDE is a subordinate command of the TSC, or by extension the ESC. The Sust Bde is a flexible, multifunctional sustainment organization, tailored and task organized according to METT-TC. It plans, prepares, executes, and assesses sustainment operations within an area of operations. It provides C2 and staff supervision of sustainment operations and distribution management.

SUST BDEs are primarily employed in a support relationship. Under certain METT-TC conditions, they may be OPCON to the Army forces commander when operating as the senior sustainment command or TACON for operational area security or other types of operations.

See chap. 2 for further discussion.

7. Medical Command (Deployment Support)

The MEDCOM (DS) serves as the senior medical command within the theater in support of the CCDR. The MEDCOM (DS) provides the medical C2 for medical units delivering health care in support of deployed forces. The MEDCOM (DS) is a regionally focused command and provides subordinate medical organizations to operate under the MEDBDE and/or MMB and forward surgical teams or other augmentation required by supported units.

8. Medical Bde (MEDBDE)

The MEDBDE provides a scalable expeditionary medical C2 capability for assigned and attached medical functional organizations task-organized for support of the BCTs and supported units at echelons above brigade (EAB).

9. Multifunctional Medical Battalion (MMB)

The MMB is designed as a multifunctional HQ. It provides medical C2, administrative assistance, medical logistics (MEDLOG) support, and technical supervision for assigned and attached medical functional organizations (companies, detachments, and teams) task-organized for support of a division and its BCTs. It can also be deployed to provide medical C2 to expeditionary forces in early entry operations and facilitate the RSOI of theater medical forces.

10. Sustainment Brigade (Special Operations)

The Sust Bde (SO) is a subordinate command of the U.S. Army Special Operations Command. Its mission is to provide limited sustainment, FHP, and signal support to Army Special Operations Forces (ARSOF). ARSOF are not logistically self-sufficient. The planning and execution of logistics support to ARSOF must be nested within the GCC's concepts of operation and support, as well as tailored to interface with theater logistics structures (see FM 3-05.140).

11. Brigade Support Battalion (BSB)

BSBs are organic components of BCT, Fires, and Maneuver Enhancement Brigades. The BSB is tailored to support the particular brigade to which it is organic. For example, the BSB of a heavy brigade combat team (HBCT) has more fuel distribution capabilities and maintenance than does a fires brigade BSB. The battalion provides supply, maintenance, motor transport, and medical support to the supported brigade. The BSB plans, prepares, and executes, logistics operations in support of brigade operations (see FM 4-90.1).

12. Aviation Support Battalion (ASB)

The ASB is the primary aviation logistics organization organic to combat aviation brigade (CAB) and the theater aviation brigade. The ASB performs the BSB mission. Combat aviation brigades typically conduct attack, reconnaissance, security, movement to contact, air assault, air movement, aero medical evacuation, personnel recovery, and C2 support missions. It provides aviation and ground field maintenance, brigade-wide satellite signal support, replenishment of all supplies, and medical support to the aviation brigade. The ASB has been optimized to support the CAB's forward support companies, aviation maintenance companies, and the brigade HQ and HQ company (see FM 4-90.23).

VI. Interagency Coordination

Interagency coordination is the coordination that occurs between elements of Department of Defense and engaged U.S. Government agencies for the purpose of achieving an objective (FM 3-0). It is an essential characteristic of unified action. Military operations must be coordinated with the activities of other agencies of the United States government, IGO, NGO, and activities of various HN agencies.

The SECDEF may determine that it is in the national interest to task U.S. military forces with missions that bring them into close contact with (if not in support of) IGOs and NGOs. In such circumstances, it is mutually beneficial to closely coordinate the activities of all participants. Unity of effort between IGOs, NGOs, and military forces should be the goal. Taskings to support IGOs and NGOs are normally for a short-term purpose due to extraordinary events. In most situations, sustainment, communications, mobility, and security are the capabilities most needed.

In a national emergency or complex contingency operation, DOD and the U.S. military often serve in a supporting role to other agencies and organizations. Commanders and their staffs should develop an understanding of how military operations and capabilities can be coordinated with those of other agencies and organizations to focus and optimize the military's contributions to accomplish the desired end state.

VII. Multinational Sustainment Operations

A major objective when Army forces participate in the sustainment of multinational deployments is to maximize operational effectiveness. Support provided and received in multinational operations must be in accordance with existing legal authorities. There are two types of multinational operations; alliances and coalitions.

1. Alliance

An alliance is the relationship that results from a formal agreement (such as a treaty) between two or more nations for broad, long-term objectives that further the common interests of the members (see JP1-02). The North Atlantic Treaty Organization (NATO) is an example of an alliance. An alliance may use an integrated staff, instead of merely augmenting the staff of one nation's organization with other national representatives. Each primary staff officer could be a different nationality and usually the deputy commander represents a major participant other than the lead nation.

2. Coalition

A coalition is an ad hoc arrangement between two or more nations for common action (see JP1-02). Many coalitions are formed under the guidance of the United Nations (UN). The UN does not have a military organization; and therefore, no preplanned formal military structures. The American, British, Canadian, Australian, and New Zealand (ABCA) Armies Program represent a coalition of English speaking nations. ABCA forces have never been employed under the program however. However, the ABCA nations have served together in ad hoc coalitions on numerous occasions to pursue common objectives (ABCA Publication 332).

In multinational operations, sustainment of forces is primarily a national responsibility. However, relations between the United States and its NATO allies have evolved to where sustainment is viewed as a collective responsibility (NATO Military Committee Decision [MCD] 319/1). In multinational operations, the multinational commander must have sufficient authority and control mechanisms over assets, resources, and forces to effectively achieve his/her mission. For each nation of an alliance or coalition to perform sustainment functions separately would be inefficient and expensive.

Multinational Logistics Options

Ref: FM 4-0, Sustainment (Aug '09), pp. 2-16 to 2-17.

The basic sustainment support options for multinational operations may range from totally integrated multinational sustainment forces to purely national support. NATO Allied Publication 4.2 provides details on the following support options.

National Support Element (NSE)

A NSE is any national organization or activity that supports national forces that are a part of a MNF. Their mission is nation-specific support to units and common support that is retained by the nation. It should also be noted that NSEs operating in the NATO commander's AO are subject to the SOFA, memorandums of agreements, and other HN arrangements.

Host Nation Support (HNS)

HNS is civil and military assistance rendered by a nation to foreign forces within its territory during peacetime, crises or emergencies, or war based on agreements mutually concluded between nations. Many HNS agreements have already been negotiated between NATO nations. Potential HNS agreements may address labor support arrangements for port and terminal operations, using available transportation assets in country, using bulk petroleum distribution and storage facilities, possible supply of Class III (Bulk) and Class IV items, and developing and using field services.

Contracting Support to Multinational Operations

A deployed force may be required to set up contractual arrangements with local (and non-local) contractors. These are normally negotiated individually with vendors to make use of whatever resources are available.

Multinational Integrated Logistics Units (MILU)

A MILU is formed when two or more nations agree, under OPCON of a NATO commander, to provide logistics support to a MNF. MILUs are designed to provide specific logistics support where national forces cannot be provided, or could be better utilized to support the commander's overall sustainment plan.

Lead Nation

A Lead Nation for Logistic Support has agreed to assume overall responsibility for coordinating and/or providing an agreed spectrum of sustainment for all or part of a MNF within a defined geographical area. This responsibility may also include procurement of goods and services with compensation and/or reimbursement.

Role Specialization

One nation may assume the responsibility for providing or procuring a particular class of supply or service for all or part of the MNF. A Role Specialist Nation's responsibilities include the provision of assets needed to deliver the supply or service. Compensation and/or reimbursement will then be subject to agreement between the parties involved.

Third Party Logistic Support Services (TPLSS)

TPLSS is the use of preplanned civilian contracting to perform selected sustainment. Its aim is to enable competent commercial partners to provide a proportion of deployed sustainment so that such support is assured for the commander and optimizes the most efficient and effective use of resources. TPLSS is most likely to be of use in a non-Article 5 Crisis Response Operation, and especially once the operational environment has become more benign.

Mutual Support Agreements (MSA)

Participating nations have the option to develop mutual support arrangements (bi- and multi-laterally) to ensure provision of logistic support to their forces. This is especially useful when nations have small force contingents collocated with the forces of another nation that have the capacity to support them.

A. NATO Logistics Options

NATO doctrine allows for the formation of a Combined Joint Force Land Component Command (CJFLCC). The CJFLCC HQ can be set at a sub-regional command level or formation level. The CJFLCC commander establishes requirements and sets priorities for support of forces in accordance with the overall direction given by the Joint Force Commander. The commander coordinates sustainment operations with all participating nations.

Merging national sustainment systems into multinational support systems requires the willingness to share the control of vital support functions with a NATO commander and requires technical interoperability of national support assets. STANAGs provide agreed policy and standards to NATO nations and contribute to the essential framework for specific support concepts, doctrine procedures, and technical designs. Non-NATO nations will be expected to comply with NATO publications while on NATO-led operations.

B. Other Sustainment Options

Chapter 138 of Title 10 USC authorizes exchanging support between U.S. services and those of other countries. It authorizes DOD acquisition from other countries by payment or replacement-in-kind, without establishing a cross-servicing agreement.

C. Acquistion Cross-Servicing Agreement (ACSA)

Under ACSA authority (Title 10 USC, sections 2341 and 2342), the SECDEF can enter into agreements for the acquisition or cross-service of logistics support, supplies, and services on a reimbursable, replacement-in-kind, or exchange-for-equal-value basis. These agreements can be with eligible nations and international organizations of which the United States is a member. An ACSA is a broad overall agreement, which is generally supplemented with an implementing agreement (IA).

III. Sustainment Command & Control (C2)

Ref: FM 4-0, Sustainment (Aug '09), chap. 3. For complete discussion of battle command and the operations process, see The Battle Staff SMARTbook or FM 5-0.

Command and Control (C2) is the exercise of authority and direction by a properly designated commander over assigned and attached forces in the accomplishment of a mission. C2 is an art and a science. Commanders combine the art of command and the science of control to accomplish missions. This section will discuss C2, mission command, and the art of battle command from a sustainment perspective. Although the processes of C2 are the same for all commands, the mission focus, knowledge, understanding, and visualization of how support will be provided is different. C2 is fundamental to any discussion of sustainment because of the importance and flexibility of the modular force. The linking of C2 systems enables effective decision making as people, materiel, and medical support moves from generating force to operating force units. It also affects how commanders visualize, describe, and direct support. Because of the uncertain and ever changing nature of operations, mission command—as opposed to detailed command—is the preferred method for exercising C2. The use of mission orders, full familiarity with the commander's intent and concept of operations, and mutual trust and understanding between commanders and subordinates are prerequisites for mission command.

Command

Command is the authority that an armed forces commander lawfully exercises over subordinates by virtue of rank or assignment. Command includes the authority and responsibility for effectively using available resources and for planning the employment of, organizing, directing, coordinating, and controlling military forces for accomplishment of assigned missions (see FM 3-0).

Command is an individual and personal function. It blends imaginative problem solving, motivational and communications skills, and a thorough understanding of the dynamics of operations. Command during operations requires understanding the complex, dynamic relationships among friendly forces, enemies, and other aspects of the operational environment (OE). This understanding helps commanders visualize and describe their intent and develop focused planning guidance. Command is a specific and legal leadership responsibility unique to the military (see FM 6-22).

Control

Control is the regulation of forces and warfighting functions (WFFs) to accomplish the mission in accordance with the commander's intent (FM 3-0). Control is fundamental to directing operations. Commanders and staff both exercise control. Commanders and staffs must understand the science of control to overcome the physical and procedural constraints under which units operate. Control also requires a realistic appreciation for time-distance factors and the time required to initiate certain actions. It demands understanding those aspects of operations that can be analyzed and measured. It relies on objectivity, facts, empirical methods, and analysis.

Control of sustainment spans the strategic to tactical level. It demands an understanding of sustainment functions and related systems that support all aspects of the distribution process. It also requires the availability of organizations, centers, and activities designed with the mission to control sustainment. As a result; Movement Control Battalions (MCB), DMCs, materiel management sections, and support operations (SPO) sections within sustainment commands are responsible this control. Brigades and battalions are primarily responsible for the control and coordination of distribution operations.

Battle Command and Sustainment

Battle command is the art and science of understanding, visualizing, describing, directing, leading, and assessing forces to impose the commander's will on a hostile, thinking, and adaptive enemy. Battle command applies leadership to translate decisions into actions—by synchronizing forces and WFFs in time, space, and purpose—to accomplish missions. Battle command is guided by the commander's professional judgment gained from experience, knowledge, education, intelligence, assessment skills, intuition, and leadership.

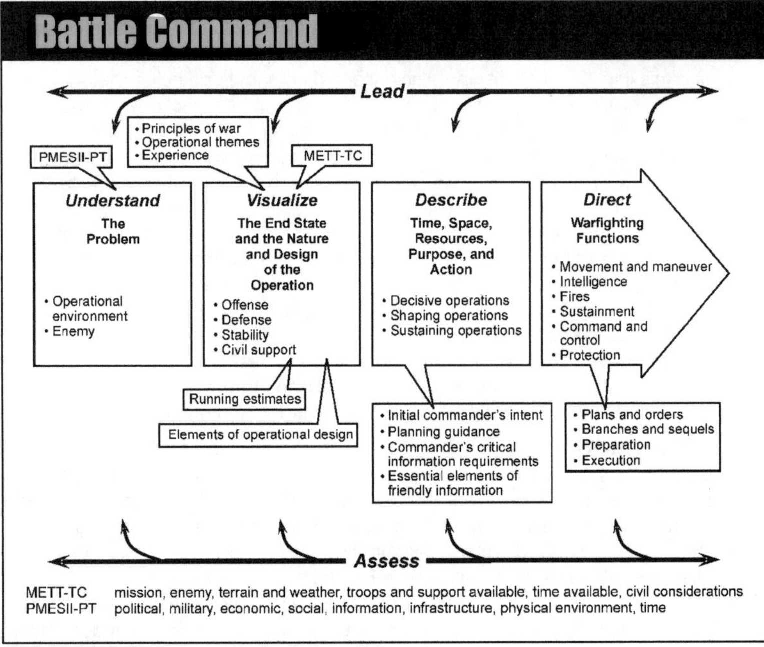

Ref: FM 4-0, Sustainment, fig. 3-1, p. 3-2.

Sustainment commanders must have broad perspective, understanding, and knowledge of sustainment activities throughout the operational area. They must share the visualization of the operational commander and then how to employ all elements of sustainment capabilities at their disposal in support of the operation. Then they must describe and direct how these capabilities are provided. The elements of battle command from the perspective of the sustainment commander are discussed below.

I. Understand

Understanding is fundamental to battle command. Sustainment commanders must first understand the supported commanders' intent and concept of operations. They understand how and what the supported commander thinks. They specifically must understand the supported commander's intent and concept of the operations then track developments and adjust plans as the operations unfold. Sustainment commanders must understand processes and procedures for the provision of sustainment, in relation to the operational environment and the resources available to them. They must understand the relationship between each of the WFFs and how sustainment impacts each. Sustainment commanders must also understand the flow of sustainment and the critical decision points at which they can effect or adjust resources based on changing mission requirements.

Understanding changes as operations progress. Understanding the changes to the operation allows the commander to choose and exploit METT-TC factors that best support the mission. Sustainment commanders build upon their understanding by collecting, storing, and sorting through information that impacts the operation. As a result, the sustainment estimate and commanders' understanding have to be reviewed and re-evaluated throughout an operation. They use a variety of tools, methods, and resources to increase their understanding. A few are discussed below.

1. Relevant Information (RI)

RI is defined as information of importance to commanders and staffs in the exercise of C2 (FM 3-0). RI provides the answers commanders and staffs need to conduct operations successfully. Effective information management helps staffs collect and store information commanders need for better understanding and thus knowledge of the mission, task, or operation. For the sustainment commander, RI drives how he/she visualizes the concept of support. It aides the sustainment commander in determining what, when, and where support is needed. It helps commanders anticipate requirements and prioritize support for current and future operations.

2. Intelligence

Intelligence Preparation of the Battlefield (IPB) is a continuous staff planning activity undertaken by the entire staff to understand the operational environment and options it presents to friendly and threat forces. It is a systematic process of analyzing and visualizing the operational environment in a specific geographic area and for each mission. By applying IPB, commanders gain the information necessary to selectively apply and maximize combat power at critical points in time and space. Understanding intelligence data is critical to sustainment operations. Analysis of intelligence information may help commanders avoid potential enemy activity and threats. Sustainment Soldiers in many ways (such as convoy operations and support to stability operations) become a valuable source for collecting intelligence data which must be processed and passed through intelligence channels.

3. Liaison

Liaison is that contact or intercommunication maintained between elements of military forces or other agencies to ensure mutual understanding and unity of purpose and action (JP 3-08). The Liaison Officer (LNO) is the personal and official representative of the sending organization commander and should be authorized direct face-to-face liaison with the supported commander. LNOs must have the commander's full confidence and the necessary rank and experience for the mission. Using an LNO conserves manpower while guaranteeing the consistent, accurate flow of information, coordination, advice, and assistance.

LNOs are essential for the sustainment mission for several reasons. Through monitoring of the supported command's mission, the sustainment LNOs provide quick information on mission changes thus enabling responsive adjustments in support of the operation. The LNO enables sustainment command staffs and supported command staffs in their planning and coordination, thereby assuring unity of effort. The LNO is an important advisor to the supported commander aiding in the employment of sustainment assets.

4. Command Visits

Another technique used to facilitate understanding is command and staff visits with supported commands. It enables commanders to determine the implications of what is happening (situational awareness) and anticipate what may happen (commander's visualization). It also establishes character, presence, and intellect (attributes of leadership) and instills competence. It enables commanders to see firsthand, the operational environment and the supported commander's mission. As a result, discussion, comparison of views, and continuous study facilitates situational understanding. Sustainment commanders and their staffs obtain a better understanding

Sustainment Information Systems

Ref: FM 4-0, Sustainment (Aug '09), app. A.

Information systems are essential for providing commanders and staffs situational understanding and building the common operational picture. FM 4-0, appendix A describes a number of the C2 and STAMIS systems used in sustainment operations. Highlighted below are key systems. However, the list is not all inclusive.

A. Human Resources (HR)

1. Defense Integrated Military Human Resources System (DIMHRS). DIMHRS, when implemented, will be a fully integrated web-based, all-Service, all-Component, military personnel and pay system that will support military personnel throughout their careers.

2. Defense Casualty Information Processing System (DCIPS). DCIPS is a single uniform casualty reporting system for use by all services.

3. Defense Casualty Information Processing System-Forward (DCIPS-Fwd). DCIPS-Fwd is an automated system to record and report casualty data.

4. Deployed Theater Accountability Software (DTAS). DTAS fills the current void within the Personnel Automation Architecture. It provides the essential personnel functionality to support a commander's tactical decision-making process.

5. Synchronized Pre-deployment Operational Tracker (SPOT). SPOT is a web accessible database designed to account for contractor personnel.

B. Financial Management

1. General Funds Enterprise Business System (GFEBS). GFEBS is the Army's core FM system to provide capabilities such as distribution and execution of appropriated funds, cost management, financial reporting, and management of real property.

2. The Corporate Electronic Document Management System (CEDMS). CEDMS is a web-based electronic file room.

3. Wide Area Work Flow (WAWF). WAWF is an e-commerce business solution for DOD and defense contractors. It allows online submission of invoices/receiving report and electronic disbursement to vendors.

4. Resource Management Tool (RMT). RMT consolidates and integrates financial and manpower data from multiple sources into a single database. RMT links unit FM information into the Standard Finance System (STANFINS) and when fully deployed, GFEBS.

5. International Treasury Services.gov. ITS.gov is an international payment and collection system used for processing international direct deposit payments to benefit recipients with both electronic and check payments for vendor pay, foreign payroll, and miscellaneous payment recipients in foreign countries.

6. Cash-Link. Cash-Link is a web-based system used to research Treasury deposits and debit transactions.

7. Paper Check Conversion (PCC). PCC is a Treasury system which converts a personal/business checks into electronic funds transfers.

8. Financial Management Tactical Platform (FMTP). FMTP is a deployable, modular local area network-configured hardware platform that supports finance and RM operations and functions across the entire spectrum of conflict.

C. Logistics

1. Global Combat Support System Army (GCSS-Army). GCSS-Army is replacing a variety of legacy tactical-level logistics information systems and automated capabilities such as the Standard Army Retail Supply System (SARSS), the Standard Army Maintenance System-Enhanced (SAMS-E), Unit Level Logistics System Aviation Enhanced (ULLS-AE), and the Property Book Unit Supply Enhanced (PBUSE). The Army Enterprise System Integration Program (AESIP) will link GCSS-Army with Logistics Modernization Program (LMP)—the Army's national-level logistics system. GCSS-Army will provide a single access point to the Single Army Logistics Enterprise (SALE) through AESIP.

2. Global Combat Support System – Engineer (GCSS-EN). The GCSS-EN is a tool used to support quantitative aspects of engineering support planning and execution.

3. Battle Command Sustainment Support System (BCS3). BCS3 is the logistics component of the Army Battle Command System (ABCS).

4. Force XXI Battle Command Brigade and Below (FBCB2). FBCB2 is the principal digital C2 system for the U.S. Army at brigade level and below.

5. Movement Tracking System (MTS). The MTS is a vehicle based tracking and messaging system using commercial satellites (L-band), two-way free text messaging, digital maps, encryption, military Global Positioning System, and RFID interrogation.

*STAMIS -Standard Army Management Info Systems

The current baseline of tactical sustainment Standard Army Management Information Systems (STAMIS) operate to support the war-fighter. These systems are fielded in Army logistics activities of the active and reserve components in virtually all TOE units and at the installation level in the Army.

1. Property Book Unit Supply Enhanced. The Property Book Unit Supply Enhanced is a web-based property accountability system that replaced the Standard Property Book System-Redesign and Unit Level Logistics Systems-S4.

2. Unit Level Logistics System Aviation (ULLS-A) (E). Company crew chiefs and unit level aviation maintenance personnel operate ULLS-A (E), a microcomputer based software system, to perform repair part supply and aviation maintenance management. This STAMIS will be integrated into the new enterprise solution, Global Combat Support System - Army (GCSS-A).

3. Standard Army Maintenance System (SAMS-E). SAMS-E consists of both SAMS-1E and SAMS-2E applications and supports sustainment Table of Organization and Equipment unit level maintenance elements and Field and Sustainment maintenance shop production activities.

4. Standard Army Retail Supply System (SARSS). SARSS supports receipt, storage, issue, and management of Class II, Class IIIP, Class IV, and Class IX items of supply.

5. Standard Army Ammunition System Modernization (SAAS-MOD). The SAAS-MOD is designed to provide centralized information management to support ammunition management functions on the battlefield and in garrison, within Army Commands, and ASCCs.

6. TC AIMS II. TC AIMS II is the Army's unit deployment and theater operations (movement control) automated system. It provides critical planning data to JOPES and execution data to the Global Transportation Network.

7. CMOS. CMOS is being fielded to IMCOM and the Installation Transportation Office to support inbound and outbound freight operations.

Combat Service Support (CSS) Automated Information Systems Interface (CAISI). Provides commercial and tactical network connections for sustainment Standard Army Management Information System (STAMIS), along with emerging systems.

8. The Installation Support Module Central Issue Facility (ISM CIF). This system is required to manage personnel clothing issue records.

D. Health Management

1. Medical Communications for Combat Casualty Care (MC4). MC4 is the Army's medical information system. As the Army component of the deployed Defense Health Information Management System (DHIMS), MC4 will provide the hardware infrastructure for the DHIMS medical functionality software, as well as software required to ensure The MC4/DHIMS systems will rely on Army communications systems for transmission of health care information.

DHIMS provides an integrated suite of software to support the military's deployed medical business practice. The theater family of systems supports complete clinical care documentation, medical supply and equipment tracking, patient movement visibility, and health surveillance in austere communications environments. A description of the theater DHIMS systems are described below.

- **Armed Forces Health Longitudinal Technology Applications Mobile (AHLTA-Mobile).** AHLTA Mobile is the first responder's handheld data capture device.
- **AHLTA Theater.** AHLTA Theater extends the sustaining-base electronic medical record capability, look, and feel operation.
- **Theater Medical Information Program Composite Health Care System Cache (TC2).** This system provides documentation for inpatient health care and ancillary services order-entry and result-reporting.
- **Theater Medical Data Store (TMDS).** Information from the theater medical systems are transferred to the TMDS which serves as the authoritative theater database for collecting, distributing, and viewing Service members' pertinent medical information.
- **Joint Medical Workstation (JMeWS).** JMeWS provides medical situational awareness, medical surveillance, and force health decision support.
- **DOD Occupational and Environmental Health Readiness System–Industrial Hygiene (DOEHRS-IH).** DOEHRS-IH supports the reduction of worksite hazards and tracking of long-term environmental exposure.
- **Defense Medical Logistics Standard System Customer Assistance Module (DCAM).** DCAM is the medical logistics ordering tools that allows operational units to order and monitor Class VIII medical supplies and replenish levels when required.

2. Theater Enterprise-Wide Logistics System (TEWLS). The TEWLS application is designed to transfer the capability for theater-level Class VIII supply chain management from TAMMIS into a Systems Applications and Products (SAP)-based enterprise architecture.

of the requirements of supported units and the operational environment in which they operate. To maintain situational understanding, commanders talk with their peers, subordinates, superiors, and with their staffs, and with community and civilian agency leaders. This assures sustainment commanders are better able to integrate sustainment into operations, anticipate support requirements, and provide responsive and continuous support.

5. Information Systems

An information system is equipment and facilities that collect, process, store, display, and disseminate information. This includes computers—hardware and software—and communications as well as policies and procedures for their use (see FM 3-0). Having access to these systems gives the commander a common operational picture (COP). The COP is a single display of RI within a commander's area of interest tailored to the user's requirements and based on common data and information shared by more than one command (FM 3-0).

See previous pages (pp. 1-4 to 1-5) and FM 4-0, appendix A for a discussion of sustainment-related information systems.

II. Visualize

Visualization follows the commanders understanding. Commander's visualization is the mental process of developing situational understanding, determining a desired end state, and envisioning the broad sequence of events by which the force will achieve that end state (see FM 3-0). Understanding helps the commander to pull all of the pieces of the puzzle together to build the picture in his/her mind. The sustainment commander's visualization requires him/her to picture current and future operations and how to employ sustainment assets and resources in support. His/Her visualization takes into account several factors such as METT-TC, defining the end state, and determining the most effective method for employing availability of sustainment resources, the principles of sustainment, and the integration of the WFFs.

See following pages (pp. 1-30 to 1-31) for discussion of the factors of Mission, Enemy, Terrain and Weather, Troops and Support Available, Time Available, and Civil Considerations (METT-TC).

End State

The end state is a set of required conditions that defines achievement of the commander's objectives (JP 1-02). For the sustainment commander, achieving the desired end state involves determining the most effective means for getting the supported commander what he/she needs, when he/she needs it, and where he/she needs it to conduct full spectrum operations.

1. Strategic End State

At the strategic level, the sustainment commanders' focus is on force readiness. The end state is the ability of a combat force to mobilize, deploy, sustain, redeploy, and reset. The key sustainment end state is continuous cycle of ensuring units are equipped, manned, and healthy to conduct operations globally. At the strategic level, the sustainment commander's visualization may include, but not be limited to, what budget requirements are needed to fund readiness initiatives, and how to modernize forces to make them more combat effective while minimizing deployment resources. It may include things such as what forces are needed to support the GCC's operation based on METT-TC and priority for employing sustainment forces to support theater operations.

2. Operational End State

At the operational level, the end state is more narrowly focused. While readiness is a critical factor, the sustainment end state at this level may be the distribution of sustainment to support the GCC mission. It is focused on continuity of support and how best to enable the operational reach of Army forces. It may also be how to be more responsive to the needs of the commander.

3. Tactical End State

At the tactical level, the end state is the uninterrupted provision of sustainment to all units to support continuous operations in an assigned area. As a result, visualizing an end state is a continuous process and requires continuous monitoring of the situation. Commanders may make adjustments as the situation may rapidly change to any combination of offense, defense, and stability support.

III. Describe

The visualization process results in commanders describing to their staffs and subordinates the shared understanding of the mission and intent. Commanders ensure subordinates understand the visualization well enough to begin planning. Commanders describe their visualization in doctrinal terms, refining and clarifying it as circumstances require. Commanders express their initial visualization in terms of:

1. Initial Commander's Intent

Commanders summarize their visualization in their initial intent statement. The purpose of the initial commander's intent is to facilitate planning while focusing on the overall operations process. The sustainment commander's intent should reflect his/her visualization for supporting the operational commander. His/Her intent must integrate elements of the operational commander's intent to ensure synchronization and unity of effort. The sustainment staffs must analyze the commander's intent to ensure supportability of the operation.

2. Planning Guidance

Planning guidance conveys the essence of the commander's visualization. It broadly describes when, where, and how the commander intends to employ combat power to accomplish the mission. Sustainment commander's guidance conveys his/her vision for sustaining combat power. His/Her guidance may include such factors as the placements of sustainment assets to best provide responsive support. It may include guidance for supply rates or evacuation requirements. His/Her guidance may also establish priorities of support based on the missions within his/her designated support area. His/Her planning guidance ensures staffs understand the broad outline of his/her visualization while allowing the latitude necessary to explore different options.

3. Commander's Critical Information Requirements (CCIR)

A commander's critical information requirement is an information requirement identified by the commander as being critical to facilitating timely decision making. One of the staff's priorities is to provide the commander with answers to CCIR. Some examples of CCIR for sustainment commanders may be: What are the consumption rates for various classes of supply? What and where are those supplies in the distribution pipeline? Where are the most likely casualties to occur and are there assets available to evacuate them? What type and where are personnel replacements needed? What is the maintenance status of critical combat equipment? While most staffs provide RI, a good staff expertly distills that information. It identifies answers to CCIR and gets them immediately to the commander. It also identifies vital information that does not answer a CCIR, but that the commander nonetheless needs to know. The two key elements are friendly force information requirements and priority intelligence requirements (JP 3-0).

4. Essential Elements of Friendly Information (EEFI)

An essential element of friendly information is a critical aspect of a friendly operation that, if known by the enemy, would subsequently compromise, lead to failure, or limit success of the operation, and therefore should be protected from enemy detection. An EEFI establishes an element of information to protect rather than one to collect. For sustainment, a few examples of EEFI may include readiness status of units or critical personnel, equipment, and/or maintenance shortfalls. Other factors may be supply routes or schedules for resupply operations and locations of essential stocks or resources.

Visualize - The Factors of METT-TC

Ref: FM 4-0, Sustainment (Aug '09), pp. 3-5 to 3-6.

Mission, Enemy, Terrain and Weather, Troops and Support Available, Time Available, & Civil Considerations (METT-TC)

The assignment of a mission provides the focus for developing the commander's visualization. Commanders use METT-TC as a means for identifying mission variables.

The Factors of METT-TC

Mission

Enemy

Terrain and Weather

Troops and Support Available

Time Available

Civil Considerations

M - Mission

The mission is the task, together with the purpose, that clearly indicates the action to be taken and the reason therefore (JP 1-02). Commanders analyze a mission in terms of specified tasks, implied tasks, and the commander's intent two echelons up. Sustainment commanders must understand the supported commander's mission. The supported commander establishes the priority of support. Since sustainment is generally provided to a designated area of operations, commanders carefully assess the operational mission to determine the types of units operating in the area and their sustainment needs. Results of that analysis yield the essential tasks that—with the purpose of the operation—clearly specify the sustainment actions required which then become the sustainment unit's mission. The sustainment commander and staff work closely with the operational staffs to ensure the integration of sustainment with the operations and mission plans.

E - Enemy

The enemy may consider such sustainment operations, as convoys and medical evacuations, as relatively soft targets to attack. In the current operational environment, the enemy has used IEDs and ambushes on convoys as one of the methods to disrupt sustainment operations. To reduce this risk, sustainment commanders may develop, alter, and/or improvise plans and actions for avoiding potential sites for attacks. Understanding when and how the enemy is most likely to attack require detailed, timely, and accurate information. Effective intelligence, surveillance, and reconnaissance information is important for identifying threat capabilities and vulnerabilities.

T - Terrain and Weather

Terrain includes both natural and man-made features such as rivers, mountains, cities, airfields, and bridges. Weather includes atmospheric conditions such as excessive heat, cold, rain, snow, and a variety of storms. Terrain and weather significantly impact sustainment. It influences the sustainment commander's decision and visualization for supporting operations. Sustainment commanders visualize the advantages and disadvantages afforded by terrain and weather. Natural terrain features may help conceal sustainment forces or operations. Urban areas may provide more access to contract capabilities, but may also serve as bottlenecks for convoys or impede medical support. Weather likewise has advantages and disadvantages. For example, cloud cover may conceal sustainment operations from aerial attack while it may also hinder aerial delivery of supplies or MEDEVAC operations.

Terrain and weather also influences the type of sustainment provided. For example, urban operations may require increased quantities of small arms and crew served ammunition versus tank or artillery ammunition for open terrain. Weather factors such as heat or cold will increase the demand on supplies such as water or cold weather equipment.

T - Troops and Support Available

Troops and support available is the number, type, capabilities, and condition of available friendly troops. These include resources from joint, interagency, multinational, host nation, commercial, and private organizations. It also includes support provided by civilians. Troops and support available falls largely within the sustainment area and encompasses much of the sustainment commander's visualization. Sustainment staffs track readiness including training, maintenance, logistics, health and welfare, and morale.

T - Time Available

Time is critical to the responsiveness of all operations. Sustainment commanders must understand the time sensitive nature of operations and maximize all available time to get commanders what they need when they need it. A key consideration for sustainment commander's visualization is the time it will take to get resources to supported commanders. There are numerous techniques sustainment commanders may use to resolve timing issues. One solution may be positioning support in proximity of the operations. Another solution is the ability to anticipate support requirements and initiate processes and procedures to begin the flow of support through the sustainment chain. In certain unique circumstances, the commander may have to improvise or take risk assuring critical support is provided. He/She may also direct by passing support nodes or aerial delivery or other means.

C - Civil Considerations

Understanding the operational environment requires understanding civil considerations. Civil considerations reflect man-made infrastructure, civilian institutions, attitudes, and activities. The civilian leaders, populations, and organizations within an area of operations influence the conduct of military operations. In instances where stability and offensive operations are concurrent, sustainment commanders may be required to provide support to civilian populations in addition to ongoing military operations, until other agencies or HNS is available. Supporting such operations places heavy demands on sustainment forces and activities. Military movements, supply distribution, contracting, and other sustainment activities may be strained. Commanders should avoid providing support to civilian populations that compete with economic factors in the community.

Sustainment Staffs

Ref: FM 4-0, Sustainment (Aug '09), pp. 3-8 to 3-11.

Staffs assist the commander in providing control over and executing timely decisions for operations. Commanders and staffs are continually alert for opportunities to streamline cumbersome or time-consuming procedures. They provide RI to help commanders achieve situational understanding. One piece of information alone may not be significant; however, when combined with other information from the COP, it may allow the commander to formulate an accurate visualization and make an appropriate decision.

Sustainment Cell Roles

Modular force theater army, corps, and division headquarters have been realigned in accordance with the WFFs of Movement and Maneuver, Fires, Intelligence, Command and Control, Protection, and Sustainment. The WFF Cell – Sustainment combines many of the functions formerly found in G1, G4, G8, and Surgeon staffs, and the Engineer Coordinator (ENCOORD). These functions are now organized into a G1 Division, G4 Division, G8, Surgeon, and Engineer Division.

Sustainment Staff Coordination

The sustainment staffs are responsible for providing staff support activities for the commander. The sustainment staff integrator monitors and coordinates sustainment functions between the sustainment staffs and other WFF staffs and advises the commander on force readiness. The sustainment staffs also coordinate with the corresponding sustainment commands and specialized functional centers (HRSC and FMC) for oversight of operations.

G-1 Staff

The G-1 staff is an element of Theater Army sustainment HQ and staff section of corps or division. At each command level, the G-1 is the senior HR advisor to the commander. The G-1's mission is to ensure HR readiness and plan HR support in compliance with the commander's priorities, intent, and policies. The HR mission is accomplished through a combination of direct communication with the USAHRC, G-1 and S-1 staffs at Theater Army, corps and division, and TSCs and HRSC.

The G-1 focus is establishing HR policy, priorities, and providing guidance and oversight for the HR functions. These functions include man the force, provide HR services, coordinate personnel support, and conduct HR planning and staff operations.

The G-1 coordinates through the TSC to the HRSC for personnel accountability (PA), reception, replacement, and return to duty, rest and recuperation, redeployment, postal, and casualty operations. The G-1 also advises the commander on the allocation and employment of HR support and units within the AOR (see FM 1-0).

G-4 Staff

The G-4 is the principal staff officer responsible for monitoring and advising on logistics within the Theater Army, corps, or division commands. At the corps and division, the G-4 also serves as the sustainment chief/integrator for the commander. The G-4 staff develops the logistics plan in support of the operational plan. The staff provides recommendations on a variety of command priorities including force structure, HNS, contracting support, materiel management, and movement control.

The G-4 staff may include divisions, branches, and specialized sections for supporting various types of operations. In addition, it may have joint and multinational capabilities for supporting requests for logistics support to joint and MNFs, U.S. Agencies, NGOs, and IGOs in the theater. These logistic requests are coordinated through the G-9 (Assistant Chief of Staff for Civil-Military Operations) or the civil-military operations center. The staff may also serve as the focal point for the coordinating LOGCAP and other contracting support.

G-8 Staff

The G-8 is the senior FM advisor to the commander at ASCC, corps, and division levels. The staff's mission is to fund the force through the coordination and synchronization of resources against unit requirements. It provides advice and guidance concerning resources commanders and staff. It also synchronizes RM operations and performs management requirements as determined by the commander.

The G-8 performs the following functions:

- Acquires, manages, & distributes funds
- Manages DOD resources
- Plans, programs, and develops budget
- Coordinates and supports accounting
- Supervises the G-8 staff
- Estimates, tracks, and reports costs for specific operations to support requests to the U.S. Congress for appropriation

The Army may be appointed the lead Service responsible for common FM support. If so, Army G-8 staffs ensure that RM, banking, and currency support are provided for joint interagency and multinational operations. The Army also provides financial analysis and recommendations to joint forces for the most efficient use of fiscal resources. FM provides the funding essential to support contracting requirements and accomplish joint programs.

Surgeon

A command surgeon is designated for all command levels. This medical officer is a special staff officer charged with advising on the AHS mission. The duties and responsibilities of command surgeons for HSS include: Advise the commander on the health of the command; plan and coordinate AHS support for units in theater; establish polices for care of non-U.S personnel; maintain HSS situational awareness; monitor troop strength of medical personnel; monitor medical logistics and blood management operations; and monitor medical regulating and patient tracking operations.

The duties of command surgeons for FHP include: coordinate for veterinary support for food safety, animal care, and veterinary preventive medicine; plan and implement FHP operations to counter health threats; recommend combat and operational stress control, behavioral health, and substance abuse control programs; advise commanders on FHP CBRN defensive actions; ensure the provision of dental services; and ensure a medical laboratory capability for the identification and confirmation of the use of suspect biological warfare and chemical agents.

The Theater Army surgeon ensures that all AHS support functions are considered and included in OPLANs and OPORDs. The surgeon is a full-time special staff officer answering directly to the Theater Army commander on matters that pertain to the health of the command. He/She coordinates for AHS support for both HSS and FHP. The Theater Army surgeon maintains a technical relationship with the MEDCOM (DS) commander and helps establish medical policy for the theater.

The corps surgeon is solely responsible for planning, coordinating, and synchronizing the AHS effort within the Corps AO. The corps surgeon maintains a technical relationship with the MEDCOM (DS) commander and helps establish medical policy for the AO.

Chaplain

The chaplain is a member of the commander's personal staff. The chaplain is responsible for religious support operations. The chaplain advises the commander on matters of religion, morals, and morale as affected by religion and on the impact of indigenous religions on military operations. No coordinating staff officer exercises responsibility over the chaplain.

Judge Advocate General

The staff judge advocate (SJA) is a member of the commander's personal staff. The HQ legal team participates in actions related to C2 of its subordinates. Command and staff functions include advice to commanders, staffs, and Soldiers on the legal aspects of command authority, command discipline, applying force, and the law of war.

Legal support to sustainment activities includes negotiating acquisition and cross-servicing agreements, SOFAs with host nations, combat contingency contracting, fiscal law, processing claims arising in an operational environment, and environmental law.

IV. Direct

Commanders direct all aspects of operations. This direction takes different forms throughout the operations process. Commanders make decisions and direct actions based on their situational understanding, which they maintain by continuous assessment. They use control measures to focus the operation on the desired end state. Sustainment commanders direct operations by:

- Assigning and adjusting missions, tasks, task organization, and control measures based on the concept of operations of the supported commander and his/her priorities
- Positioning units to maximize support, anticipate combat operations and changes to combat intensity, or create or preserve maneuver options
- Positioning key leaders to ensure observation and supervision at critical times and places
- Adjusting support operations to execute changed priorities based on exploiting opportunities or avoiding threats

Sustainment commanders direct forces by issuing orders to subordinates. The orders issued may be verbal, written, or electronically produced using matrices or overlays. These orders may be of several types (see FM 5-0).

1. Operation Order (OPORD)

An operation order (OPORD) is a directive issued by a commander to subordinate commanders for the purpose of effecting the coordinated execution of an operation (FM 5-0).

2. Warning Order (WARNO)

The warning order (WARNO) is a preliminary notice of an order or action, which is to follow. WARNOs help subordinate units and staffs prepare for new missions.

3. Fragmentary Order (FRAGO)

A fragmentary order (FRAGO) is an abbreviated form of an operation order (verbal, written, or digital) usually issued on a day-to-day basis that eliminates the need for restating information contained in a basic operation order.

For complete discussion of battle command and the operations process, see The Battle Staff SMARTbook or FM 5-0.

IV. Integrating into Operations

Ref: FM 4-0, Sustainment (Aug '09), chap. 4.

Integration is a key principle of sustainment. Effective integration of sustainment sets the conditions to ensure mission success and extend strategic and operational reach. Integration begins with the operations process—planning, preparation, execution, and continuous assessment. It is conducted simultaneously and in synchronization with the operations plan development. Sustainment must be integrated across each level of war and with joint and multinational operations.

Integrating Sustainment into Operations

The operations process consists of the major C2 activities performed during operations: planning, preparing, executing, and continuously assessing the operations (FM 3-0). Integrating sustainment with the operations process across each level of war is vital for ensuring the synchronization of sustainment with the WFFs and unity of effort during operations. Integrating sustainment with joint and multinational operations allow forces to conduct operations using mutual support capabilities while reducing redundancy and competition for limited resources. Commanders and staffs at every level must make all efforts to integrate sustainment with the operations process.

The Operations Process

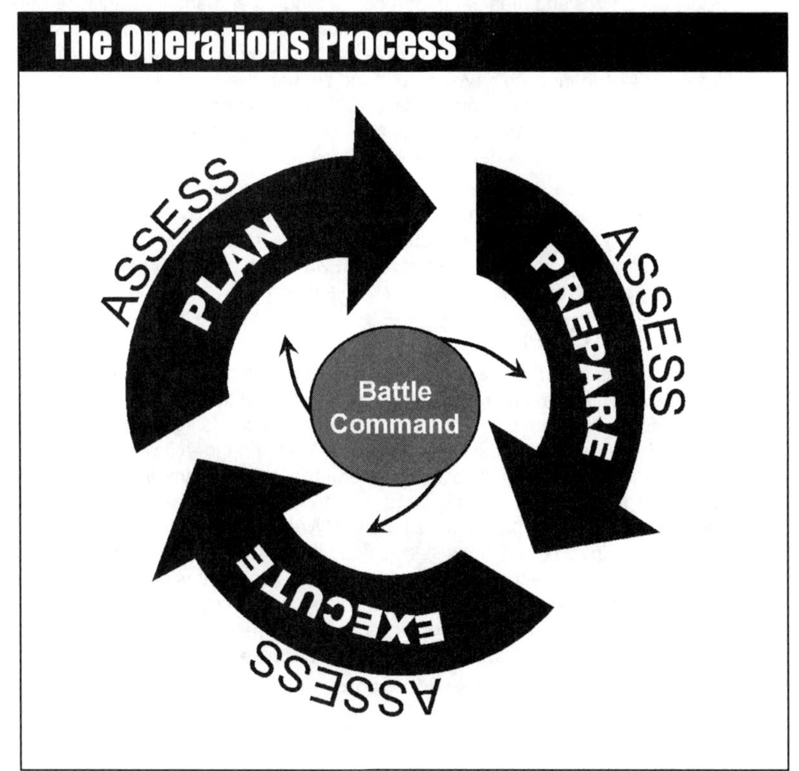

Ref: FM 4-0, Sustainment, fig. 4-1, p. 4-1.

I. Planning the Sustainment of Operations

The previous section covered many of the C2 planning functions (such as battle command, the determination of end state, and sustainment staff roles). This section will focus on more specific tools and planning considerations sustainment commanders and staffs use in planning for sustainment of full spectrum operations.

Planning begins with analysis and assessment of the conditions in the operational environment with emphasis on the enemy. It involves understanding and framing the problem and envisioning the set of conditions that represent the desired end state (FM 3-0). Sustainment planning indirectly focuses on the enemy but more specifically on sustaining friendly forces to the degree that the Army as a whole accomplishes the desired end state.

See chap. 4, Sustainment Planning (pp. 3-1 to 3-36), for further discussion.

II. Preparing for Sustainment of Operations

Preparation for the sustainment of operations consists of activities performed by units to improve their ability to execute an operation. Preparation includes but is not limited to plan refinement, rehearsals, intelligence, surveillance and reconnaissance, coordination, inspections, and movements (FM 3-0). For sustainment to be effective, several actions and activities are performed across the levels of war to properly prepare forces for operations.

A. Negotiations and Agreements

Negotiating HNS and theater support contracting agreements may include pre-positioning of supplies and equipment, civilian support contracts, OCONUS training programs, and humanitarian and civil assistance programs. These agreements are designed to enhance the development and cooperative solidarity of the host nation and provide infrastructure compensation should deployment of forces to the target country be required. The pre-arrangement of these agreements reduces planning times in relation to contingency plans and operations.

Negotiation of agreements enables access to HNS resources identified in the requirements determination phase of planning. This negotiation process may facilitate force tailoring by identifying available resources (such as infrastructure, transportation, warehousing, and other requirements) which if not available would require deploying additional sustainment assets to support.

B. Theater Opening (TO)

Theater opening (TO) is the ability to rapidly establish and initially operate ports of debarkation (air, sea, and rail), to establish the distribution system and sustainment bases, and to facilitate port throughput for the reception, staging, and onward movement of forces within a theater of operations. Preparing for efficient and effective TO operations requires unity of effort among the various commands and a seamless strategic-to-tactical interface. It is a complex joint process involving the GCC and strategic and joint partners such as USTRANSCOM, its components, and DLA. Working together, TO functions set the conditions for effective support and lay the groundwork for subsequent expansion of the theater distribution system. It comprises many of the sustainment functions including, human resources (including Military Mail Terminal Team), FM, HSS, engineering, movement (air/land/water transport, inland terminal operations), materiel management, maintenance, and contracting.

See pp. 2-19 to 2-21 for further discussion.

C. Operational Contract Support

Ref: FM 4-0, Sustainment (Aug '09), pp. 4-7 to 4-8. See also pp. 2-10 to 2-11.

Operational contract support plays an ever increasing role in the sustainment of operations and is an integral part of the overall process of obtaining support. Contract support is used to augment other support capabilities by providing an additional source for required supplies and services. Because of the importance and unique challenges of operational contract support, commanders and staffs need to fully understand their role in managing contract support in the AO.

The requiring activity (normally brigade through ASCC level units), in close coordination with the supporting contracting unit/office or Team LOGCAP-Forward, must be able to describe what is needed to fulfill the minimum acceptable standard for the government.

An important capability for the commander is to incorporate contract support with operational reach. The major challenge is ensuring that theater support and external support contracts are integrated with the overall sustainment plan. It is imperative that the TSC/ESC SPO and the ASCC G-4 coordinate with the supporting CSB. The CSB assists the Theater Army G-4 to develop the contracting support integration plans. The CSB commands contracting deployed units to support those plans.

1. Theater Support Contracts

Theater support contracts assist deployed operational forces under prearranged contracts or contracts awarded in the AO by contracting officers under the C2 of the CSB. Theater-support contractors acquire goods, services, and minor construction support, usually from the local commercial sources, to meet the immediate needs of operational commanders. Theater support contracts are typically associated with contingency contracting. When this support involves a service contract, the unit must be prepared to provide a contracting officer representative.

2. External Support Contracts

External support contracts provide a variety of support functionalities to deployed forces. External support contracts may be prearranged contracts or contracts awarded during the contingency itself to support the mission and may include a mix of U.S. citizens, third-country nationals, and local national subcontractor employees. The largest and most commonly used external support contract is LOGCAP. This Army program is commonly used to provide life support, transportation support, and other support functions to deployed Army forces and other elements of the joint force as well. Depending on METT-TC factors, the TSC will often serve as the requiring activity for mission related LOGCAP support requirements. If designated by the Army forces as the priority unit for LOGCAP support, the TSC would normally be augmented by an USAMC logistics support officer from Team LOGCAP-Forward.

3. System Support Contracts

System support contracts are pre-arranged contracts by the USAMC LCMCs and separate ASA(ALT) program executive and product/project management offices. The AFSB coordinates the administration and execution of system contracts within an AO in coordination with LCMC and separate ASA(ALT) program executive and product/project management offices. Supported systems include, but are not limited to, newly fielded weapon systems, C2 infrastructure (such as the Army Battle Command Systems (ABCS) and standard Army management information systems (STAMIS)), and communications equipment. System contractors, made up mostly of U.S. citizens, provide support in garrison and may deploy with the force to both training and real-world operations. They may provide either temporary support during the initial fielding of a system, called interim contracted support, or long-term support for selected materiel systems, often referred to as contractor logistics support.

D. Army Pre-Positioned Stocks (APS)

Ref: FM 4-0, Sustainment (Aug '09), pp. 4-8 to 4-9.

The APS program is a key Army strategic program. APS is essential in facilitating strategic and operational reach. USAMC executes the APS program and provides accountability, storage, maintenance, and transfer (issue and receipt) of all equipment and stocks (except medical supplies and subsistence items) (FM 1-01). Medical APS stocks are managed by U.S. Army Medical Materiel Agency for the Office of the Surgeon General and subsistence items are managed for the Army by DLA. The reserve stocks are intended to provide support essential to sustain operations until resupply lines of communication can be established. Prepositioning of stocks in potential theaters provides the capability to rapidly resupply forces until air and sea lines of communication are established. Army prepositioned stocks are located at or near the point of planned use or at other designated locations. This reduces the initial amount of strategic lift required for power projection, to sustain the war fight until the LOC with CONUS is established, and industrial base surge capacity is achieved (FM 3-35.1):

1. Prepositioned Unit Sets

Prepositioned Unit Sets consist of prepositioned organizational equipment (end items, supplies, and secondary items) stored in unit configurations to reduce force deployment response time. Materiel is prepositioned ashore and afloat to meet the Army's global prepositioning strategic requirements of more than one contingency in more than one theater of operations.

2. Operational Projects Stocks

Operational projects stocks are materiel above normal table of organization and equipment (TOE), table of distribution and allowances (TDA), and common table of allowance (CTA) authorizations, tailored to key strategic capabilities essential to the Army's ability to execute force projection. They authorize supplies and equipment above normal modified TOE authorizations to support one or more Army operation, plan, or contingency. They are primarily positioned in CONUS, with tailored portions or packages prepositioned overseas and afloat. The operational projects stocks include aerial delivery, MA, and Force Provider (FP) base camp modules.

3. Army War Reserve Sustainment Stocks

Army War reserve sustainment stocks are acquired in peacetime to meet increased wartime requirements. They consist of major and secondary materiel aligned and designated to satisfy wartime sustainment requirements. The major items replace battle losses and the secondary items provide minimum essential supply support to contingency operations. Stocks are prepositioned in or near a theater of operations to reduce dependence on strategic lift in the initial stages of a contingency. They are intended to last until resupply at wartime rates or emergency rates are established.

4. War Reserve Stocks for Allies

War Reserve Stocks for Allies (WRSA) is an Office of the Secretary of Defense (OSD)–directed program that ensures U.S. preparedness to assist designated allies in case of war. The United States owns and finances WRSA assets and prepositions them in the appropriate theater.

Army Pre-positioned Afloat (APA)

Army prepositioned afloat (APA) is the expanded reserve of equipment for an infantry brigade combat team (IBCT), theater-opening sustainment units, port-opening capabilities, and sustainment stocks aboard forward-deployed prepositioned afloat ships. APA operations are predicated on the concept of airlifting an Army IBCT with sustainment elements into a theater to link up with its equipment and supplies prepositioned aboard APA ships (see FM 3-35.1)

E. Joint Deployment Distribution Operations Center (JDDOC)

Also critical to the TO effort is the JDDOC. The JDDOC mission is to improve in-transit visibility and to support the geographic CCDR's operational objectives. The operational objective is accomplished by synchronizing and optimizing the interface of intertheater and intratheater distribution to integrate the proper flow of forces, equipment, and supplies. The JDDOC, under the control and direction of the GCC, plans and coordinates deployment and redeployment and strategic distribution operations. The JDDOC is an integral component of the GCC staff, normally under the direction of the GCC Director of Logistics (J4). However, GCC's can place the JDDOC at any location required or under the OPCON of another entity in the GCC area of responsibility. The JDDOC will coordinate with the TSC/ESC. However, on small scale operations, the JDDOC may coordinate directly with a Sust Bde operating as the senior Army LOG C2 HQ in the theater of operations. The JDDOC is directly linked to USTRANSCOM and provides strategic visibility. *See also p. 2-7.*

F. Port Opening

Port opening and port operations are critical components for preparing TO. Commanders and staffs coordinate with the HN to ensure sea ports and aerial ports possess sufficient capabilities to support arriving vessels and aircraft. USTRANSCOM is the port manager for deploying U.S. forces (see FM 55-50 and 55-60).

Joint Task Force Port Opening (JTF-PO)

The JTF-PO is a joint capability designed to rapidly deploy and initially operate aerial and sea ports of debarkation, establish a distribution node, and facilitate port throughput within a theater of operations. The JTF-PO is a standing task force that is a jointly trained, ready set of forces constituted as a joint task force at the time of need.

The JTF-PO facilitates joint RSOI and theater distribution by providing an effective interface with the theater JDDOC and the Sust Bde for initial aerial port of debarkation (APOD) operations. The JTF-PO is designed to deploy and operate for up to 60 days. As follow-on theater logistic capabilities arrive, the JTF-PO will begin the process of transferring mission responsibilities to arriving Sust Bde forces or contracted capabilities to ensure the seamless continuation of airfield and distribution operations.

See also p. 2-6. See pp. 2-19 to 2-21 for discussion of theater opening, and pp. 6-29 to 6-40 for discussion of RSOI.

G. Warehouse and Billeting and Other Support

Warehousing, billeting, and other infrastructure capabilities must be identified at each port of debarkation (POD) prior to the arrival of forces in theater. Any limitations influence the efficiency of the entire sustainment system. Host nation infrastructure such as electrical power grids, sanitation, bulk petroleum, oil, and lubricant (POL) availability, POL 'Tank Farms', and potable water sources and facilities are important to the successful employment and deployment of forces.

Force Provider

Force provider is one system for providing life support for transient forces deploying to operations. Force provider can be configured in a 600 person life support capability. It can be configured for transport in 150 person increments that provide environmentally controlled billeting, feeding, and field hygiene (laundry, shower, and latrine) capabilities. Add on capabilities include: cold weather kit; prime power connection kit; and morale, welfare, and recreation kit.

H. Medical Logistics Support

Medical units must be capable of operations immediately upon arrival and initial entry of forces. Therefore, medical logistics support must be included in planning for port opening and early entry operations. Medical logistics support to arriving forces includes Class VIII sustainment of primary medical care (sick call), including support to combat units so that organic medical supply levels are not depleted during RSOI. Medical logistics also includes management of special medical materiel, such as medical chemical defense materiel, special vaccines, and other medical materiel under the control of the ASCC surgeon. Port operations may also include the issue of medical unit sets from APS and the integration of potency and dated pharmaceuticals, refrigerated, and controlled substances with those assemblages (see FM 4-02.1).

I. Rehearsals and Training

Rehearsals are a vital component of preparing for operations. Large rehearsals require considerable resources, but provide the most planning, preparation, and training benefit. Depending on circumstances, units may conduct a reduced force or full dress rehearsal. The integration of sustainment and operational rehearsals are preparation activities. FM 5-0 describes the following rehearsals:

1. Rock Drill

Rock drills allow key leaders to rehearse operational concepts prior to execution.

2. Full Dress

Full rehearsals help Soldiers to clearly understand what is expected of them and gain confidence in their ability to accomplish the mission.

3. Support Rehearsals

Support rehearsals complement preparations for the operation. They may be conducted separately and then combined into full dress rehearsals.

4. Battle Drills / Standing Operating Procedures (SOP)

A battle drill or SOP rehearsal ensures that all participants understand a technique or a specific set of procedures.

III. Executing Sustainment Operations

Execution means putting a plan into action by applying combat power to accomplish the mission and using situational understanding to assess progress and make adjustments (FM 3-0). It focuses on concerted actions to seize, retain, and exploit the initiative. Execution of sustainment operations includes supporting force projection, basing, distribution, and reconstitution of forces. The provision of sustainment maintains combat power and prolongs endurance.

A. Strategic and Operational Reach and Endurance

Strategic reach is the distance a Nation can project decisive military power against complex, adaptive threats operating anywhere. Operational reach is the distance and duration across which a unit can successfully employ military capabilities. The ability to conduct strategic and operational reach combines joint military capabilities—air, land, maritime, space, special operations, and information systems with those of the other instruments of national power.

Sustainment enables strategic and operational reach. It provides joint forces with the lift, materiel, supplies, health support, and other support functions necessary to sustain operations for extended periods of time. Army forces require strategic sustainment capabilities and global distribution systems to deploy, maintain, and conduct operations anywhere with little or no advanced notice.

See also p. 3-2.

B. Basing

Ref: FM 4-0, Sustainment (Aug '09), p. 4-15.

A base is a locality from which operations are projected or supported (JP 1-02). The base includes installations and facilities that provide sustainment. Bases may be joint or single Service areas. Commanders often designate a specific area as a base and assign responsibility for protection and terrain management with the base to a single commander. Units located within the base are under the tactical control of the base commander, primarily for the purpose of facilitating local base defense. Within large bases, controlling cdrs may designate base clusters for mutual protection and C2.

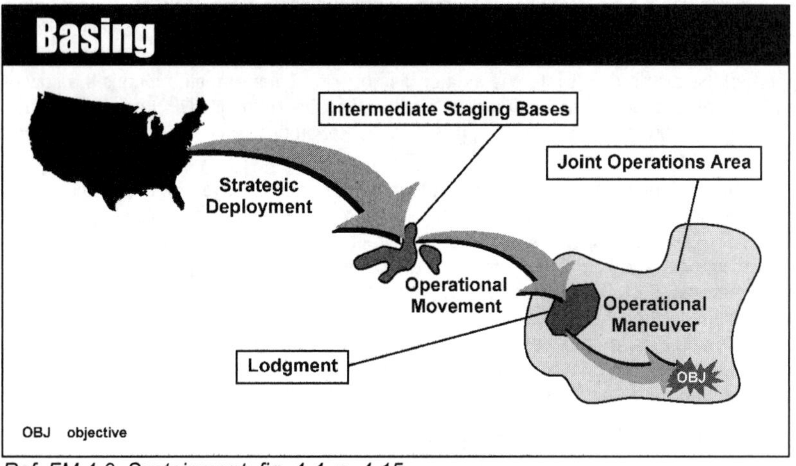

Ref: FM 4-0, Sustainment, fig. 4-4, p. 4-15.

1. Intermediate Staging Bases (ISB)

An ISB is a secure base established near, but not in, the AOR through which forces and equipment deploy (FM 3-0). While not a requirement in all situations, the ISB may provide a secure, high-throughput facility when circumstances warrant. The commander may use an ISB as a temporary staging area en route to a joint operation, as a long-term secure forward support base, and/or secure staging areas for redeploying units, and noncombatant evacuation operations (NEO). It may provide life support to staging forces in transit to operations or serve as a support base supporting the theater distribution plan.

As a support base, an ISB may serve as a transportation node that allows the switch from strategic to intratheater modes of transportation. ISB personnel may perform limited sustainment functions, such as materiel management and selected sustainment maintenance functions.

2. Forward Operating Bases (FOB)

Forward operating bases extend and maintain the operational reach by providing secure locations from which to conduct and sustain operations. They not only enable extending operations in time and space; they also contribute to the overall endurance of the force. Forward operating bases allow forward deployed forces to reduce operational risk, maintain momentum, and avoid culmination.

C. Distribution

Ref: FM 4-0, Sustainment (Aug '09), pp. 4-16 to 4-18. See also pp. 2-22 to 2-25 and 2-61.

Distribution is the key component for executing sustainment. It is based on a distribution system defined as that complex of facilities, installations, methods, and procedures designed to receive, store, maintain, distribute, and control the flow of military materiel between point of receipt into the military system and point of issue to using activities and units.

The Joint segment of the distribution system is referred to as global distribution. It is defined as the process that synchronizes and integrates the fulfillment of joint requirements with the employment of joint forces. It provides national resources (personnel and materiel) to support the execution of joint operations. The ultimate objective of the process is the effective and efficient accomplishment of joint operations. The Army segment of the distribution system is theater distribution. Theater distribution is the flow of equipment, personnel, and materiel within theater to meet the CCDR's mission. The theater segment extends from the ports of debarkation or source of supply (in theater) to the points of need (Soldier) (FM 4-01.4).

Theater distribution is enabled by a distribution management system. Distribution management is the function of synchronizing and coordinating a complex of networks (physical, communications, information, and resources) and the sustainment WFF (logistics, personnel services, and HSS) to achieve responsive support to operational requirements. Distribution management includes the management of transportation and movement control, warehousing, inventory control, materiel handling, order administration, site and location analysis, packaging, data processing, accountability for people and equipment, and communications. It involves activities related to the movement of materiel and personnel from source to end user, as well as retrograde operations.

Distribution Management Center (DMC)

Theater distribution management is conducted by the DMCs located within the support operations (SPO) section of the TSC and ESC. The DMC develops the theater distribution plan and monitors distribution performance in coordination with strategic distribution process owners and the support operations staffs in Sust Bdes and BSBs, This coordination ensures timely movement and retrograde of sustainment within the CCDR's area of responsibility. The DMC coordinates distribution with the HRSC and ASCC G-4/G-1/G-8 to ensure personnel and resources are linked. It exercises staff supervision of movement control units in a theater.

The DMC orchestrates the distribution of all classes of supply and manages all aspects of theater distribution by maintaining visibility of requirements, managing the capacity of the system, and controlling the execution of distribution operations. The DMC considers the impact of unit movement requirements on the distribution system. It provides current information on location of mode assets and movement of critical supplies along main supply routes. They provide staff recommendations to direct, redirect, retrograde, and cross-level resources to meet the distribution mission and user mission requirements.

The distribution management of medical materiel is accomplished by a support team from the MLMC.

In-Transit Visibility (ITV)

ITV is the ability to track the identity, status, and location of DOD units, and non-unit cargo (excluding bulk petroleum, oils, and lubricants) and passengers; patients and personal property from origin to consignee, or destination across the range of military operations (JP 3-35). This includes force tracking and visibility of convoys, containers/ pallets, transportation assets, other cargo, and distribution resources within the activities of a distribution node.

Intransit Visibility Architecture

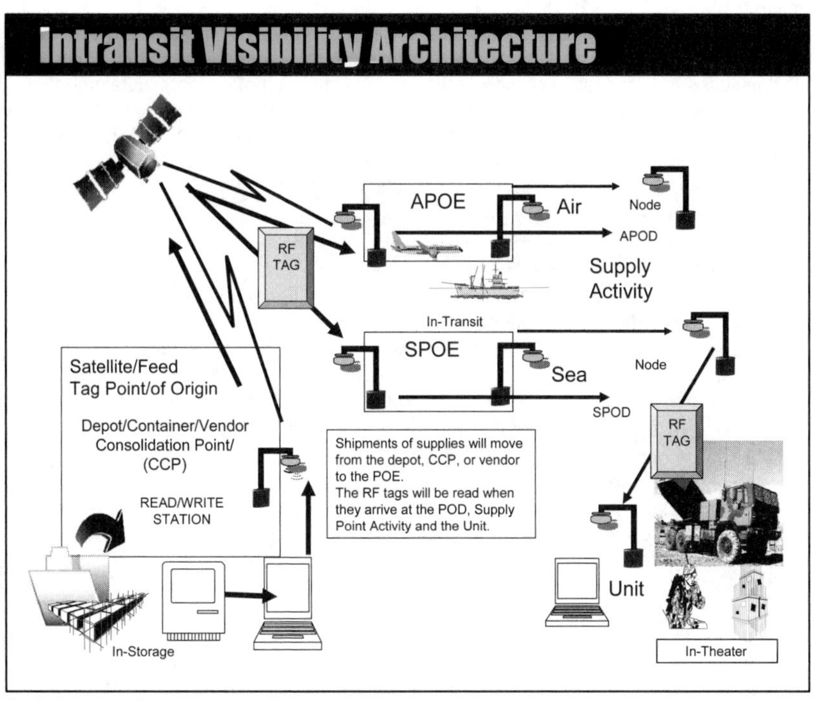

Ref: FM 4-0, Sustainment, fig. 4-5, p. 4-17.

Retrograde of Materiel

Retrograde of materiel is the return of materiel from the owning/using unit back through the distribution system to the source of supply, directed ship-to location and/or point of disposal (FM 4-01.4). Retrograde includes turn-in/classification, preparation, packing, transporting, and shipping. To ensure these functions are properly executed, commanders must enforce supply accountability and discipline and utilize the proper packing materials. Retrograde of materiel can take place as part of theater distribution operations and as part of redeployment operations.

Early retrograde planning is essential and necessary to preclude the loss of materiel assets, minimize environmental impact, and maximize use of transportation capabilities. Planners must consider environmental issues when retrograding hazardous materiel.

Contractor or HNS may be used in the retrograde of materiel. This support is planned and negotiated early in the operation. HNS must be identified early enough to ensure they are properly screened and present no security risk.

The theater distribution system provides the ASCC the ability to monitor and manage retrograde materiel through the system. Retrograde materiel flows through the distribution system in the reverse order from the tactical to strategic levels. Retrograde materiel is consolidated at the lowest supply support activity (SSA) and reported up through the support operations for distribution instructions.

An approved military customs inspection program must be in place prior to redeployment to pre-clear not only redeployment materiel but also the shipment of battle damaged equipment out of theater. The ASCC is responsible for establishing the customs inspection program to perform U.S. customs pre-clearance and United States Department of Agriculture inspection and wash down on all materiel retrograded to the United States in accordance with DOD 4500.9-R.

D. Force Projection

Force projection is the military element of national power that systemically and rapidly moves military forces in response to requirements across the spectrum of conflict. It includes the processes of mobilization, deployment, employment, sustainment, and redeployment of forces. Sustainment to force projection operations is a complex process involving the GCC, strategic and joint partners such as USTRANSCOM, and transportation component commands like AMC, MSC, SDDC, USAMC, DLA, Service Component Commands, and Army generating forces.

See p. 6-3 for further discussion of force projection processes.

E. In-theater Reconstitution

In-theater reconstitution is extraordinary actions that commanders take to restore a degraded unit to combat effectiveness commensurate with mission requirements and available resources. In-theater reconstitution should be considered when the operational tempo, mission, or time, does not allow for replacements by an available unit. Reconstitution requires both generating and operating force involvement. Generally it should be conducted in a relatively low stress environment.

The combat readiness of the unit, mission requirements, risk, and the availability of a replacement unit are the keys for considering reconstitution operations. Commanders must closely evaluate the combat worthiness of a unit to determine whether a reconstitution operation should be ordered. He/She must also decide what type of reconstitution effort would be best for the organization based on METT-TC factors. In-theater reconstitution includes reorganization, regeneration, and rehabilitation.

F. Internment/Resettlement (I/R) Operations

The Army is DOD's EA for all detainee operations. Additionally, the Army is DOD's EA for long-term confinement of U.S. military prisoners. I/R operations are defined as operations that take or keep selected individuals in custody or control as a result of military operations to control their movement, restrict their activity, provide safety, and/or gain intelligence (FM 3-19.40).

I/R operations comprise those measures necessary to guard, protect, sustain, and account for people that are captured, detained, confined, or evacuated from their homes by the U.S. armed forces. I/R operations require detailed advanced planning to prevent the degradation of operational momentum while providing a safe and secure environment for prisoners. U.S. policy mandates that all individuals captured, interned, evacuated, or held by U.S. armed forces are treated humanely. This policy applies from the moment they are under the control of U.S. armed forces until they are released, repatriated, or resettled.

Sustainment to I/R operations involves a wide range of support including logistics, personnel services, and medical treatment to detained persons. It encompasses providing all classes of supplies and materiel, health and personnel services, and general engineering support. General engineering provides horizontal and vertical construction, as well as repair and maintenance of the infrastructure.

See FM 3-34.400 and p. 1-3 for further discussion.

IV. Assessing the Sustainment of Operations

Assessment is the continuous process that occurs throughout the operations process. Sustainment commanders and staffs monitor and evaluate the current situation and the progress of the operation and compare it with the concept of support, mission, and commander's intent. Based on their assessment, commanders direct adjustments to sustainment operations, ensuring that they remain focused on the mission and commander's intent.

The primary tools for assessing are the staff running estimates, see pp. 4-20 to 4-23 for further discussion.

V. Sustainment: Logistics Sub-Function

Ref: FM 4-0, Sustainment (Aug '09), chap. 5.

Logistics is the planning and executing the movement and support of forces. It includes those aspects of military operations that deal with: design and development, acquisition, storage, movement, distribution, maintenance, evacuation, and disposition of materiel; movement, evacuation, and hospitalization of personnel; acquisition or construction, maintenance, operation, and disposition of facilities; and acquisition or furnishing of services (JP 4-0).

See pp. 1-2 to 1-3 for discussion of logistics as a sub-function of the sustainment warfighting function and pp. 2-31 to 2-62 for discussion of logistics support to the warfighter (sustainment brigade operations).

Logisitics Sub-Function

- I Supply
- II Field Services
- III Transportation
- IV Maintenance
- V Distribution
- VI Operational Contract Support
- VII General Engineering Support

Ref: FM 4-0, Sustainment, chap. 5.

I. Supply Operations

Supply operations include the requisitioning, receipt, storage, issue, distribution, protection, maintenance, retrograde, and redistribution of supplies. Levels of supply are broadly classified under the levels of war as tactical, operational, and strategic.

See pp. 2-25 to 2-49 for discussion of supply operations in support of the warfighter.

A. Tactical-Level Supplies

Tactical level supplies are those items provided to and carried within each maneuver or support brigade to sustain operational endurance. They also consist of those supplies held by Sust Bdes to provide area support.

B. Operational-Level Supplies

Operational supplies are theater stocks positioned to replenish tactical stocks, when strategic replenishment is not feasible.

C. Strategic-Level Supplies

Strategic supplies are items under the control of strategic managers and are available for worldwide materiel release. These supplies are considered inventory in motion and part of the distribution system.

Supply operations with total asset visibility enablers merge the tactical, operational, and strategic levels into a seamless supply system. The automated management systems allow units to place their requests and assists sustainment units in providing responsive support in a timely manner.

While munitions is a class of supply, it is unique due to the complexities of activities associated with its handling. Munitions are a dominant factor in determining the outcome of full spectrum operations. Munitions provide the means to defeat and destroy the enemy. Planning munitions support is considered and synchronized from strategic to tactical levels. The results of planning and integrating munitions operations is to ensure munitions arrive in the right quantities and proper types where and when needed.

D. Ammunition Support

The ASCC has overall responsibility for in-theater receipts, accountability, and management of munitions stocks. The ASCC is also responsible for establishing a Theater Support Area and Ammunition Supply Points (ASPs). It is also responsible for coordinating distribution between storage sites, forward Ammunition Transfer and Holding Point (ATHP), and direct issue to using units on an area support basis.

Ammunition Supply Point (ASP)

The ASP is run by an ordnance company assigned to a sustainment brigade. ASPs receive, store, issue, and maintain a one- to three-day supply of ammunition to meet a routine surge and emergency requirements for supported units. ASP stockage levels are based on tactical plans, availability of ammunition, and the threat to the supply operations.

Ammunition Transfer and Holding Point (ATHP)

ATHPs are the most mobile and responsive of all ASAs. Each BCT and selected support brigades are authorized an ATHP. It is located within the brigade support area (BSA) and is manned and operated by the ATHP section of the BSB distribution company.

See FM 4-30.13 for more details of munitions support.

Classes of Supplies

Ref: FM 4-0, Sustainment (Aug '09), table 5-1. p. 5-2.

The Army divides supply into ten classes for administrative and management purposes.

Class	Symbol	Description
Class I		Subsistence, including health and welfare items.
Class II		Clothing, individual equipment, tentage, organizational tool sets and kits, hand tools, administrative and housekeeping supplies and equipment (including maps).
Class III		POL, petroleum and solid fuels, including bulk and packaged fuels, lubricating oils and lubricants, petroleum specialty products; solid fuels, coal, and related products.
Class IV		Construction materials, to include installed equipment and all fortification/barrier materials.
Class V		Ammunition of all types (including chemical, radiological, and special weapons), bombs, explosives, mines, fuses, detonators, pyrotechnics, missiles, rockets, propellants, associated items.
Class VI		Personal demand items (nonmilitary sales items).
Class VII		Major items: A final combination of end products which is ready for its intended use.
Class VIII		Medical material, including medical peculiar repair parts.
Class IX		Repair parts and components, including kits, assemblies and subassemblies, reparable and non repairable, required for maintenance support of all equipment.
Class X	CA	Material to support nonmilitary programs; such as, agricultural and economic development, not included in Class I through Class IX.

II. Field Services

Field services provide quality of life for Soldiers. Field service provides life support functions, including field laundry, showers, light textile repair, FP, MA, aerial delivery support, food services, billeting, and sanitation as discussed below.

See pp. 2-50 to 2-51 for discussion of field services in support of the warfighter.

A. Shower and Laundry

Shower and laundry capabilities resident within the Field Services Company are provided from the Sust Bdes for supported units as far forward as possible. The mission is to provide Soldiers a minimum of one weekly shower and up to 15 pounds of laundered clothing each week (comprising two uniform sets, undergarments, socks, and two towels). Shower and Laundry Clothing Repair Teams from the modular Quartermaster Field Services Company can be moved forward to provide field services for the BCT. The laundry and shower function does not include laundry decontamination support (see FM 3-11.5).

B. Food Preparation

Food preparation is a basic unit function and one of the most important factors in Soldier health, morale, and welfare. Virtually every type of unit in the force structure has some organic food service personnel.

The field feeding system assumes use of Meals Ready to Eat (MREs) for the first several days following deployment, followed by transition to prepared group feeding rations. The theater initially transitions from MREs to Unitized Group Rations. Then, as the operational situation permits, A-rations (fresh foods) are introduced into theater. This requires extensive sustainment expansion since it requires refrigeration, storage, distribution, and ice making. The standard is to provide Soldiers at all echelons three quality meals per day. Proper refuse and waste disposal is important to avoid unit signature trails and maintain field sanitation standards.

See FM 4-20.2 for more details.

C. Water Production and Distribution

Water is an essential commodity. It is necessary for hydration, sanitation, food preparation, medical treatment, hygiene, construction, and decontamination. Support activities, such as helicopter maintenance, FP, and operation of medical facilities, consume large volumes of water. Classification of the water function is somewhat different from other commodities; it is both a field service and a supply function. Water purification is a field service. Quartermaster supply units normally perform purification in conjunction with storage and distribution of potable water which are supply functions. It is the users' responsibility to determine potable water requirements and submit them through supply channels.

Water supply units perform routine testing. However, water quality monitoring is primarily the responsibility of preventive medicine personnel of the MEDCOM (DS). The command surgeon ensures the performance of tests associated with water source approval, monitoring and interpreting test results. Each service provides its own water resource support. Typically, the Army, as directed by the JFC, provides support in a joint operation. AR 700-136 details the responsibilities of Army elements for water support.

Engineers play a major role in providing water to Army forces. They are responsible for finding subsurface water, drilling wells, and constructing, repairing, or maintaining water facilities. Geospatial engineers generate, manage, and analyze hydrologic data and work together with ground-survey teams and well-drilling teams to locate water sources.

The quantity of water required depends on the regional climate and the type and scope of operations. Temperate, tropic, and arctic environments normally have enough fresh surface and subsurface water sources to meet raw water requirements for the force. In arid regions, providing water takes on significantly greater dimensions. Soldiers must drink more water. Water requirements are significantly greater in areas, where demand is heavy for aircraft and vehicle washing, medical treatment, laundry and shower facilities, and where construction projects are conducted.

I/R operations may require a large amount of unanticipated bulk water consumption. Units must consider the potential absence of water capability in enemy units and the requirement for on-site sanitation, shower, delousing, and medical support for in-coming detainees. Since water is a critical commodity in arid regions, managers must strictly control its use. Commanders must establish priorities.

Because of the scarcity of potable water in some contingency areas, water support equipment may be prepositioned afloat. This allows initial support to a contingency force. Additional water equipment is available in CONUS depots to sustain operations. Most of this equipment is packaged for tactical transportability. Its configuration allows for throughput to the user with minimal handling in the AO.

D. Clothing and Light Textile Repair

Clean, serviceable clothing is essential for hygiene, discipline, and morale purposes. During peacetime, fixed facilities or field expedient methods normally provide clothing repair for short-duration exercises. During combat operations, they are provided as far forward as the brigade area.

Forces receive clothing support from a combination of units, HNS, and contractors. In low levels of hostilities, HNS and contractors may provide much of this support. LOGCAP offers considerable capability during the early deployment stages. A field service company provides direct support at the tactical level. The company has the modular capability of sending small teams as far forward as desired by the supported commander.

E. Aerial Delivery

Aerial delivery equipment and systems include parachute packing, air item maintenance, and rigging of supplies and equipment. This function supports both airborne insertions and airdrop/airland resupply. Airborne insertion involves the delivery of fighting forces, along with their supplies and equipment, to an objective area by parachute. Airdrop resupply operations apply to all Army forces. The airdrop function supports the movement of personnel, equipment, and supplies. It is a vital link in the distribution system and provides the capability of supplying the force even when land LOCs have been disrupted or terrain is too hostile, thus adding flexibility to the distribution system.

USAMC manages most airdrop equipment and systems (ADES). It maintains the national inventory control point and national maintenance point for ADES. At the operational level, the airdrop equipment repair and supply company provides supply and maintenance support to airdrop supply companies.

Aerial delivery support (ADS) companies provide airdrop resupply support in the corps/division area. They also provide personnel parachute support to units such as airborne and long range surveillance units. If the corps cannot support an airdrop request, it passes the request to a Sust Bde at theater level. Heavy Airdrop Supply Companies provide theater level support. Most of the supplies used for rigging by the ADS Company come directly from strategic level, bypassing the Airdrop Equipment Repair and Supply (AERS) Company. The AERS Company provides airdrop equipment supply and maintenance for the Heavy Airdrop Supply Company.

Airdrop resupply support must be flexible. Certain contingencies may require airdrop resupply from the beginning of hostilities. However, the requisite airdrop support structure is not likely to be in place due to other deployment priorities. In such cases, the operational commander should consider having a portion of the airdrop supply company deploy to the depot responsible for supply support to the contingency area. If forces require airdrop resupply before airdrop support units deploy to the theater, the units may rig supplies for airdrop at the depot. Supplies can then be flown directly to the airdrop location.

F. Mortuary Affairs (MA)

The MA program is a broadly based military program to provide for the necessary care and disposition of deceased personnel. Each service has the responsibility for returning remains and personal effects to CONUS. The Army is designated as the EA for the Joint MA Program. It maintains a Central Joint MA Office and provides general support to other services when requirements exceed their capabilities. The MA Program is divided into three subprograms:

1. Current Death Program

The Current Death Program provides mortuary supplies and associated services for permanently disposing remains and personal effects of persons for whom the Army is or becomes responsible.

2. Graves Registration Program

The Graves Registration Program provides search, recovery, initial identification, and temporary burial of deceased personnel in temporary burial sites. It also provides for the maintenance of burial sites and for the handling and disposing of personal effects.

3. Concurrent Return Program

The Concurrent Return Program is a combination of the Current Death and Graves Registration Programs. This program provides for the search, recovery, and evacuation of remains to collection points and further evacuation to a mortuary. It provides for identification and preparation of remains at the mortuary and shipment to final destination as directed by next of kin.

MA units operate theater collection points, evacuation points, and personal effects depots. MA personnel initially process remains in theater. A MA Decontamination Collection Point may become operational whenever the threat of CBRN warfare exists. They then arrange to evacuate remains and personal effects, usually by air, to a CONUS POD mortuary. CONUS POD mortuaries provide a positive identification of the remains and prepare them for release in accordance with the desires of next of kin. MA processing points include mortuary affairs collection points (MACPs), theater mortuary evacuation points, mortuary affairs decontamination collection points, temporary interment sites, ID laboratories, and a port mortuary.

When directed by the CCDR, MA units establish cemeteries and provide for temporary interment of remains. MA units may also operate in-theater mortuaries, but they require personnel and equipment augmentation or HNS for embalming and other procedures (see JP 4-06 and FM 4-20.64).

A process and location for evacuation of personnel remains and equipment must be established. The responsibility for manning and running this activity must be done by the service responsible for the theater and have coordination and automation capability to process actions in the personnel automation systems as well as logistical systems. All personnel have clothing and other issue documents that must be cleared as well as personal effects that must be inventoried, cleaned, disposed of, and entered back into the system for issue. This is normally done by units, but also done at the Joint Personnel Effects Depot for those killed in action as well as wounded in action that have been evacuated and separated from their equipment.

III. Transportation

Army transportation units play a key role in facilitating force projection and sustainment. Army transportation ensures that Army and joint forces that are projected globally are able to be sustained in operations. Transportation operations encompass the wide range of capabilities provided by transportation units and Soldiers. In joint operations, Army transportation units provide the full range of capabilities needed to allow joint and Army commanders to achieve operational objectives.

See pp. 2-52 to 2-62 for discussion of transportation operations in support of the warfighter.

Transportation Management

The Army transports personnel, cargo, and equipment by motor, rail, air, and water with organic or contract assets. While each situation may not be conducive to using a particular mode, the Army must be able to manage, operate, and supervise these modes of transport. Mode platforms include trucks, trains, containers, flatracks, watercraft, aircraft, and host nation assets. To successfully execute force projection operations and extend operational reach and endurance, Army transportation units must execute the following functions: movement control, terminal operations, and mode operations (see FM 55-1).

A. Movement Control

Movement control is the planning, routing, scheduling, controlling, coordination, and in-transit visibility of personnel, units, equipment, and supplies moving over line(s) of communication (LOC) and the commitment of allocated transportation assets according to command planning directives. It is a continuum that involves synchronizing and integrating sustainment efforts with other programs that span the spectrum of military operations. Movement control is a tool used to help allocate resources based on the CCDR's priorities and to balance requirements against capabilities (FM 4-01.30).

The five basic principles of movement control provide a basis for all transportation operations. These principles are:

1. Centralized Control and Decentralized Execution

Centralized Control means that a focal point for transportation planning and resource allocation exists at each level of command involved in an operation. Decentralized execution of transportation missions means terminal and mode operators remain free to assign and control the specific transportation assets that will meet the requirement.

2. Regulated Movements

Movement control authorities regulate moves to prevent terminal and route congestion and scheduling conflicts among Service components.

3. Fluid and Flexible Movements

Transportation systems must provide the uninterrupted movement of personnel, supplies, and services. To do this, the system must be capable of rerouting and diverting traffic.

4. Effective Use of Carrying Capacity

This principle is simple: keep transportation assets fully loaded and moving as much as the tactical situation permits.

5. Support Forward

Support forward is throughput, which means rapid delivery of supplies and personnel as far forward as possible. It is dependent on fast, reliable transportation to move supplies and personnel as far forward as the tactical situation requires and permits.

B. Mode Operations

Ref: FM 4-0, Sustainment (Aug '09), p. 5-6.

Mode operations and movement control elements working together match up the correct asset capability depending on cargo characteristics and required delivery time. Movement control sections coordinate transportation assets. When allocated, Army aviation assets for sustainment support direct coordination between the MCB and the Aviation Brigade is vital in providing responsive support. Requests for use of Air Force fixed winged aircraft for sustainment resupply requires coordination between the MCB and the theater airlift liaison officer. Airlift providers may be the Army, Navy, Air Force, MNFs, host nation military, or commercial aircraft.

1. Motor

Army motor transportation provides essential distribution capabilities for Army organizations. Army transportation units are the single largest provider of land surface movement within joint forces. Motor transportation includes organic, host nation, and contracted resources.

2. Rail

Rail is potentially the most efficient ground transportation method for hauling large tonnages. The Army has limited railway operating, construction, and repair capabilities. Rail capability may be provided through HNS. The Army augments HNS by providing personnel resources.

3. Air

Airlift is a mode of transportation. Wide-ranging sustainment needs within a theater require Air Force and Army airlift assets to support. Army utility helicopters provide support at the through movement control channels in response to mission requirements and the commander's priorities. Likewise, the U.S. Air Force provides intratheater airlift to all services within a theater through an allocation process on a routine basis or provides immediate support to operational requirements. While airlift is the preferred method of delivery, airdrop is a field service that can provide additional flexibility. It makes possible rapid resupply of critical items over extended distances directly to or near forward units.

4. Water

Army watercraft is a component of intratheater transportation. It can augment other modes when integrated with appropriate terminal operations or may be the primary means of transport in specific areas in a theater. Army watercraft move materiel and equipment over inland waterways, along theater coastlines, and within marine terminals. Their primary role is to support cargo discharge from strategic lift assets, conduct onward movement, and provide distribution of cargo and equipment from the SPOD to inland terminals and austere delivery points or retrograde materiel from those areas.

Watercraft can perform utility missions including patrolling, salvage, ship-to-shore transport of personnel, and harbor master duties. Although not an Army watercraft mission, they can perform limited docking and undocking services for strategic transport vessels when required. The watercraft fleet consists of a variety of vessels such as landing craft, tug boats, floating cranes, barges, causeways, and associated equipment. Army watercraft are organized into companies and detachments which can operate under a variety of command relationships (such as attached to a Transportation Terminal Battalion or SDDC units (see FM 55-80).

C. Terminal Operations

Ref: FM 4-0, Sustainment (Aug '09), pp. 5-6 to 5-8.

Terminal operations are key elements in Army force projection operations and support endurance and reach operations. They provide loading, unloading, and handling of materiel, cargo, and personnel between various transportation modes. When linked by the modes of transport (air, rail, and sea), they define the physical network for distribution operations.

Well established terminal operations are essential in supporting deployment, reception, staging, and onward movement and sustainment of the force. Crucial to the successful execution of the terminal/nodal operation is the assignment of the right personnel, cargo, and material handling equipment at each terminal. ITV of materiel moving through the transportation system also provides the GCC with information pertaining to location and destination of all cargo and equipment. There are two types of terminal/nodal operations: marine and inland.

1. Marine Terminals

The type, size, number, and location of military marine terminals selected for use, dictate the number and types of units needed to sustain theater support requirements. Using small or geographically dispersed terminals may be necessary for flexibility and survivability. However, this creates a greater need for C2 organizations. A fixed-port facility operated by a HN under contract may only require a contract supervision team. A similar facility operated as a military marine terminal may require a terminal battalion. Fixed-port facilities are designed for oceangoing vessel discharge operations and port clearance. These facilities have sufficient water depth and pier length to accommodate deep-draft vessels. They also have highly sophisticated facilities, equipment, and organization to effectively support cargo discharge and port clearance operations (see FM 55-60). Marine terminals consist of three types of facilities :

- Fixed-port facilities
- Unimproved port facilities
- Bare beach facilities

2. Inland Terminals

Inland terminals provide cargo transfer facilities. These include air, motor transport, inland waterway, and rail:

- **Air.** Air cargo transfer takes place at common-use APODs and service controlled airfields and landing strips throughout the theater. A capability assessment should be conducted for each airfield to determine the maximum aircraft on the ground (MOG) that can be parked (called parking MOG) and the number and type of aircraft that can be worked (called working MOG) with available personnel, MHE, and ramp space.
- **Motor transport.** Distribution terminals of Centralized Receiving and Shipping Points are normally located at both ends of a line-haul operation. They form the connecting link between local hauls and the line-haul service. They may also be located at intermediate points along the line-haul route where terrain necessitates a change in type of carrier.
- **Inland waterway.** Inland waterway terminals are limited by the size and configuration of the terminal, types of watercraft, and capabilities of the unit's cargo handling equipment.
- **Rail**. Rail terminals may include yard tracks, repair and servicing facilities, train crew accommodations, and railheads. They are located at originating and terminating points and at sites that mark the limits of rail operations.

D. Container Management

USTRANSCOM has designated SDDC as the global container manager (GCM), to include the authority over execution of container policy across Services as coordinated with a GCC's concept of operations and support. SDDC issues, numbers, and maintains the register of all USAMC DOD-owned inter-modal containers and ISO-configured shelters by DOD Activity Address Code and type container (see FM 4-01.52). *See also p. 2-52.*

SDDC provides a theater container database that monitors the inventory, management, and accountability of all containers via the Container Management Element. It also uses automated information systems to monitor movement of containers throughout the theater. The GCM manages, monitors, reports, and provides asset visibility of DOD-owned, leased, and commercial inter-modal surface shipping platforms and containers while in the Defense Transportation System. They provide data expertise to the Army for determining container and container handling equipment requirements to support Army and joint forces contingency, exercises, and peacetime operations.

Containerization

Containerization facilitates and optimizes cargo carrying capabilities via multiple modes of transport (sea, highway, rail, and air) without intermediate handling of the container's contents. This decrease in time, MHE, personnel, and handling ensures rapid deployment and cargo integrity during shipment. The standardization of the container has facilitated the ease of handling associated with the ISO container and associated MHE. This method of cargo distribution provides fast and flexible preparation, employment, deployment, and sustainment of forces in a theater of operations and extends operational reach. Containerization provides minimum obstructions to the deployment throughput and facilitates unit integrity and cargo security, while enabling container tracking and cargo ITV.

Service components must plan for theater reception, staging, onward movement, and integration plans. They must include in their plans, methods for container and pallet management and control. When planning to use DOD-owned, Service-owned, or leased containers, the following factors must be considered:

- Availability and location of containers
- Time and resources required
- Original out load capability
- Theater infrastructure/Force structure
- Availability of MHE at shipping point and at destination
- Tracking capability, labeling and marking of owner/addressee and destination
- Method of securing container (lock or serial band)

Initial Entry Container Control Site

A critical node for containerized cargo is the initial entry container control site. This site may be a sea port, aerial port, or rail head. Another critical area of cargo transfer ashore is during LOTS, the shore operations at the beach. At both locations, the container control site will receive, identify, and direct inland distribution and retrograde of containers. The ability of control site personnel to rapidly identify the sender address and the receiver address is the primary enabler of a rapid and successful mission for the force.

To avoid having large quantities of government-owned containers on hand, the strategy requires partnering with commercial container leasing companies to ensure that leased containers are made available to support military missions which are then staged at depots and power projection platforms in accordance with specific timelines. As operations stabilize, a trans-load operation can commence when directed by the Joint Task Force or CCDR. This would allow government-owned and government-leased containers being used for storage to remain in place, while using ocean carrier provided containers to resupply and sustain operations. These carrier-provided containers would then be unloaded and returned to the carrier within the allotted free time. It is anticipated that these containers will also be used for retrograde operations. Deployment, redeployment, and force rotation requirements will continually be met primarily with government-owned or government-leased containers.

Current operations resulted in containers being used for non-traditional/non-commercial transportation purposes, such as long-term storage. Future contingencies are expected to be characterized by similar indefinite durations in austere environments. The commercial practice of using ocean carrier provided containers that must be unloaded and returned within a specific time period does not support combat operations and results in high detention costs. Container usage in future contingencies must be addressed in the planning stage in order to minimize cost while supporting mission requirements.

IV. Maintenance

Maintenance is performed at the tactical through strategic levels of war. The Army's two levels of maintenance are field maintenance and sustainment maintenance (see FM 4-30.3).

See pp. 2-31 to 2-34 for discussion of maintenance support to the warfighter.

Fundamentals of Army Maintenance

Field maintenance is repair and return to user and is generally characterized by on-(or near) system maintenance, often utilizing line replaceable unit, component replacement, battle damage assessment, repair, and recovery. Field level maintenance is not limited to remove and replace but also provides adjustment, alignment, and fault/failure diagnoses. Included in field maintenance is the scheduled service/condition based maintenance required on equipment in accordance with the specified technical manual, to include preventative maintenance checks and services. Field maintenance is performed at all levels of the Army and most units have at least some organic field level maintenance capability. Sustainment maintenance is characterized by “off system” component repair and/or ”repair and return to supply system.” The sustainment maintenance function can be employed at any point in the integrated logistics chain. The intent of this level is to perform commodity-oriented repairs on all supported items to one standard that provides a consistent and measurable level of reliability.

A. Field Maintenance

Field maintenance is focused on returning a system to an operational status. The field maintenance level accomplishes this mission by fault isolating and replacing the failed component, assembly, or module. The field maintenance level consists of the maintenance functions of inspection, test, service, adjust, align, remove, replace, and repair. Field maintenance also includes battlefield damage and repair tasks performed by either the crew or support personnel to maintain system in an operational state. Within the BCT, the Field Maintenance Company of the BSB provides:

- Automotive, armament, recovery, ground support, missile and electronic maintenance, and maintenance to brigade base elements (HQ, BSB, and Special Troops Battalion)
- Maintenance advice and management to the brigade
- Low density equipment support to the field maintenance platoon (FMP)

In the BCT, each maneuver battalion will have a forward support company (FSC) that performs field maintenance. Each FSC has a maintenance platoon that provides recovery support, automotive and tracked vehicle repair, and ground support equipment repair to the battalion. Field maintenance teams deploy with each maneuver company and provide automotive and track vehicle repair support. Ground-support equipment repairs are conducted at the FSC located with the maneuver battalion HQ. The FMP is organized with a maintenance control section to provide maintenance management for the battalion. In EAB, the support maintenance company (SMC) of the Sust Bde provides field maintenance support on an area basis (see FM 4-30.3).

B. Sustainment Maintenance

Sustainment maintenance is generally characterized as "off system" and "repair rear". The intent of this level is to perform commodity-oriented repairs on all supported items to one standard that provides a consistent and measurable level of reliability. Off-system maintenance consists of overhaul and remanufacturing activities designed to return components, modules, assemblies, and end items to the supply system or to units, resulting in extended or improved operational life expectancies.

In sustainment maintenance, component repair work is coordinated by the USAMC National Maintenance Office to a single standard that provides consistent and measurable level of reliability. End item repair may be performed by either military or civilian technicians at a sustainment maintenance activity. Repair rear is synonymous with the term "off-system" and "sustainment maintenance". The intent is to repair components, assemblies, or end items and return them to the supply system for redistribution.

Component repair companies (CRC) provide sustainment level support at the operational level. The CRC may be attached to the CSSB to facilitate overall maintenance support. CRC units may be employed in any location along the distribution system. These units can be pushed forward into the AO as needed to repair and return components, modules, and assemblies to the supply system.

Collection and Classification companies establish and operate collection and classification facilities for the receipt, inspection, segregation, disassembly, preservation, and disposition of serviceable and unserviceable Class VII and Class IX materiel and similar foreign materiel. It also operates a cannibalization point when authorized by higher HQ. It supports distribution hub teams that perform vital maintenance inspection functions at distribution hubs along the distribution system.

V. Distribution

Distribution is defined as the operational process of synchronizing all elements of the logistics system to deliver the right things to the right place and right time to support the CCDR. It is a diverse process incorporating distribution management and asset visibility.

See pp.1-42 to 1-43, 2-22 to 2-25, and 2-61 for further discussion of distribution.

VI. Operational Contract Support

Operational contract support is the process of planning for and obtaining supplies, services, and construction from commercial sources in support of operations along with the associated contractor management functions.

See p. 1-37 for further discussion of operational contract support.

VII. General Engineering Support

General engineering activities modify, maintain, or protect the physical environment (see FM 3-34.400).

Fundamentals of General Engineering

General engineering capabilities are applied to establish and maintain the infrastructure necessary for sustaining military operations in theater. At times, the military operation may extend general engineering support to restore facilities, power, and life-support systems within the infrastructure of the AO. This effort aids in the recovery and the transition to pre-conflict conditions or may be the objective of stability or civil support operations (see FM 3-34 and FM 3-34.400).

General engineering capabilities employed in an operation will include a broad array of joint, multinational, contract, and other construction and engineering resources. The U.S. Army Corps of Engineers (USACE) provides and coordinates significant engineering resources to enable general engineering support. USACE is the Army's Direct Reporting Unit assigned responsibility to execute Army and DOD military construction, real estate acquisition, and development of the nation's infrastructure through the civil works program. USACE, through its field force engineering (FFE) and reach back assets, provides for technical and contract engineering support and a means to integrate capabilities of other Services and other sources of engineering-related support.

A. General Engineering

General engineering sustainment requirements will compete for priority in any operation with general engineering requirements related to protection, enabling operational movement, as augmentation to combat engineering, and supporting the other WFFs. Within the sustainment WFF, general engineer applications are primarily linked to providing logistics support. General engineer support includes:

- Restore Damaged Areas
- Construct and Maintain Sustainment Lines of Communications
- Provide Engineer Construction (to include pipeline) Support
- Supply Mobile Electric Power
- Provide Facilities Engineering Support

Sustainment of stability and civil support operations involves a shift to the establishment of services that support civilian agencies in addition to the normal support of U.S. forces. Stability operations tend to be of a long duration compared to the other elements in full spectrum operations. As such, the general engineering level of effort, including FFE support from USACE, is very high at the onset and gradually decreases as the theater matures. As the AO matures, the general engineering effort may transfer to theater or external support contracts such as LOGCAP.

B. Engineer Coordination

The senior engineer staff officer at each echelon HQ, designated the ENCOORD, is responsible for coordinating sustainment related general engineer support. An engineer brigade or theater engineer command will typically be task organized with those general engineering capabilities not provided to subordinate BCTs or other brigades. The engineer brigade or theater engineer command focuses general engineering efforts on priorities established by the JFC.

Engineering priorities will typically include sustainment related general engineer support. The engineer support commander may align engineer assets to provide general support on an area basis. If assets are available and priorities support a more direct relationship, the commander may place an engineer brigade in DS to the TSC, with subordinate engineer elements DS to the support brigades or CSSBs as required.

C. Real Estate Planning and Acquisition

The JFC is responsible for the coordination of planning, programming, and construction of facilities to meet the requirements of assigned forces. Facility requirements are consistent with operational requirements, duration of need, and forces to be supported. Engineer planners coordinate with sustainment and other planners to identify facility requirements for contingency operations.

USACE theater elements provide technical real estate guidance and advice to the JFC. In addition to recommending real estate policies and operational procedures, they acquire, manage, dispose of, administer payment for rent and damages, handle claims, and prepare records and reports for real estate used within the theater.

D. Real Property Maintenance

The JFC has overall responsibility for real property maintenance activities (RPMA). The JFC normally delegates authority to the ASCC/ARFOR. The TSC and installation commanders (in most cases a CSSB) normally provide the needed RPMA support. RPMA in an AO includes operation, repair, and maintenance of facilities and utilities; fire prevention and protection; and refuse collection and disposal. RPMA requirements that exceed the organization's capabilities are forwarded to the local engineer commander (in most cases, the engineer group providing support to a CSSB on an area basis) or USACE element for execution. The TSC provides technical RPMA guidance to subordinate units.

E. Base Camp Construction

A base camp is an evolving military facility that supports the military operations of a deployed unit and provides the necessary support and services for sustained operations. It is a grouping of facilities collocated within a contiguous area of land, or within close proximity to each other, for the purpose of supporting an assigned mission, be it tactical, operational, or logistical. Base camps may be located near a key piece of real estate such as a port, an airfield, a railroad, or other major LOCs. Base camps support the tenants and their equipment; and while they are not installations, they have many of the same facilities and attributes the longer they are in existence.

The CCDR specifies the construction standards for facilities in the theater to minimize the engineer effort expended on any given facility while assuring that the facilities are adequate for health, safety, and mission accomplishment. Typically, the CCDR will develop the base camp construction standards for use within the theater, utilizing the guidelines provided in JP 3-34 and facilities standards handbooks developed by the specific combatant command. The engineer must recommend the most feasible solutions to each requirement based on construction guidelines and other planning factors.

F. Environmental Considerations

The ENCOORD also advises the commander on environmental issues as the staff proponent for environmental considerations. The ENCOORD coordinates with other staff offices to determine the impact of operations on the environment and helps the commander integrate environmental considerations into the decision making process. Environmental considerations include:

- Policies and responsibilities to protect and preserve the environment
- Certification of local water sources by appropriate medical personnel
- Solid, liquid, and hazardous waste management, including dumping and burning, and disposal of gray water, pesticides, human waste, and hazardous materials
- Protection of indigenous animals and vegetables
- Archaeological and historical preservation
- Contingency spill plans

VI. Sustainment: Personnel Services

Ref: FM 4-0, Sustainment (Aug '09), chap. 5.

Personnel services include HR support, religious support, financial management, legal support, and band support. Personnel services are those sustainment functions maintaining Soldier and Family readiness and fighting qualities of the Army force. Personnel services complement logistics by planning for and coordinating efforts that provide and sustain personnel (FM 3-0).

See pp. 1-2 to 1-3 for discussion of personnel services as a sub-function of the sustainment warfighting function and pp. 2-63 to 2-66 for discussion of personnel services support to the warfighter (sustainment brigade operations).

Personnel Services Sub-Function

I Human Resources (HR) Support

II Financial Management (FM) Operations

III Legal Support

IV Religious Services

V Band Support

Ref: FM 4-0, Sustainment, chap. 5.

I. Human Resources Support (HRS)

HRS is the aggregate of systems and services designed to provide and support Soldiers. HRS is important to maximizing operational reach and endurance. HRS encompasses four major categories: manning the force, HR services, personnel support, and HR planning and staff operations. Each includes major functional elements and all are covered below (see FM 1-0, FMI 1-0.1, and FMI 1-0.2).

A. Manning the Force

Manning the force involves personnel readiness of the force, maintaining accountability of the force, and management of personnel information. The manning challenge is getting the right Soldier to the right place, at the right time, with the right capabilities so that commanders have the required personnel to accomplish their mission. Manning combines anticipation, movement, and skillful positioning of personnel assets. It relies on the secure, robust, and survivable communications and digital information systems of emerging technologies that provide the common operational picture, asset visibility, predictive modeling, and exception reporting.

1. Personnel Readiness Management (PRM)

The purpose of PRM is to distribute Soldiers to units based on documented requirements, authorizations, and predictive analysis to maximize mission preparedness and provide the manpower needed to support full spectrum operations. This process involves analyzing personnel strength data to determine current mission capabilities and project future requirements. It compares an organization's personnel strength to its requirements and results in a personnel readiness assessment and allocation decision.

2. Personnel Accountability (PA)

PA plays a critical role in deployed operations and relies on timely, accurate, and complete duty status and location of personnel at all times. PA is the process for recording by-name data on Soldiers when they arrive and depart from units; when their location or duty status changes (such as from duty to hospital); or when their grade changes. PA will be accomplished primarily through the database of record and web enabled processes that facilitate personnel support from home station or abroad. Personnel Accounting Teams manage or administer all HR support activities of processing, tracking, and coordinating personnel moves into, through, or out of a deployed organization or theater. These activities include the reception of personnel, the assignment and tracking of replacements, return-to-duty and rest and recuperation personnel, and redeployment operations.

3. Strength Reporting

Strength reporting is a numerical end product of the accounting process, achieved by comparing the by-name data obtained during the personnel accountability process (faces) against specified authorizations (spaces or in some cases requirements) to determine a percentage of fill. It starts with strength-related data submitted at unit level and ends with an updated database visible at all echelons. Similar to PA, strength reporting relies on timely, accurate, and complete personnel information into the database of record. It is also a command function conducted by the G-1/S-1 to enable them to provide a method of measuring the effectiveness of combat power. Standard reports available from the personnel accounting system include:

- Personnel status report
- Personnel summary
- Personnel requirements report
- Task force personnel summary

4. Personnel Information Management (PIM)

PIM encompasses the collecting, processing, storing, displaying, and disseminating of relevant information about Soldiers, units, and civilians. PIM is the foundation for conducting or executing all HR functions and tasks. HR managers and technicians at all levels of command use a personnel information database when performing their missions. The DIMHRS, when implemented, will be the HR enterprise database for all military personnel.

B. Provide HR Services

HR Services encompass casualty operations and EPS to maintain Soldier readiness and sustain the human dimension of the force. The following is a discussion of casualty operations, EPS, and the elements of personnel support.

1. Casualty Operations Management

The casualty operations management process includes the recording, reporting, verifying, and processing of information from unit level to HQ, Department of the Army. It also involves notifying appropriate individuals and assisting family members. The process collects casualty information from a number of sources and then collates, analyzes, and determines the appropriate action. Accuracy and timeliness are critical components of casualty management and depend on satellite communications and reliable access to personnel information.

Casualties can occur on the first day of an operation. Therefore, casualty managers from each echelon of command need to deploy early. Units report all casualties, to include civilians, contract, and military personnel from Army, other services, and MNFs. Casualty operations require 100 percent personnel accounting reconciliation. The unit verifies casualty information against the database and emergency data in an individual's deployment packet. Casualty liaison teams (CLT) provide an interface between medical facilities, MACP collection points, and human resources elements.

2. Essential Personnel Services (EPS)

EPS provide Soldiers and units timely and accurate personnel services that efficiently update Soldier status, readiness and quality of life, and allow Army leadership to effectively manage the force. EPS includes actions supporting individual career advancement and development, proper identification documents for security and benefits entitlements, recognition of achievements, and service performance. It also includes personal actions such as promotions, reductions, evaluations, military pay, leave and pass, separations, and line-of-duty investigations.

C. Personnel Support

Personnel Support encompasses command interest/human resources programs, MWR, and retention functions. Personnel Support also includes substance abuse and prevention programs, enhances unit cohesion, and sustains the morale of the force.

1. Postal Operations

Postal operations and services have a significant impact on Soldiers, civilians, and their families. The Military Postal Service serves as an extension of the U.S. Postal Services; therefore, its services are regulated by public law and federal regulation. Efficient postal operations are necessary and require significant logistics and planning for issues such as air and ground transportation, specialized equipment, secured facilities, palletization crews, mail handlers, and others. Postal services also include selling stamps; cashing and selling money orders; providing registered (including classified up to secret), insured and certified mail services; and processing postal claims and inquiries.

2. Morale, Welfare, and Recreation (MWR) and Community Support

MWR and community support provide Soldiers, Army civilians, and other authorized personnel with recreational and fitness activities, goods, and services. The MWR network provides unit recreation, library books, sports programs, and rest areas for brigade-sized and larger units. Community support programs include the American Red Cross, AAFES, and family support system.

The MWR system becomes an immediate outlet for Soldiers to reduce stress, which is critical to sustaining the readiness of the force, particularly as the speed and intensity of operations escalate. The MWR system relies on FP packages and recreation specialists. It capitalizes on using cellular, e-mail, and video-teleconference technologies to provide links between Soldiers and their families. Soldiers are also entertained through the latest in visual and audio entertainment over satellite, worldwide web, and virtual reality technologies. The human dimension of the Soldier is critical to the strength of Army forces. The human resource element of sustainment to the fighting force contributes to both the National will and the will of the Soldier to fight.

Human Resources Planning and Staff Operations

Human Resources Planning and Staff Operations are the means by which the HR provider envisions a desired HR end state in support of the operational commander's mission requirements. HR planning addresses the effective ways of achieving success, communicates to subordinate HR providers and HR unit leaders the intent, expected requirements, and outcomes to be achieved, and provides the support OPLANs, OPORDs, or Planning Annex.

Planning and staff operations are also the process of tracking current and near-term (future) execution of the planned HR support to ensure effective support to the operational commander through the military decision process. Effective planning includes ensuring HR C2 nodes are established, operated, and that connectivity to HR data and voice communications nodes is maintained.

II. Financial Management (FM)

The FM mission is to analyze the commander's tasks and priorities to ensure that proper financial resources are available to accomplish the mission and to provide recommendations to the commander on the best allocation of scarce resources. FM support enhances the commander's ability to manage and apply available resources at the right time and place in a fiscally responsible manner. FM provides the capability for full spectrum finance and RM operations across the theater to include all unified operations.

FM is comprised of two core functions: finance operations and RM. These two processes are similar and mutually supporting in organizational structure and focus. The ASCC G-8, in consultation with the FMC Director, is the principal adviser to the ASCC in all matters relating to FM operations. The integration of finance and RM under one entity offers the commander a single focal point for FM operations (see FM 1-06).

The Army may be appointed the EA responsible for FM operations. The EA for FM normally will fund multi-Service contract costs, unique joint force operational costs, special programs, joint force HQ operational costs, and any other designated support costs. The EA also provides financial analysis and recommendations to joint forces for the most efficient use of fiscal resources (see JP 1-06, Financial Management Support in Joint Operations, for further information).

Regardless of the scale or scope of sustainment operations, finance and RM operations play a key role in providing responsive agile support to deployed forces across the spectrum of conflict (see FM 1-06).

A. Finance Operations

Finance operations must be responsive to the demands of the unit commanders at all levels, requiring FM leaders to anticipate and initiate the finance support needed. This section summarizes finance operations during all operational stages. It presents a stable body of technical principles rooted in actual military experience from which commanders can guide their actions in support of national objectives.

See facing page (p. 1-63) for further discussion.

Finance Operations

Ref: FM 4-0, Sustainment (Aug '09), pp. 5-14 to 5-15.

Finance operations must be responsive to the demands of the unit commanders at all levels, requiring FM leaders to anticipate and initiate the finance support needed. This section summarizes finance operations during all operational stages. It presents a stable body of technical principles rooted in actual military experience from which commanders can guide their actions in support of national objectives.

1. Procurement Support

The success of all operations depends on the support provided to the sustainment system and to contingency contracting efforts. A large percentage of the FM mission is to support the procurement process and provide oversight. Oversight is critical in preventing improper or illegal payments. By coordinating with the contracting officer and the SJA regarding local business practices, financial managers greatly reduce the probability of improper or illegal payments. Procurement support includes two areas: contracting support and commercial vendor services support.

- **Contracting Support.** Contracting support involves payment to vendors for goods and services. This includes all classes of supply, laundry operations, bath operations, transportation, and maintenance. Financial managers are crucial to successful contracting operations.
- **Commercial Vendor Services (CVS).** CVS provides for the immediate needs of the force. These are needs the standard logistics systems cannot support. This usually includes payments of cash (U.S. or local currency). Cash payments are usually for day laborers, Class I supplements (not otherwise on contract) and the purchase of construction material not readily available through the contract or supply system.

2. Limited Pay Support

FM units provide limited travel support, casual payments, check cashing and currency exchanges to Soldiers and civilians in permanent change of station and temporary duty status, NEO travel advances, and non-US pay support (EPW, civilian internee, host nation employees, and day laborers).

3. Disbursing Support

Disbursing support includes training and funding paying agents in support of local procurement, administering the Stored Value Card (SVC), supporting rewards programs, and making condolence and solatium payments. Individual support is provided to Soldiers and/or civilians through check cashing, foreign currency conversions, receiving collections (such as Savings Deposit Program), making payments on prepared and certified vouchers, funding FM units, determining the need for currency (U.S. and foreign) and its replenishment, and receiving and processing all captured currencies and precious metals.

4. Accounting Support

Accounting support includes ensuring proper financial resources are available to the commander by supporting the fiscal triad (Contracting, RM, and Finance) in reconciling expenditures and thus providing the most accurate and timely financial data.

5. Banking and Currency Support

Banking relationships and procedures are established with any banking industry of a host nation to include establishing local depository accounts, limited depository accounts for current contract payments, and foreign currency resupply.

B. Resource Management (RM) Support

RM operations are a critical enabler at all levels of the Army's chain of command. The RM operations mission is to analyze tasks and priorities and to identify the resource requirements that will enable the commander to accomplish the mission. In advising the commander, financial managers perform the tasks shown below.

1. Identify, Acquire, Distribute and Control Funds

Financial managers identify the sources of funds available from various DOD and other Federal agencies. They also acquire the funds and distribute funds to subordinate elements to support the mission and commander's intent.

2. Develop Resource Requirements

Determining what resources are required and available to support the mission and commander's intent includes, but is not limited to, contracting, transportation, multinational support, support to other agencies and international organizations, foreign humanitarian assistance, and force sustainment. Developing and determining resource requirements also includes:

- Preparing the FM annex to operations plan and order
- Determining and validating costs to accomplish the mission
- Determining when resources are needed throughout the fiscal year(s)
- Making resources available at the time and amount needed
- Developing budgets
- Coordinating fiscal issues associated with all unified action operations

3. Track, Analyze, and Report Budget Execution

Procedures are established to track costs in order to determine obligation rates and conduct analysis on use of funds in support of the mission and to identify trends to foresee resourcing challenges. Reports are submitted as required by policy. Tracking, analyzing, and reporting budget execution include the following:

- Analyze RM and accounting reports
- Establish procedures to track costs
- Establish management internal control processes

4. Accounting Support

Financial managers ensure official accounting records are accurate, properly supported by source documentation, and resolve accounting issues in a timely manner.

III. Legal Support

Members of The Judge Advocate General's Corps (JAGC) provide proactive legal support on all issues affecting the Army and the Joint Force and deliver quality legal services to Soldiers, retirees, and their families. Legal support centers on six core disciplines across full-spectrum operations. The six core disciplines are: military justice, international and operational law, contracts and fiscal law, administrative and civil law, claims, and legal assistance. Each discipline is described on the facing page (see FM 1-04).

See facing page (p. 1-65) for further discussion.

Units at the BCT Level and Echelons Above

Units at the BCT level and echelons above have organic legal elements to support the mission. At the BCT, the brigade legal section (BLS) is responsible for providing services in all core legal disciplines that extend across the full spectrum of operations. The BLS mission is dictated primarily by the brigade commander's guidance and direction and the brigade judge advocate's professional judgment.

Legal Support

Ref: FM 4-0, Sustainment (Aug '09), pp. 5-16 to 5-17.

1. Military Justice

Military justice is the administration of the Uniform Code of Military Justice (UCMJ). The Judge Advocate General is responsible for the overall supervision and administration of military justice within the Army. Commanders are responsible for the administration of military justice in their units and must communicate directly with their servicing SJAs about military justice matters (AR 27-10).

2. International and Operational Law

International law is the application of international agreements, U.S. and international law, and customs related to military operations and activities. Within the Army, the practice of international law includes the interpretation and application of foreign law, comparative law, martial law, and domestic law affecting overseas activities, intelligence, security assistance, counter-drug, stability operations, and rule of law activities.

Operational Law (OPLAW) is that body of domestic, foreign, and international law that directly affects the conduct of military operations. OPLAW encompasses the law of war, but goes beyond the traditional international law concerns to incorporate all relevant aspects of military law that affect the conduct of operations. The OPLAW attorney supports the commander's military decision-making process by performing mission analysis, preparing legal estimates, designing the operational legal support architecture, war gaming, writing legal annexes, assisting in the development and training of rules of engagement (ROE), and reviewing plans and orders.

3. Administrative and Civil Law

The practice of administrative law includes advice to commanders and litigation on behalf of the Army involving many specialized legal areas, including military personnel law, government information practices, investigations, relationships with private organizations, labor relations, civilian employment law, military installations, regulatory law, intellectual property law, and government ethics.

Civil law is the body of law containing the statutes, regulations, and judicial decisions that govern the rights and duties of military organizations and installations with regard to civil authorities. The practice of civil law includes environmental law.

4. Contract and Fiscal Law

The practice of contract law includes battlefield acquisition, contingency contracting, bid protests and contract dispute litigation, procurement fraud oversight, commercial activities, and acquisition and cross-servicing agreements. The SJA's contract law responsibilities include furnishing legal advice and assistance to procurement officials during all phases of the contracting process, overseeing an effective procurement fraud abatement program, and providing legal advice to the command concerning battlefield acquisition, contingency contracting, use of LOGCAP, ACSAs, the commercial activities program, and overseas real estate and construction.

Fiscal law is the application of domestic statutes and regulations to the funding of military operations and support to non-federal agencies and organizations. Fiscal law applies to the method of paying for obligations created by procurements.

5. Claims

The Army Claims Program investigates, processes, adjudicates, and settles claims on behalf of and against the United States world-wide under the authority conferred by statutes, regulations, international and interagency agreements, and DOD Directives.

6. Legal Assistance

Legal assistance is the provision of personal civil legal services to Soldiers, their family members, and other eligible personnel.

The level of service that a brigade legal section is able to provide will depend on a number of factors, including:

- Operational tempo
- Deployment status
- Experience level of the BLS
- Availability of additional judge advocate or paralegal support during "surge" periods and the existence of actual conflicts of interest

When faced with situations where the BLS is unable to provide the proper level of service, the brigade judge advocate should use the brigade chain of command and the JAGC technical channels to address the shortfalls.

Units at Division Level and higher

Units at division level and higher receive legal support from an Office of the Staff Judge Advocate which is responsible for the provision of legal services across all core legal disciplines to the appropriate commander and General Court Martial Convening Authority. The division SJA is a personal staff officer with direct access to the commander. In addition, SJAs typically possess logistical and professional capabilities which allow them to enhance the resources of a subordinate legal section.

IV. Religious Support

Religious support facilitates the free exercise of religion, provides religious activities, and advises commands on matters of morals and morale. The First Amendment of the U.S. Constitution and Army Regulation (AR) 165-1 guarantees every American the right to the free exercise of religion. Commanders are responsible for fostering religious freedoms. Chaplains and chaplain assistants functioning as Unit Ministry Teams (UMT) perform and provide RS in the Army to ensure the free exercise of religion (see FM 1-05).

Religious support to the Army is guided by historical precedence. The three broad functions of religious support include nurturing the living, caring for the wounded, and honoring the dead. There are several additional aspects of religious support. They include:

- Facilitating individual freedom of worship and observation of holy days in accordance with Army regulations and mission requirements
- Advising the command on morals and morale as affected by religion and the impact of indigenous religions
- Advising the command on the ethical impact of command decisions, policies, and procedures
- Resolution of medical treatment religious and ethical issues, religious apparel issues, and religious dietary restrictions in accordance with AR 600-20
- Respect for the constitutional, statutory, and regulatory requirements ensuring freedom of religion for every Soldier, family member, and command authorized civilian

See facing page (p. 1-67) for further discussion.

Fundamentals of Religious Service

Ref: FM 4-0, Sustainment (Aug '09), pp. 5-18 to 5-19.

Religious support to the Army is guided by historical precedence. The three broad functions of religious support include nurturing the living, caring for the wounded, and honoring the dead. These include all other specific activities carried out by chaplains and chaplain's assistants, as discussed below.

1. Nurture the Living

In preparation for missions that span the spectrum of conflict, UMTs develop and provide religious support activities to strengthen and sustain the spiritual resilience of Soldiers and family members. During the operation, UMTs bring hope and strength to those who have been wounded and traumatized in body, mind, and spirit, thus assisting the healing process.

2. Caring for the Wounded

UMTs provide religious support, spiritual care, comfort, and hope to the wounded. This focus of RS affirms the sanctity of life, which is at the heart of the chaplaincy. Through prayer and presence, the UMT provides the Soldiers and their families with courage and comfort in the face of death.

3. Honoring the Dead

Our Nation reveres those who have died in military service. RS honors the dead. Funerals, memorial ceremonies, and services reflect the emphasis the American people place on the worth and value of the individual. Chaplains conduct ceremonies and services, fulfilling a vital role in rendering tribute to America's sons and daughters who paid the ultimate price serving the Nation in the defense of freedom.

Religious Services

Chaplains are obligated to provide for those religious services or practices that they cannot personally perform. Chaplains perform religious services when their actions are in accordance with the tenets or beliefs of their faith group. Chaplain assistants assist the chaplain in providing or performing this religious support.

- **Unit Support** is provided to the unit to which the UMT is assigned or attached. The team normally gives priority to this mission.
- **Area Support** is provided to Soldiers, members of other services, and authorized civilians who are not a part of the unit, but operate within a unit's AO.
- **Denominational Support** is given to Soldiers and other authorized persons of the chaplain's denomination or distinctive faith group. Availability of assets may limit the availability of denominational support provided.

Staff Integration and Coordination

Chaplains personally deliver religious support. They have two technical roles: religious leader and religious staff advisor. The chaplain as a religious leader executes the religious support mission which ensures the free exercise of religion for Soldiers and authorized personnel. The chaplain is a non-combatant and will not bear arms (see AR 165-1).

Chaplain assistants are enlisted personnel and are combatants. As combatants, they integrate UMTs into tactical formations for security and survivability. Chaplain assistants also possess specific technical and staff competencies to support administering the Command Master Religious Program on behalf of the commander.

The chaplain is a personal staff officer responsible for coordinating the religious support assets and activities within the command. The chaplain is a confidential advisor to the commander for religious matters. A chaplain is located at every echelon of command from battalion through Army Service component command (see FM 1-05).

V. Band Support

Army bands provide critical support to the force by tailoring music support throughout military operations. Music instills in Soldiers the will to fight and win, foster the support of our citizens, and promote America's interests at home and abroad. (See FM 1-0 and FM 1-19)

Fundamentals of Band Support

Army bands sustain the operational Army throughout the full spectrum of operations through the provision of tailored music that enhances Warrior morale, supports Army recruiting efforts, and maintains a connection with the American public. Music serves as a useful tool to reinforce relations with host nation populations and favorably shapes the civil situation throughout the peace building process. Inherently capable of providing a climate for international relations, bands serve as ambassadors in multi-national operations or to the host nation population.

Army bands are modular units designed to support Army, Joint, and Multinational formations. Army bands are organized into four types: Small, Medium, Large, and Special. Army Bands Small and Medium are further subdivided based upon their operational capabilities with regard to the deployment in support of ongoing operations. Special bands have unique responsibilities in support of the Military District of Washington, Headquarters, Department of the Army Public Affairs, or the United States Military Academy.

The modular structure of Army bands, with "plug-in" augmentation, enables split operations in support of musical mission requirements. Home station missions among Army bands include music support of Soldier and family, wounded warrior outreach, and community relations. Deployment operations of Army bands include the music support of morale-building events among Soldiers, joint-services, and MNFs as well as building alliances or shaping the civil situation with host nation and/or regional populations. Army bands can also provide music support of nondenominational religious activity in accordance with AR 220-90, both the at home station and during deployment operations.

VII. Sustainment: Health Services Support (HSS)

Ref: FM 4-0, Sustainment (Aug '09), chap. 5; FM 3-0, Operations (Feb '08), chap. 4; and FM 4-02.1, Combat Health Logistics (Sept '01), chap. 1.

Health services support is all support and services performed, provided, and arranged by the AMEDD to promote, improve, conserve, or restore the mental and physical well being of personnel in the Army and, as directed in other Services, agencies and organizations. Army Health System (AHS) support includes both HSS and force health protection (FHP). The HSS mission is a part of the sustainment WFF. The FHP mission falls under the protection WFF, but will be included to provide an accurate description of AHS support.

See pp. 1-2 to 1-3 for discussion of health services support as a sub-function of the sustainment warfighting function and pp. 2-67 to 2-70 for discussion of health services support to the warfighter (sustainment brigade operations). See also pp. 3-31 to 3-36 for discussion of HSS planning considerations.

HSS Sub-Function

HSS includes casualty care, which involves all Army Medical Department functions, to include:

- **Organic and area medical support**
- **Hospitalization**
- **Dental care**
- **Behavioral health**
- **Clinical laboratory services**
- **Medical evaluation**
- **Medical logistics**

Ref: FM 4-0, Sustainment, chap. 5.

I. Army Health System (AHS) Support

AHS support includes all mission support services performed, provided, or arranged by the AMEDD to support HSS and FHP mission requirements for the Army and as directed, for joint, intergovernmental agencies, coalitions, and MNFs (see FM 4-02).

AHS support involves the delineation of support responsibilities by capabilities (roles of care) and geographical area (area support). The AHS support encompasses the promotion of wellness and preventive, curative, and rehabilitative medical services:

- Provide prompt medical treatment consisting of those measures necessary to locate, acquire, resuscitate, stabilize, document, and prepare patients for evacuation to the next role of care and/or return to duty (RTD)
- Employ standardized air and ground medical evacuation units/resources, in conjunction with the aviation brigades for air ambulances
- Provide flexible, responsive, and deployable medical support designed and structured to sustain a force projection Army and its varied missions. This capability includes hospitalization resources to provide essential care to all patients who cannot recover within the theater evacuation policy and are stabilized and evacuated out of theater and definitive care to those Soldiers capable of returning to duty (see FM 4-02.10)
- Provide a medical logistics system (to include blood management) that is anticipatory and tailored to continuously support missions throughout full spectrum operations (see FM 4-02.1)
- Provide dental services to maximize the RTD of dental patients by providing operational dental care and maintaining the dental fitness of theater forces
- Provide medical laboratory functions in medical operations
- Provide blood management services
- Provide preventive dentistry activities
- Provide combat and operational stress control and behavioral health (BH) preventive services

Army Health Support Fundamentals

AHS support is guided by six fundamentals, consistent with JP 4-02:

1. OPLAN Conformance

By ensuring that Force Health Projection support conforms to the tactical commander's OPLAN, the AHS support planner can determine support requirements and plan for the support needed to prevent non-battle injuries and to effectively clear the battlefield of the ill, injured, and wounded.

2. Surgeon technical direction

Technical direction/guidance and staff supervision of AHS support activities must remain with the appropriate cmd-level surgeon.

3. Continuity of care

The AHS support staff must maintain continuity of care since an interruption of treatment may cause an increase in morbidity and mortality. No patient is evacuated farther to the rear than his/her medical condition or the tactical situation dictate.

4. Proximity

The proximity of AHS support assets to the supported forces is dictated by the tactical situation METT-TC.

5. Flexibility

The AHS support plan must be flexible to enhance the capability of reallocating AHS support resources to meet changing requirements.

6. Mobility

The mobility and survivability of medical units/medical platforms must be equal to the forces supported.

II. Combat Health Logistics (CHL)

Ref: FM 4-02.1, Combat Health Logistics (Sept '01), pp. 1-5 to 1-6.

Logistics support may be executed by strategic, operational, or tactical logistics systems. These three levels of logistics support correlate to the three levels of war (FM 3-0):

1. Strategic Logistics

Strategic logistics supports the attainment of broad goals and objectives established by the National Command Authorities in national security policies. It includes special activities under DA control and the national inventory control points; national maintenance points; and depots, arsenals, data banks, plants, and factories associated with the US Army Materiel Command (AMC). Strategic functions are performed in CONUS and in the rear of the theater.

2. Operational Logistics

Operational logistics support the commander's plan in either a mature or immature theater. Operational logistics link strategic logistics to tactical logistics on the battlefield, ensuring support and success at the tactical level. Operational support attempts to balance the strategic planning requirements with the needs of tactical operations in joint and combined campaigns, major operations, and other military operations within an AO. Operational logistics are conducted by echelons above corps (EAC) and corps and below organizations to support tactical logistics.

3. Tactical Logistics

Tactical logistics support the commander's plan at the operational level of military operations. At this level, the essential functions of supply, maintenance, transportation, technical assistance, human resources support (HRS), CHS, and field services are delivered to soldiers to permit them to accomplish their mission. The medical logistician focuses on CHL to support and sustain the soldier.

CHL Functional Areas

Combat health logistics encompasses functional areas that are all tied together as a subsystem of the multifunctional CHS system. Combat health logistics support is characterized by goals, policies, procedures, and organizational structures and is directly related to the overall CHS system. It interfaces as a facilitating-type subsystem responsive first and foremost to patient care and secondly to the Army's logistical system. The functional areas include:

- Materiel procurement
- Materiel management (receiving, shipping, storage, and property accounting)
- Medical equipment maintenance and repair support
- Prescription optical lens fabrication
- Blood storage and distribution
- Arranging contract support

The materiel system has long recognized that certain commodities possess peculiarities or characteristics that make them sufficiently distinctive so that they must be managed by specifically trained personnel. Class III and Class IV are typical examples, as is medical materiel. In this regard, in their decision of 20 July 1967, the Joint Chiefs of Staff directed that medical materiel be removed from Class III and Class IV and be designated as a separate class of supply (Class VIII).

III. Levels of Medical Care

Ref: FM 4-02, Force Health Protection (Feb '03), pp. 2-5 to 2-7.

Health service support is arranged in levels of medical care. They extend rearward throughout the theater to the CONUS support-base. Each level reflects an increase in capability, with the functions of each lower level being within the capabilities of higher level.

A basic characteristic of organizing modern HSS is the distribution of medical resources and capabilities to facilities at various levels of location and capability, which are referred to as levels. Echelonment is a matter of principle, practice, and organizational pattern; not a matter of rigid prescription. Scopes and functions may be expanded or contracted on sound indication. As a general rule, no level will be bypassed except on grounds of efficiency or battlefield expediency. The rationale for this rule is to ensure the stabilization/survivability of the patient through advanced trauma management (ATM) and far forward resuscitative surgery prior to movement between MTFs (Levels I through III).

Level I

The first medical care a soldier receives occurs at Level I. It is provided by the trauma specialist/special operations forces combat medics (assisted by self-aid, buddy aid, and combat lifesaver skills, and at the battalion aid station [BAS] by the physician and physician assistant). This level of care includes immediate lifesaving measures, prevention of DNBI, COSC preventive measures, patient collection, and medical evacuation to supported medical treatment elements.

Level II

Medical companies and troops of brigades, divisions, separate brigades, armored cavalry regiments, and area support medical battalions (ASMBs) render care at Level II. They examine and evaluate the casualty's wounds and general status to determine treatment and evacuation precedence. This level of care duplicates Level I and expands services available by adding limited dental, laboratory, optometry, preventive medicine, health service logistics, COSC/mental health services, and patient-holding capabilities. When required to provide far-forward surgical intervention, the medical company may be augmented with a forward surgical team (FST) to provide initial wound surgery. The FST is organic to airborne and air assault divisions.

Level III

Level III is the first level of care with hospital facilities. Within the combat zone, the combat support hospital (CSH) provides resuscitation, initial wound surgery, and post-operative treatment. At the CSH, personnel treat patients for return to duty (RTD) or stabilize patients for continued evacuation. Those patients expected to RTD within the theater evacuation policy are regulated to an echelon above corps (EAC) CSH.

Level IV

At Level IV, the patient is treated at an EAC CSH. Those patients not expected to RTD within the theater evacuation policy are stabilized and evacuated to a Level V facility.

Level V

Definitive care to all categories of patients characterizes Level V (primarily CONUS-based) care. The Department of Defense (DOD) and Department of Veteran's Affairs (VA) hospitals provide this care. During mobilization, the National Disaster Medical System (NDMS) may be activated. Under this system, civilian hospitals care for patients beyond the capabilities of the DOD and VA hospitals.

I. Sustainment Brigade Operations

Ref: FMI 4-93.2, Sustainment Brigade (Feb '09), chap. 1.

In response to the challenge of transforming into an expeditionary Army, the modular force was designed. To compliment the modular force, the Modular Force Logistics Concept was developed to provide commensurate increased operational flex and unity of command. For the logistician, this involved streamlining traditional systems for command and control (C2), theater opening, and theater distribution. Logisticians today must be prepared to conduct a wide-ranging array of concurrent operations to support deployment, employment, sustainment, redeployment, and reconstitution.

I. Theater Operational Environment

In the recent past, the nature of operational environments changed significantly. This section addresses these changes and how the Army has adapted to accommodate them. One key feature is a distribution system that relies on visibility and flexibility instead of mass. The sustainment brigade is a key element in providing support and services to war-fighting units to ensure freedom of action, extended operational reach, and prolonged endurance.

A. Changes in the Operational Environment

Commanders must be prepared and able to conduct operations in permissive, uncertain, and hostile environments. These environments are likely to comprise difficult terrain, adverse climates, and adaptive enemies. Within the context of social, physical, and economic conditions characteristic of failed states and fractured societies, commanders may expect rampant crime with international linkages as well as religious and ethnic tension. This environment is further characterized by:

- A complex, noncontiguous battlefield, where boundaries will not be clearly defined
- A threat scenario in which potential adversaries are not readily identifiable
- Simultaneous, geographically dispersed operations that will result in extremely long and potentially unsecured lines of communications (LOC's)
- The prevalence of joint organizations at the operational level and single service organizations operating in a collaborative or interdependent joint environment at the tactical level
- A significant degree of joint and single service interaction with other governmental and nongovernmental organizations (NGO's), multinational forces, and contractors

B. Theater of Operations

A theater of operations is a geographical area for which a Geographic Combatant Commander (GCC) is assigned military responsibility. The command views a theater of operations from a strategic perspective and assesses the level of international military cooperation available with the degree of dedicated US military resources necessary. These factors influence prospective Army operations in each theater of operations or GCC area of responsibility (AOR).

C. Designation of the Area of Operations

To conduct operations within its geographic area of responsibility, the GCC may designate a specific area within the AOR as a theater of war, theater of operations, or a joint operations area (JOA). Commanders may use these terms independently or in conjunction with one another, depending on the needs of the operation. If used in conjunction, the theater of war would encompass the larger area with smaller theaters of operation and JOA's within it. Joint Publication (JP) 3-0 describes the criteria for each designation in more detail. This manual uses the more generic term area of operations (AO) to refer to any area where the commander may deploy a sustainment brigade to conduct operations. The GCC (or subordinate combatant commander) maintains responsibility for the operations of US forces in an AOR or designates a joint task force (JTF) to command forces in a designated area. The Army Service Component Commander (ASCC) provides Army forces to the joint force commander (JFC)/JTF to support those operations.

II. Command and Support Relationships

A. Command Relationships

1. Theater Sustainment Command (TSC)

The TSC is the central Army logistics C2 headquarters (HQ) in a theater of operations and the senior Army logistics HQ for the Theater Army (for example, USAREUR-7th Army, United States Army, Pacific Command--8th Army) or a JFC. The TSC consolidates many of the functions previously performed by Corps Support Commands and Theater Support Commands into a central operational echelon that is responsible for C2 of theater opening (TO), theater distribution (TD), and sustainment operations conducted in support of Army and, on order, joint, interagency, and multinational forces. The TSC is regionally focused and globally employable. Its modular design provides the TSC commander with the operational flexibility to adapt C2 as requirements develop, including deploying an Expeditionary Sustainment Command (ESC) to provide an additional measure of responsiveness, agility, and flexibility for employment. *See pp. 2-71 to 2-78 for further discussion.*

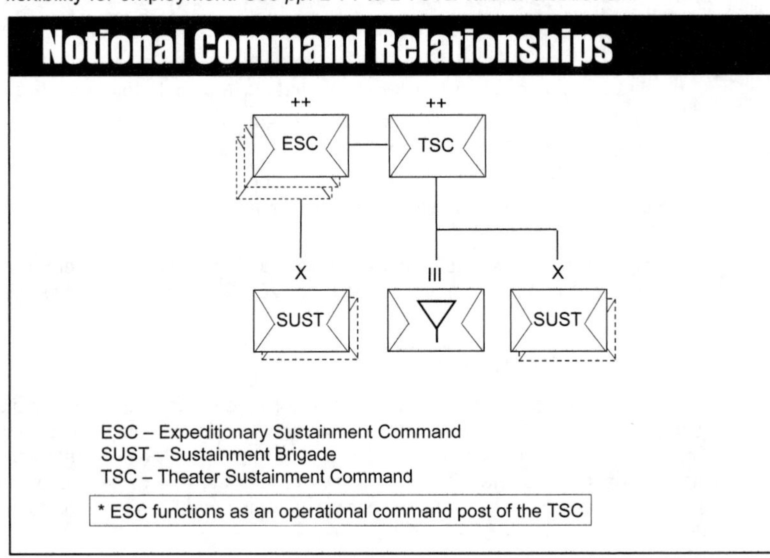

Ref: FM 4-93.2, The Sustainment Brigade, fig. 1-1, p. 1-3.

2. Expeditionary Sustainment Command (ESC)

The Expeditionary Sustainment Command (ESC), attached to a Theater Sustainment Command (TSC), provides command and control (C2) for attached units in an area of operation as defined by the TSC. As a deployable command post for the TSC, the ESC provides operational reach and span of control. The ESC plans and executes sustainment, distribution, theater opening and reception, staging, and onward movement for Army forces in full spectrum operations. It may serve as the basis for an expeditionary joint sustainment command when directed by the Combatant Commander or his designated Coalition/Joint Task Force Commander. The TSC establishes C2 of operational level theater opening, sustainment, distribution, and redistribution in specific areas of operation by employing one or more ESC. Each ESC provides a rapidly deployable, regionally focused, control and synchronization capability, mirroring, on a smaller scale, the organizational structure of the TSC. By design, the ESC provides C2 for operations that are limited in scale and scope; employing reach capabilities to provide augmented support where practical. The ESC also oversees TO, TD, and sustainment operations in accordance with TSC plans, policies, programs, and mission guidance.

The TSC may operate from a command center located in sanctuary, employing one or multiple ESC to establish a forward presence to control and direct deployed units.

See FM 4-93.4 and also pp. 2-71 to 2-78 for further discussion.

3. Sustainment Brigades

The sustainment brigades consolidate selected functions previously performed by corps and division support commands and area support groups into a central operational echelon and provide C2 of TO, TD, and sustainment operations. Combat Sustainment Support Battalions (CSSB) are the building blocks of the sustainment brigades. Their designs are standardized and can consist of up to eight companies. CSSB's are modular and task organized to support TO, TD, area sustainment, or life support missions.

Notional Support Operations in a DTO

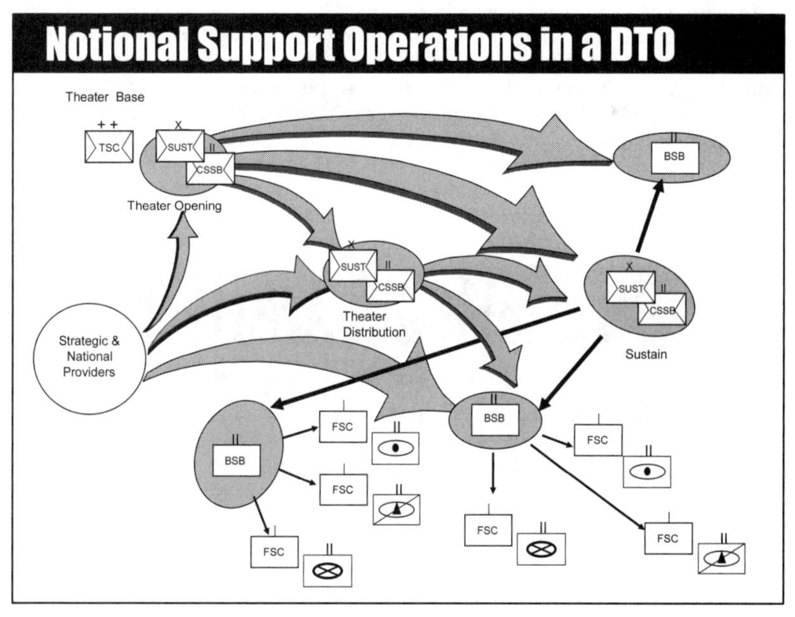

Ref: FM 4-93.2, The Sustainment Brigade, fig. 1-2, p. 1-4.

B. Support Relationships

Ref: FMI 4-93.2, Sustainment Brigade (Feb '09), pp. 1-4 to 1-6. See also p. 2-73.

In the Army, support is a specified relationship whereas in JP 3-0 support is a joint command relationship. Support is the action of a force that aids, protects, complements, or sustains another force in accordance with a directive requiring such action. The primary purpose of the support relationship is to indicate which commanders are providing support to a designated command/organization. Designation of a support relationship does not provide authority to organize and employ commands and forces, nor does it include authoritative direction for administrative and logistics support.

- The support is more effective when the supporting unit is controlled by a commander with the requisite technical and tactical expertise
- The echelon of the supporting unit is the same as or higher than that of the supported unit. For example, the supporting unit may be a brigade and the supported unit may be a battalion. It would be inappropriate for the brigade to be subordinated to the battalion, hence the use of an Army support relationship.
- The supporting unit may provide support to several units simultaneously. Prioritization of requirements is an essential function of command, as assigning support relationships is an aspect of mission command.

1. Echelons Above Brigade (EAB) Sustainment Units

The parent organization of the TSC is the Theater Army. The ESC is intended to be an operational command post of the TSC, but may be employed as a separate echelon of command. When employed as a separate command, the ESC may need augmentation from the TSC as determined through mission analysis. The sustainment brigade is assigned to the TSC, or the ESC when employed as a separate command. Any of these EAB support units may be placed under tactical control (TACON) of a combined arms HQ for a specific purpose such as protection or METT-TC. CSSB's and functional battalions are assigned to the sustainment brigade. The TSC is assigned to support a particular theater of operations. Sustainment brigades and their CSSB's provide support on an area basis and may also be assigned specific units to support. In a small operation and in the absence of an ESC, a sustainment brigade may be the senior logistical C2 HQ in a JOA and may be TACON to the senior maneuver HQ. During theater opening operations, the sustainment brigade may also provide C2 for all units in their AO (such as MPs, engineer, or chemical) as directed by an OPORD and in the absence of their brigade HQ. The TSC, ESC, sustainment brigade, and CSSB are in a general support relationship to the ARFOR, Corps as JTF, Division, brigade combat teams (BCT), and functional Brigades (see also JP 3-0 and FM 3-0).

Unit[9]	Parent Org	ARFOR in JOA	Corps as JTF	Division	BCT & Support BDE
TSC [2]	Theater Army	GS [2/3]	GS	GS	GS
TSC/TSC (-) [4]	Theater Army	GS [4/5]	GS [6]	GS	GS
ESC [6/7]	TSC	GS	GS	GS	GS
SUST	TSC	GS	GS	GS [8]	GS
CSSB / Functional Bn	SUS BDE	GS	GS	GS	GS

All of the logistics units listed here are assigned to the TSC. This chart describes the support relationships that typically exist within a theater of operations.

2. Relationships For Brigades and Below

The heavy BCT, infantry BCT, and the fires brigade have organic Brigade Support Battalions (BSB's) and Forward Support Companies (FSC's). The FSC's are assigned to the BSB and can be in direct support, operational control (OPCON), attached, or assigned to the supported battalions. Combat aviation brigades also have organic BSB's and FSC's, but the FSC's are assigned to the supported aviation battalions. The Striker Brigade Combat Team (SBCT) has an organic BSB, but no FSC's. The SBCT task organizes support packages into forward logistics elements to provide support similar to that of an FSC. The Maneuver Enhancement Brigade (MEB) has organic BSB's and FSC's. Within the MEB, engineer battalions have FSC's, but military police (MP) and chemical battalions do not. The engineer battalions are the parent of the FSC's. Support to the Battlefield Surveillance Brigade (BFSB) comes from a brigade support company assigned directly to the BFSB. The sustainment brigades have neither BSB's nor FSC's.

Bde Type	organic BSB[1]	organic FSC[1]	FSC'S Parent[2]	FSC to BSB Relationship[3]	FSC to SPT'd BN Relationship[4]
H/IBCT	Yes	Yes	BSB	Assigned	DS/OPCON/Attached
SBCT	Yes	No[5]	N/A	N/A	N/A
SUST	No	No	N/A	N/A	N/A
FIRES	Yes	Yes	BSB	Assigned	DS/OPCON/Attached
AVN Bde	Yes	Yes	AV Bn	DS[6]	Assigned
MEB	Yes	Yes[7]	EN Bn	DS[6]	Assigned
BFSB	No	BSC	N/A[8]	N/A[8]	N/A

1– These 2 columns address which BDEs have BSBs and FSCs
2– Addresses what unit the FSC is assigned to.
3 – Addresses the relationship between the FSC and the BSB.
4 – Defines the likely command or support relationship that may exist between the FSC and the unit it supports.
5 – The SBCT task organizes support packages into forward logistics elements (FLEs).
6 – 'DS' means that the BSB provides replenishment to the FSC, but no formal relationship exists.
7 – Engineers battalions have FSCs, but MP and chemical battalions do not.
8 – Support to BFSB comes from a Bde Support Company assigned directly to BFSB; there is no BSB or FSC.

Priorities of Support and Unity of Effort

From the President, Secretary of Defense, and GCC's, on down to divisions; commanders communicate their requirements and priorities for support through commander's intent, orders, the planning process, and briefings/conferences. While doing so, they also ensure that coordination occurs not only with subordinate units, but also with their higher HQ and laterally to units which may be called upon to perform in a supporting role. Commanders at all levels continually ensure cohesiveness and unity of effort. Under the concept of centralized logistics C2, the TSC/ESC supports the GCC or JTF commander by ensuring that all actions throughout the theater of operations or JOA, for which the TSC/ESC is responsible, continually support unified action and reinforce the commander's intent. Commanders at all levels must ensure that the supported commander has confidence in the concept of support and that supporting plans enable the objectives of the supported commanders. They do this by continuous coordination, to include attending battlefield update briefings and commanders' conferences of both the supported commander and their own higher HQ (or sending appropriate command representation).

III. Strategic Level Support Organizations

A. US Transportation Command (USTRANSCOM)

USTRANSCOM provides air, land, and sea transportation for the Department of Defense (DOD), both in time of peace and in time of war through its component commands: the Air Force's Air Mobility Command, the Army's Military Surface Deployment and Distribution Command, and the Navy's Military Sealift Command. The command maintains the capability to rapidly open and manage common-use aerial ports and seaports for the GCC. The contributions of USTRANSCOM's component commands are discussed below.

1. Air Mobility Command (AMC)

AMC provides strategic and tactical airlift, air refueling, and aero-medical evacuation services for deploying, sustaining, and redeploying US forces, including rapidly employing aerial ports around the world. The AMC also contracts with commercial air carriers through Civil Reserve Air Fleet and other programs for movement of DOD passengers and cargo. The sustainment brigade will receive airlift schedules from and coordinate strategic air movements with representatives of AMC.

2. Military Sealift Command (MSC)

MSC provides sealift transportation services to deploy, sustain, and redeploy US forces around the globe with a fleet of government-owned and chartered US-flagged ships. MSC executes Voluntary Inter-modal Sealift Agreement contracts for chartered vessels. Sealift ships principally move unit equipment from the US to theaters of operation all over the world. In addition to sealift ships, MSC operates a fleet of pre-positioned ships strategically placed around the world and loaded with equipment and supplies to sustain Army, Navy, Marine Corps, Air Force, and DLA operations. These ships remain at sea, ready to deploy on short notice, which significantly reduces the response time for the delivery of urgently needed equipment and supplies to a theater of operation.

3. Military Surface Deployment and Distribution Command (SDDC)

SDDC provides ocean terminal, commercial ocean liner service, and traffic management services to deploy, sustain, and redeploy US forces on a global basis. The command is responsible for surface transportation and is the interface between DOD shippers and the commercial transportation carrier industry. The command also provides transportation for troops and materiel to ports of embarkation in the US and overseas and manages seaports worldwide, coordinating for onward movement with the sustainment brigade.

4. Joint Task Force – Port Opening (JTF-PO)

The JTF-PO facilitates joint reception, staging, onward movement, and integration and theater distribution by providing an effective interface with the theater JDDOC and the sustainment brigade for initial aerial port of debarkation (APOD) operations. Its capabilities include:

- APOD assessment
- APOD opening and initial operation
- Movement control including coordination for onward movement of arriving cargo and passengers
- Establishment of joint in-transit visibility and radio frequency identification
- Moving cargo up to 10 miles to a designated location for handover to the sustainment brigade for onward movement

The JTF-PO is designed to deploy and operate for 45 to 60 days. As follow-on theater logistic capabilities arrive, the JTF-PO will begin the process of transferring mission responsibilities to arriving sustainment brigade forces or contracted capabilities. *See also p. 1-39.*

5. Joint Deployment Distribution Operations Center (JDDOC)

The JDDOC is a joint capability designed to support GCC operational objectives by assisting in the synchronization of strategic and theater resources to maximize distribution, force deployment, and sustainment. Its goal is to maximize GCC combat effectiveness through improved total asset visibility.

The JDDOC, under the control and direction of the GCC, directs, coordinates, and synchronizes forces' deployment and redeployment execution and strategic distribution operations to enhance the GCC's ability to effectively and efficiently build and sustain combat power. The JDDOC is an integral component of the GCC staff, normally under the direction of the GCC Director of Logistics (J4). However, GCC's can place the JDDOC at any location required or under the operational control of another entity in the GCC area of responsibility. The JDDOC will normally be coordinating with the TSC/ESC, but on small-scale operations, may be coordinating directly with a sustainment brigade operating as the senior Army LOG C2 HQ in the theater of operations. *See also p. 1-39.*

B. Defense Logistics Agency (DLA)

DLA is the DOD's primary strategic-level logistics provider and is responsible for providing a variety of logistics support to the military services. DLA has the capability of providing a forward presence in the operational area via its DLA contingency support teams (DCST's). DLA normally provides a DCST to each major joint operation to serve as the focal point for coordinating DLA support. The DLA DCST's may either collocate with the joint force commander J-4, the TSC Distribution Management Center (DMC), or the AFSB when the Army is the lead Service for significant common user logistics (CUL) support. The in-theater DCST integrates strategic to operational level materiel management support of DLA common commodities such as subsistence, clothing and other general supplies, Class IV construction/barrier materiel, package/bulk petroleum, and medical materiel.

DLA also provides in-theater defense reutilization and marketing services. DLA assists the joint force commander J-4; establishes theater specific procedures for the reuse, demilitarization, or disposal of facilities, equipment, and supplies, to include hazardous materiel (HAZMAT) and waste. Initially, salvage and excess materiel is collected in the main theater distribution point and/or the brigade support areas as the situation permits. As the theater of operations matures, this materiel is evacuated to collection points for inspection and classification. The TSC or sustainment brigade coordinates with Defense Reutilization and Marketing Service to perform distribution management operations for the Army forces. *See also p. 1-12.*

C. US Army Materiel Command (USAMC)

USAMC support to deployed Army forces is coordinated via the Army Sustainment Command (ASC) and is executed in the operational area by the Army Field Support Brigade (AFSB) and Contracting Support Brigade (CSB). The AFSB, when deployed, will be OPCON to the TSC and responsible for planning and controlling all USAMC and other acquisition logistics and technology (ALT) functions in the operational area less theater support contracting. The CSB, when deployed, will also be OPCON to the TSC and responsible for theater support contracting.

1. Operational Contract and Other Acquisition, Logistics, and Technology Support

Acquisition, logistics, and technology (ALT) support consists of a numerous unique support functions in a deployed AO to include rapid equipment fielding support, technical equipment support, pre-positioned stock support, and contracting support. Most of this ALT support is conducted through two O-6 level USAMC units: the Army Field

Support Brigade (AFSB) and the Contracting Support Brigade (CSB). AFSB and CSB provide support to the sustainment brigade under the C2 of the TSC or ESC. The exception to this doctrinal rule of thumb is when the sustainment brigade is the senior Army LOG C2 HQ in the AO, when the ARFOR commander may choose to OPCON AFSB and CSB elements to the sustainment brigade. *See also p. 1-15.*

2. The Army Field Support Brigade (AFSB)

The AFSB is the primary ALT unit for the Army and is responsible to control all ALT functions, less theater support contracting and Logistic Civil Augmentation Program (LOGCAP) support, in the AO. The AFSB is a small table of organization and equipment (TOE) and augmented table of distribution and allowances (TDA) HQ that leverages reach (for technical support) and call-forward procedures to bring the requisite USAMC and Assistant Secretary of the Army for Acquisition, Logistics, and Technology (ASA[ALT]) program executive officers (PEOs) (such as, PEO Ground Combat Systems) and their subordinate product/project managers (PM's) (such as, PM Stryker Brigade Combat Team) capabilities forward to the AO. *See also p. 1-18.*

3. Contracting Support Brigade (CSB)

The Army has recently consolidated its theater support contracting capabilities into separate TOE units. These units include the CSB, Contingency Contracting Battalions (CCBN), Senior Contingency Contracting Teams (SCCT), and Contingency Contracting Teams (CCT's). These consolidated units are made up of primarily 51C military occupational specialty officers and noncommissioned officers. Like the AFSB, these units are currently assigned to the ASC, but are scheduled to come under the command of the new, US Army Contracting Command, and its subordinate, Expeditionary Contracting Command, in the near future. In major operations, the CSB will normally be OPCON to the TSC or a separate joint theater support contracting command. In these situations, theater support contracting actions will provide general support to the sustainment brigade. In smaller operations, a CCBN could be placed OPCON to the sustainment brigade, if the sustainment brigade is the senior sustainment command in the AO and the CSB is not deployed.

See following pages (pp. 2-10 to 2-11) for further discussion of managing contracting support. See also pp. 1-15 to 1-16.

IV. Support of Military, Civilian, Joint, and Multinational Organizations

The Army may operate in a joint coalition or multinational environment.

Organizing the Joint Force

The JFC can organize forces in several different ways. The JFC directs operations through service component commanders or establishes functional commands. Such functional commands may include a joint forces land component to provide centralized direction and control of all land operations, including other Services land forces. The combatant commander or JFC may assign a lead Service to provide CUL to avoid redundancy and achieve greater efficiency.

A. Interagency Support

The DOD performs both supported and supporting roles with other agencies. During combat or in humanitarian assistance operations, the DOD may likely be the lead agency and supported by other agencies. When the Army (through the DOD) is tasked to provide military support to civil authorities, its forces perform in a supporting role. The sustainment brigade may be tasked to support the local head of another agency, such as an ambassador, or may employ the resources of other US Government agencies or even private firms. Whether supported or supporting, close coordination is the key to efficient and effective interagency operations.

Sustainment Brigade's Role in Support of Joint and/or Multinational Operations

Ref: FMI 4-93.2, Sustainment Brigade (Feb '09), pp. 1-4 to 1-6.

The TSC and/or Sustainment Brigade may be called upon to support joint and/or multinational operations. As such, the roles and missions of the Sustainment Brigade may expand to include support to other services in accordance with Title 10, and/or support to forces of other nations in accordance with bilateral and multilateral agreements.

1. Setting the Theater Base

A critical component of the TSC's effort to set the theater base is its focus on organizing and positioning of capabilities to support the conduct of future operations. The sustainment brigade performing theater opening has OPCON of units processing through reception, staging, and onward movement until their HQ arrives and is operational. Key activities include: ensuring the timely arrival of deploying units in the theater of operations; coordinating support from national/strategic partners; establishing theater personnel accountability and fiscal oversight; and establishing visibility of the distribution network. The sustainment brigade must also consider and integrate the Finance Management Center and the Human Resources Sustainment Center (HRSC), including a Reception, Replacement, Rest and Recuperation, Return to Duty, and Redeployment (R5) element as appropriate.

2. Building the Theater Base

The TSC expands its operational capabilities and capacity to meet the future operational requirements of the JFC. A key consideration is the establishment of interdependencies between the Services and government agencies to minimize competition for limited strategic and theater resources. Critical tasks for the sustainment brigade (TO) include:

- Provide C2 for theater opening capabilities
- Maintain connectivity with national/strategic partners
- Conduct reception, staging, and onward movement operations
- Update/finalize distribution plans to ensure the JFC's concept is supportable

3. Setting Conditions For Follow-On Operations

The TSC sets the conditions for follow-on operations by providing for the integrated support of land component forces throughout the theater of operations. It accomplishes this through the continued synchronization of distribution operations, executed by a network of sustainment brigades. It capitalizes on the inter-dependencies created between the Services and government agencies to achieve greater operational with fewer redundant capabilities.

4. Facilitating the Redeployment Process

The sustainment brigade facilitates the redeployment and the retrograde process through its C2 of the distribution system, maintaining situational awareness of system capacity and leveraging joint capabilities. While many of the procedures used to deploy forces, draw pre-positioned stocks, conduct reception staging and onward movement (RSO), and distribute supplies within the theater of operations apply to the redeployment process, two factors in particular complicate redeployment operations.

The challenge for the sustainment brigade is effective coordination and synchronization, vertically and horizontally, to ensure responsive simultaneous support to not only on going distributed operations, but also redeployment. R5 elements are critical to the coordination and synchronization effort.

Sustainment Brigade's Role in Planning & Managing Operational Contract Support

Ref: FMI 4-93.2, Sustainment Brigade (Feb '09), pp. 1-13 to 1-14. See also p. 1-37.

Contracting is a key source of support for deployed armed forces across full spectrum operations. Because of the importance and unique challenges of operational contract support, the sustainment brigade commander and staff need to fully understand their role in planning for and managing contracted support in the AO. Current doctrine describes three broad types of contracted support: theater support, external support, and systems support.

1. Theater Support Contracts

These contracts support deployed operational forces under prearranged contracts, or contracts awarded from the mission area, by contracting officers under the C2 of the CSB. Theater support contractors are employed to acquire goods, services, and minor construction support, usually from local commercial sources, to meet the immediate needs of operational commanders. Theater support contracts are the type of contract typically associated with contingency contracting. Sustainment brigades will often be the requiring activity for theater support contract support actions related to both internal and external missions. Theater support contracts in support of the sustainment brigade's missions are normally executed through a general support CCT or regional contracting office.

2. External Support Contracts

These types of contracts provide a variety of support to deployed forces. External support contracts may be prearranged contracts or contracts awarded during the contingency itself to support the mission and may include a mix of US citizens, third-country nationals, and local national subcontractor employees. The largest and most commonly used external support contract is LOGCAP. This Army program is commonly used to provide life support, transportation support, and other support functions to deployed Army forces and other elements of the joint force as well. Depending on METT-TC factors, the sustainment brigade may or may not serve as a major requiring activity (the unit responsible to develop and assist in managing specific contract support requirements) for LOGCAP support, but in almost all operations will serve as supported unit relative to LOGCAP support. If designated by the ARFOR and/or TSC commander as the lead requiring activity for any significant LOGCAP task order support actions, the sustainment brigade would normally be augmented by an USAMC logistic support officer (LSO) from Team LOGCAP Forward (TLF).

3. System Support Contracts

System support contracts are pre-arranged contracts by the USAMC LCMC's and separate ASA(ALT) PEO and PM offices. Supported systems include, but are not limited to, newly fielded weapon systems, C2 infrastructure, such as the Army Battle Command Systems (ABCS) and standard Army management information system (STAMIS), and communications equipment. System contractors, made up mostly of US citizens, provide support in garrison and may deploy with the force to both training and real world operations. They may provide either temporary support during the initial fielding of a system, called interim contracted support or long-term support for selected materiel systems, often referred to as contractor logistic support. The sustainment brigade does not normally have a significant role to play in planning for or coordinating system support contracts other than coordinating and executing support of system support contract related personnel.

For the sustainment brigade, the major challenge is ensuring theater support and external contract support (primarily LOGCAP related support) actions are properly incorporated and synchronized with the overall sustainment brigade support effort. It is imperative that the sustainment brigade SPO, with or without LSO augmentation, closely work with the TSC/ESC SPO, the ARFOR G-4, the CSB, and the supporting TLF. It is also important to understand the sustainment brigades do not have any dedicated contingency contracting officers on their staff and this support will be provided on a GS basis through the supporting CSB. Because of these new modular force contracting support arrangements, it is imperative for the sustainment brigade staff be trained on their role in the operational contract support planning and execution process as described below:

4. Contract Planning

The sustainment brigade must be prepared to develop "acquisition ready" requirement packets for submission to the supporting contracting activity. The packets must include a detailed performance work statement (PWS) (previously referred to as a statement of work or SOW) for service requirements or detailed item description(s)/capability for a commodity requirement. In addition to the PWS, packets must include a Letter of Justification (LOJ) or Letter of Technical Direction (LOTD) for service requirements. Depending upon command policies, certain items or specific dollar amount requests may require formal acquisition review board packet review.

5. Contract Management

The sustainment brigade plays a key role in theater support contract and LOGCAP task order management. One of the most important sustainment brigade tasks in this process is to nominate and track contract officer representatives (COR's) (sometimes referred to as contract officer technical representatives or COTR's) for every service contract and LOGCAP task order as directed. It will also nominate a receiving official for all supply contracts. Quality COR and receiving official support is key to ensuring contractors provide the service or item in accordance with the contract. The sustainment brigade must also manage funding for each contract and request funds in advance of depletion of current funds or all contract work will stop until adequate funds are available. Finally, in some operations, the sustainment brigade will be required to provide formal input to LOGCAP award fee and performance evaluation boards.

6. Contract Close Out

The sustainment brigade is responsible for completing receiving reports: certifying that the contracted goods or services were received by the Army. The contracting officer shall receive a copy of the receiving report from the sustainment brigade so the contract can be closed out and the contractor can be paid.

In some major operations, the operational contract support tasks discussed above can be a major challenge for a sustainment brigade. When faced with major operational contract support management tasks, it is imperative the sustainment brigade commander organize some type of contract management cell within the S4 and/or SPO shop to ensure these tasks are properly accomplished. In long-term operations, as seen in OIF and OEF, the sustainment brigade will need to ensure direct coordination and transfer of operational contract support related information prior to relief in place/transfer of authority (RIP/TOA). Additionally, when the advance party arrives in the AO, it is essential specifically designated unit personnel actively seek out current information on local contract support capabilities, policies, and procedures. This individual must be prepared to coordinate the formal hand over of existing contract management responsibilities from the redeploying unit. It is critical to know when recurring service contracts will be ending, because it generally takes 30 to 60 days to obtain funding approval. If the unit waits until the contract is about to expire before additional funds are requested, the sustainment brigade could lose the contracted service until funds are available.

The GCC controls and is accountable for military operations within a specified area of responsibility. The commander defines the parameters, requests the right capabilities, tasks supporting DOD components, coordinates with the appropriate Federal agencies, and develops a plan to achieve the common goal. Sustainment brigades may be responsible for providing sustainment to other government agencies within their assigned AO's. During the early stages of military operations in austere environments or in small operations, the sustainment brigade may be the only organization capable of providing supplies, services, and life support to other Federal elements.

Nongovernmental organizations (NGOs) do not operate within either the military or the governmental hierarchy. Their relationship with the Armed Forces is neither supported nor supporting. An associate or partnership relationship exists between military forces and engaged NGOs. If formed, the focal point where US military forces provide coordinated support to NGOs would be the civil-military operations center of a JTF HQ.

The sustainment brigade might be required to support civil operations. The commander and staff must understand roles and responsibilities in such operations. The Army, defines the Army's role during civil support operations. Civil support operations address the consequences of manmade or natural accidents and incidents beyond the capabilities of civilian authorities. Under US law, the federal and state governments are responsible for those tasks normally associated with stability and reconstruction operations. Army forces conduct civil support operations when requested, providing expertise and capabilities to lead agency authorities. During civil support operations, the Army performs a supporting role and is called upon regularly to provide assistance to civil authorities to deal with natural disasters (such as, hurricanes, floods, and fires), as well as manmade incidents (such as, riots and drug trafficking).

B. Host Nation Support (HNS)

The sustainment brigade will coordinate for HNS (negotiated by the US Department of State) or for the contracting of resources and materiel as required in support of its mission. HNS and local procurement may provide a full range of sustainment, operational, and tactical support. HNS agreements fulfilling the command requirements for support need to be pre-negotiated. HNS may include functional or area support and use of host nation facilities, government agencies, civilians, or military units. Pre-established arrangements for HNS can reduce the requirement for early deployment of US assets and can offset requirements for early strategic lift by reducing requirements for moving resources to the theater of operations.

C. Multinational Support

Multinational support may consist of CUL support provided from one multinational partner to another. One or more of the following organizational and/or management options facilitates multinational support:

- National support elements provide national support
- Individual acquisition and cross-servicing agreements provide limited support
- A lead nation provides specific support to other contributing nation military forces
- A role-specialist nation provides a specific common supply item or service
- A multinational integrated logistics unit provides limited common supply and service support
- A multinational joint logistics center manages CUL support

In all cases, the multinational force commander directs specific multinational CUL support within the applicable laws and regulations of the HN. When operating within a formal alliance, the sustainment brigade executes CUL support in accordance with applicable standardization agreements or quadripartite standardization agreements.

Chap 2

II. Sustainment Brigade Mission & Tasks

Ref: FMI 4-93.2, Sustainment Brigade (Feb '09), chap. 2.

The sustainment brigades are subordinate commands of the TSC, designed to provide C2 of theater opening, distribution, and sustainment within an assigned area of operation. The level of assignment and mission assigned to the sustainment brigade determine the mix of functional and multifunctional subordinate battalions under the control of the brigade. The brigades provide C2 and staff supervision of replenishment, life support activities, human resources and financial management support, and distribution management.

Missions and Tasks

The major missions performed by the sustainment brigades are theater opening (TO), theater distribution (TD), and sustainment. These missions are interrelated and, throughout the course of an operation, a sustainment brigade will likely perform more than one of these functions simultaneously.

The Sustainment Brigade Mission

Plans, synchronizes, monitors, and executes logistics operations. Conducts Theater Opening and/or Theater Distribution operations when directed. Provide support to joint, interagency, and multifunctional forces as directed.

Sustainment

Theater Opening (TO)

Theater Distribution (TD)

The sustainment brigade is a flexible, tailorable organization. All sustainment brigade HQ are identical in organizational structure and capabilities. Organic to the sustainment brigade are the brigade HQ and a special troops battalion (STB). The core competency of the sustainment brigade is C2 of sustainment operations, providing C2 and staff supervision of life support activities and distribution management to include movement control as an integral component of the theater distribution system. During the ARFORGEN process, functional and multifunctional subordinate battalions are task organized to the sustainment brigade to enable it to accomplish its role and missions.

Sustainment Brigade Operational Capabilities

Ref: FMI 4-93.2, Sustainment Brigade (Feb '09), pp. 2-1 to 2-3.

The sustainment brigade, attached to an ESC/TSC, provides command and control for all subordinate units, and provides sustainment in an area of operations as defined by the ESC/TSC. The sustainment brigade plans and executes sustainment, distribution, theater opening and reception, staging, and onward movement of Army forces in full spectrum operations as directed by the ESC/TSC.

Capabilities Augmentation Possibilities

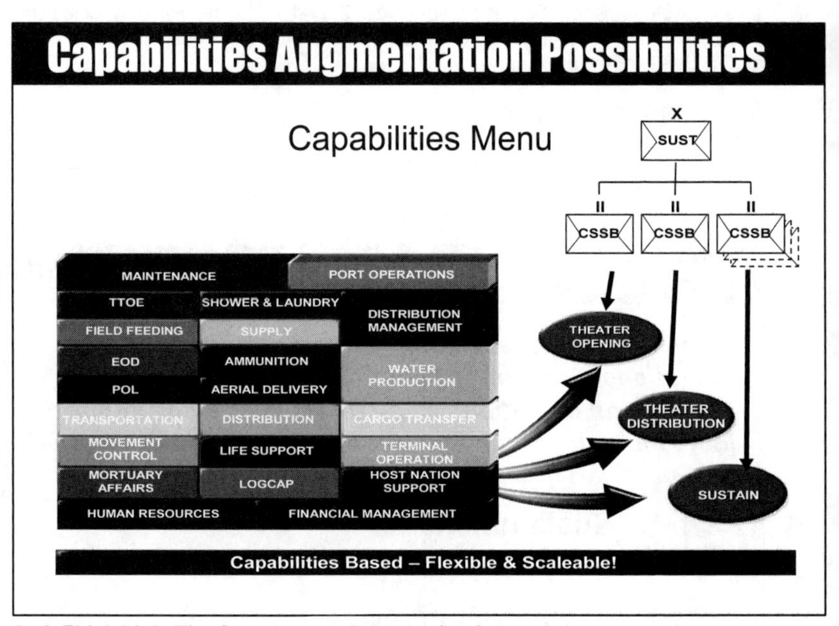

Ref: FM 4-93.2, The Sustainment Brigade, fig. 2-2, p. 2-3.

All sustainment brigades have the same general capability to manage theater opening, theater distribution, and sustainment operations. Each sustainment brigade is a multifunctional organization, tailored and task organized to provide support for multiple brigade-sized or smaller units using its' subordinate battalions, companies, platoons, and teams to perform specific sustainment functions. In the sustainment role, the brigade is primarily concerned with the continuous management and distribution of stocks, human resources support, execution of financial management support, and allocation of maintenance in the AO to provide operational reach to maneuver commanders.

The sustainment brigade management tasks are coordinating and integrating personnel, equipment, supplies, facilities, communications, and procedures to support the maneuver commander's intent. The sustainment brigade may require augmentation in those areas where it lacks staff expertise and/or functional support capabilities. For example, the TSC commander may augment the sustainment brigade with transportation units to enable it to oversee and execute port clearance and terminal operations if the sustainment brigade is given the theater-opening mission. Likewise a sustainment brigade may serve as the senior joint logistics HQ in an AO when provided augmentation commensurate to the mission. Under the modular construct, organizational designs incorporate multifunctional, self-reliant sustainment capabilities within the BCT and BSB, providing much greater self-reliance at this echelon. Each BCT, for example, carries three combat loads on organic transportation assets minimizing the need for external support, which is most likely provided by a sustainment brigade during replenishment operations. Therefore, if the sustainment brigade is supporting units which do not have this robust capability whether they are US Army units which have not transformed or are multi-national forces, the sustainment brigade and subordinates will need to be tailored to provide greater support.

During periods where only Special Operations Forces are operating in a theater, support operations may be executed under the C2 of the Sustainment Brigade (Special Operations) (Airborne)(SB [SO][A]), which has a modification table of organization and equipment-deployable organization and is assigned to USASOC. When deployed, the SB (SO)(A) acts as the logistics HQ for a joint special operations task force (JSOTF).

Sustainment Brigade Operational Capabilities

- Provides supplies, field services, field and selected sustainment level maintenance, recovery, and field feeding for itself and its assigned subordinates
- Plans and conducts base and base cluster self-defense. Defends against level I threats, assists in destruction of level II threats and escapes or evades against level III threats.
- Capable of operating as part of an Army or joint force
- Coordinates host nation support (HNS) established by the Department of State
- Capable of deploying an advance party to support early entry operations
- Provides theater opening, theater distribution, and sustainment management information and advice to commanders and staff within its AO and the TSC.
- Exercises technical supervision over operations for all assigned units
- Combat service support automation management office (CSSAMO) provides logistics STAMIS management, plans, policies, and procedures for logistics automations functions/systems to the sustainment brigade, area support and backup support to the BSB CSSAMO's in the support footprint
- Provides limited materiel management for internal stocks, Class I, II (including unclassified map stocks), Class III and water, Classes IV, V, VII, and IX and maintenance management of internal assets
- Provides a liaison team to augment other HQs as necessary
- Manages and maintains the sustainment brigade property records
- Maintains data in support of the Army equipment status reporting database and the Army equipment status reporting system
- Provides human resources, financial management, legal services, and religious support to assigned units and authorized personnel within the SB's AOR
- Appoints contracting officer representatives (COR) to monitor contractor performance, certify receipt of services, and act as liaisons

I. Sustainment Mission

Sustainment is the provision of the logistics, personnel services, and health service support necessary to maintain operations until mission accomplishment. The endurance of Army forces is primarily a function of their sustainment. Sustainment determines the depth to which Army forces can conduct decisive operations, allowing the commander to seize, retain, and exploit the initiative. Endurance is the ability to employ combat power anywhere for protracted periods. Endurance stems from the ability to generate, protect, and sustain a force. It involves anticipating requirements and making the most effective, efficient use of available resources. Sustainment also enables strategic and operational reach. Army forces require strategic and operational reach to deploy and immediately conduct operations anywhere with little or no advanced notice.

The sustainment warfighting function consists of three sub-functions: logistics, personnel services, and health service support. *See pp. 1-2 to 1-3.*

The Sustainment Warfighting Function

The Sustainment WFF is related tasks and systems that provide support and services to ensure freedom of action, extend operational reach, and prolong endurance (FM 3-0):

Logistics

Personnel Services

Health Services Support (HSS)

Ref: FM 4-0, Sustainment, chap. 1.

See pp. 2-31 to 2-62 for discussion of the logistics sub-function.

See pp. 2-63 to 2-66 for discussion of the personnel services sub-function.

See pp. 2-67 to 2-70 for discussion of the health service support sub-function.

Critical Tasks

The sustainment brigade performs sustainment operations from the operational to tactical levels. Operational sustainment focuses on theater operations that involve force deployment/reception, movement distribution, sustainment, and reconstitution. The initial focus is on generating a force ready to conduct operations. Support begins during force generation but becomes the primary focus once operations begin. Key Army functions associated with operational level sustainment include the following:

- Coordinating supply of arms, munitions, and equipment
- Synchronizing supply and distribution of fuel and water

Replenishment Operations (RO)

Ref: FMI 4-93.2, Sustainment Brigade (Feb '09), pp. 2-10 to 2-11.

Replenishment operations (RO's) are preplanned operations that allow combat forces to replenish routinely. An RO is a deliberate, time sensitive operation to replace used stocks within a BCT or support brigade. These operations, which may be, but are not normally, augmented with assets from the sustainment brigade, are quick and in-stride with the supported commander's battle rhythm.

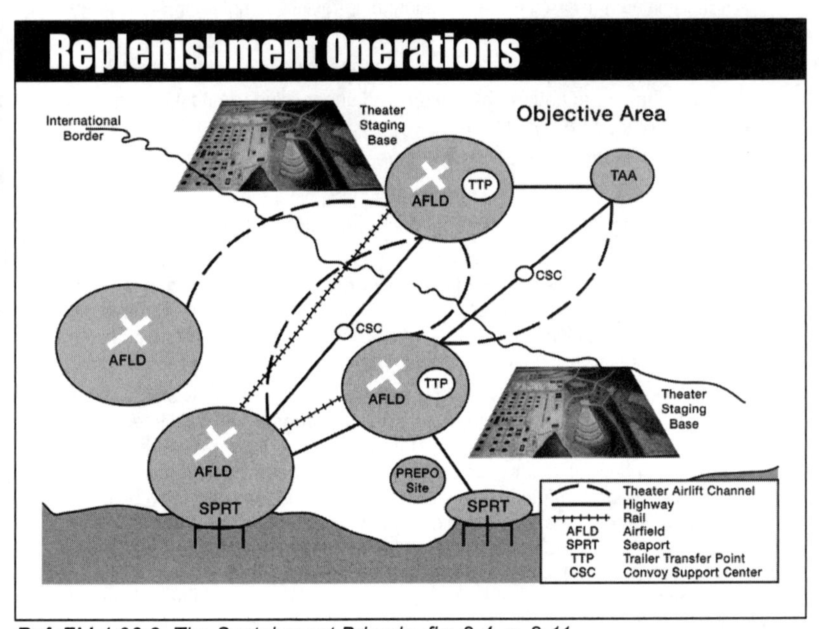

Ref: FM 4-93.2, The Sustainment Brigade, fig. 2-4, p. 2-11.

The purpose of the RO is to replace stocks used by a brigade. It may be either deliberate or hasty if circumstances allow. Typical activities that take place during the RO include rearming, refueling, maintaining, medical support, and essential personnel replacement to meet immediate needs. The BSB conducts RO to its FSC's and the FSC's also conduct RO to the combat loads of individual Soldiers and weapons platforms. The BSB will need to coordinate the timing of the RO with the sustainment brigade to ensure that the delivery from the sustainment brigade supports both the required stockade levels and the timing of its RO.

- Ensuring effective cross-leveling of supplies and efficient retrograde and redeployment of equipment, personnel, and supplies
- Maintaining equipment and stocks that support the supply system
- Coordinating support of forces, to include: personnel, human resources, supply, equipment, field services, health, religious support, financial management, and legal services
- Managing materiel, controlling movement, and managing distribution
- Providing lead service CUL to other services, multinational partners, and civilian agencies on order
- Establishing, managing, and maintaining facilities, including storage areas and maintenance areas
- Providing direct support field maintenance to units in its assigned AO
- Planning, coordinating, managing, and supervising the positioning and security of sustainment activities
- Coordinate through CSB for contingency contracting
- Maintaining visibility of customer locations
- Maintaining information regarding support relationships

The sustainment brigades are assigned multifunctional battalions and functional battalions and companies tailored and task organized to the specific mission. The sustainment brigade will normally have multiple CSSB's assigned to provide distribution and supplies to BCTs and supporting brigades operating within its assigned AO and other forces operating in or transiting its AO.

These supported brigades may be in one division or multiple divisions. The sustainment brigade (or logistics task force in a joint environment) establishes a base(s) within the AO to conduct operations. The MEB may be responsible for the terrain assignment and establishing secure movement corridors. The sustainment brigade base will be integrated into area terrain management and protection plans based on established C2 relationships and the physical space occupied.

Sustainment at the Operational Level

Key elements of the Army support structure at the operational level include APS, dedicated transportation, general support supply, sustainment maintenance, and human resources elements. Many of the stocks to support the AO are stored by Army logistics units, allowing tactical-level logistics units to remain mobile. Support at the operational level includes CUL support to joint and multinational forces, as required. Many different sources contribute to these support functions, including contractors, DA and DOD civilians, US Army and joint services, and multinational military organizations, and host nation resources.

Sustainment at the Tactical Level

Sustainment at the tactical level encompasses those activities that maintain and supply forces. The two ways that sustainment brigades provide support for a deployed Army force at this level are:

- The organic support battalions and Forward Support Companies (FSC's) of the BCT's and other brigades
- Replenishment furnished to EAB organizations on an area basis

See previous page (p. 2-17) for discussion of replenishment operations.

II. Theater Opening (TO) Mission

Theater opening (TO) is the ability to rapidly establish and initially operate ports of debarkation (air, sea, and rail) to establish sustainment bases and to facilitate port throughput for the reception, staging, and onward movement of forces within a theater of operations. Although port operations are a critical component of the theater opening function, theater opening is comprised of much more: communications, intelligence, civil-military operations, services, human resources, financial management, force health protection, engineering, movement (air/land/water transport, inland terminal operations), materiel management, maintenance, and contracting. A sustainment brigade will be one of the first organizations into a theater of operations.

When given the mission to conduct theater opening, the sustainment brigade is designated a Sustainment Brigade (TO) and a mix of functional battalions and multifunctional CSSB's are assigned based on mission requirements. The sustainment brigade HQ staff may be augmented with a Transportation Theater Opening Element (TTOE) to assist in managing the theater-opening mission. The augmentation element provides the sustainment brigade with additional manpower and expertise to C2 TO functions, to conduct transportation planning and provides additional staff management capability for oversight of reception, staging, onward movement, and integration (RSOI) operations, port operations, node and mode management, intermodal operations, and movement control.

Unity of Effort

Conducting efficient and effective theater opening operations requires unity of effort among the various commands and a seamless strategic-to-tactical interface. Theater opening is a complex joint process involving the GCC; strategic and joint partners such as USTRANSCOM, USAMC, and DLA. Also critical to the theater opening effort is the JDDOC which has the mission to improve in-transit visibility and to support the geographic combatant commander's operational objectives by synchronizing and optimizing the interface of inter-theater and intratheater distribution to integrate the proper flow of forces, equipment, and supplies. In coordination with the supporting AFSB and CSB CDR or principal assistant responsible for contracting (PARC), the sustainment brigade will participate in assessing and acquiring available host nation (HN) infrastructure capabilities and contracted support.

Critical Tasks

The critical tasks for theater opening include: C2, reach and in-transit visibility; transportation management; theater RSOI; distribution and distribution management; movement control and movement Missions and Organizations management; life support; contracting support; and initial theater sustainment. Given the mission of theater opening, a sustainment brigade, together with the TTOE, should have capabilities to conduct the following:

- Establishing the theater base and infrastructure necessary for the joint force to expand to fully developed theater distribution and support operations
- Establishing and managing initial theater distribution operations. This includes distribution management, movement control and materiel management, and both surface and aerial re-supply operations.
- Soldier support, to include all life support services needed to support theater-opening operations
- Establish the Military Mail Terminal (MMT), Theater Gateway Reception, Replacement, Return To Duty, Rest And Recuperation, and Redeployment (TG R5) Personnel Processing Center (PPC), and initial Casualty Assistance Center during theater opening operations prior to the beginning of personnel flow, if tactically feasible
- Financial management

- Establishing the required elements of the Army distribution system
- Establishing and operating staging areas and/or bases
- Establishing and operating supply support areas for staging sustainment stocks
- Clearing incoming personnel and cargo at APOD's and seaports of debarkation (SPOD's), working with USTRANSCOM elements, SDDC and AMC
- Designating and preparing routes for onward movement (in conjunction with the respective engineer coordinator)
- Coordinating movement within the theater of operations with the Movement Control Battalion (MCB)
- Identifying and occupying the real estate needed for marshaling areas and the theater staging bases
- Establish initial financial management support operations in order to provide commercial vendor services support, banking and currency support, paying agent support, and limited pay support
- Establishing C2 links to provide in-theater connectivity among tactical, joint, and strategic support systems for integrating initial Army force deployments
- Planning operational support area RSOI nodes with the ARFOR staff
- Assist in developing and executing the Army portions of the joint movement program developed by the Joint Movement Center
- Establishing connectivity with Global Transportation Network that receives data from the Worldwide Port System (WPS) and Consolidated Aerial Port System to monitor the movement of forces and cargo, as well as the movement of military and commercial airlift, sealift, and surface assets
- Establishing and conducting rail operations and Army terminal operations at SPOD's and APOD's, to include forming arrival/departure airfield control groups
- Establish and manage Joint Logistics Over-the-Shore (JLOTS) operations
- Air terminal operations include movement control, cargo transfer, unit and cargo marshaling, cargo documentation, and port clearance
- Establishing and operating initial Army force provider facilities and arranging for required commercial or HNS-operated theater-staging bases
- Providing equipment de-processing and property transfers for pre-positioned unit equipment
- Establishing areas for staging of Army Pre-positioned Stocks (APS) materiel, transitioning personnel, and supplies
- Establishing and operating in-theater force and materiel tracking systems
- Validating air, sea, rail, and highway deployment rates for the force
- Establishing movement management activities and conducting movement control operations
- Coordinating port clearance and inland theater mvmt of forces and materiel

When circumstances warrant, the SB (SO)(A) may be tasked to oversee early entry theater opening in an operation expanding from SOF to conventional forces. When assigned, the SB (SO)(A) is usually OPCON to the TSC. When assigned tailored CSSB's and functional sustainment companies, the SB (SO)(A) can—

- Receive forces and provide C2 of theater opening, theater distribution, signal, financial management, personnel, ammunition, transportation, maintenance, supply and services, human resources, and religious and other logistics elements
- Support Army special operations task force early entry operations under C2 of the TSC until relieved by a conventional sustainment brigade HQ
- Establish and manage initial theater-opening operations, to include RSO functions and the establishment of the theater base
- Establish and manage initial theater distribution operations

Transportation Theater Opening Element (TTOE)

Ref: FMI 4-93.2, Sustainment Brigade (Feb '09), pp. 2-10 to 2-11.

The TTOE is assigned to a TSC and attached to a sustainment brigade (TO). The element provides command, control, administration, logistics, and supervision of the operating elements of the unit in the performance of mission tasks. The TTOE provides staff augmentation to the sustainment brigade (TO). It augments a sustainment brigade support operations section by giving it the capabilities required to provide staff oversight of select TO operations. This includes establishing the initial distribution network and providing support to assigned customers; conducting minimum essential early entry operations prior to employment of full theater opening capabilities; and C2 of employed units. Once expansion of the theater is largely complete or as conditions warrant, the TTOE will likely move to augment the ESC/TSC to manage the theater-wide movements and transportation mission. When the TTOE is resident with the sustainment brigade and the ESC/TSC has not yet arrived, the TTOE will have the mission to coordinate with all organizations conducting distribution functions impacting the JOA theater-opening operations, regardless of whether the function occurs at the strategic, operational, or tactical levels. The TTOE will remain with the sustainment brigade, providing theater-level transportation staff augmentation, when the sustainment brigade is the senior sustainment HQ in the JOA.

The TTOE, which by design integrates into the brigade SPO section, includes the following:

1. Terminal Operations Branch

Advises on the use and implementation of assigned, attached, contracted, and HN terminal and water craft operations; provides terminal infrastructure assessment; monitors and coordinates operation and positioning of terminal operations, including motor, rail, inter-modal, air, and sea; monitors and maintains status of terminal assets to ensure proper employment and appropriate tasking; and provides advice and expertise to the plans division in matters concerning terminal operations

2. Transportation Branch

Advises on the use and implementation of assigned, attached, contracted, and HN motor transport assets; provides guidance on positioning of motor transport, air, and rail assets; monitors and maintains the status of all modal transportation assets in the AO Missions and Organizations and ensures proper tasking; and provides advice and expertise to the plans division in matters concerning mode transportation operations

3. Movements Branch

Implements and monitors movements programs and commits transportation assets in support of RSOI operations; maintains operational status; provides information and guidance on transportation operations to subordinate groups and battalions; maintains ITV; conducts transportation planning; plans support for contingency operations; and conducts exceptional movement requirements; coordinates the evacuation of civilian refugees and US civilians with proper authority; provides the senior sustainment HQ the required personnel to conduct theater-level (Executive Agent) movement control/management.

III. Theater Distribution (TD) Mission

Theater distribution (TD) is a decisive element of multifunctional support operations that include the following:

- Air, land, and sea operations
- Management of materiel
- Management of assets
- Developing requirements and priorities
- Synchronization with the capability to perform retrograde functions critical to the repair of vehicles, equipment, weapons and components

Critical Tasks

Critical tasks include the following:

- Synchronizing multi-nodal, multi-modal distribution operations across an asymmetric operational environment in support of JFC requirements
- Maintaining visibility of the distribution system
- Performing distribution management

The sustainment brigade performing TD operations will coordinate with the JDDOC, the ESC/TSC DMC, MCB/movement control teams (MCT), the J-4 of the Joint Force Land Component Commander, and the logistics staffs or SPO's of supported organizations. The theater JDDOC supports theater distribution by ensuring end-to-end visibility, managing strategic transportation assets, and synchronizing priorities.

The sustainment brigade (TD) is assigned and operates the ground transportation assets and manages selected aerial re-supply assets, as well as the theater ground distribution network (nodes, rest halts, and distribution hubs) from the theater base distribution hub to other sustainment brigades, the BSB's, or to Centralized Receiving and Shipping Points (CRSP's). It is designed to provide C2 to assigned and attached units for the purpose of conducting distribution operations in the AO. Distribution operations include: receive, store, issue, distribute, trans-load, configure, reconfigure, classify, and collect stocks and unit equipment. It also includes the reception and transportation of units and replacement personnel.

When task organized to provide theater distribution the sustainment brigade may have capabilities, which include:

- Configuring/reconfiguring loads as required. Distributing to and retrograding from BCTs, other brigades, and other forces operating in the AO (if directed)
- Storing bulk supplies and authorized stockage list (ASL) items for distribution and internal consumption
- Managing and integrating surface distribution, ILAR assets (for example Air Land, Airdrop, Helicopter Sling Load), and rail operations
- As directed by the TSC, providing the distribution of all supplies and services for which the sustainment brigade is responsible
- Planning and controlling the use of surface transport for missions within assigned AO
- Organizing the movements of subordinate units within its AO, which requires coordination with the supported maneuver elements concerning current and proposed locations and movement of units
- Providing guidance and assistance to units in the AO on matters relating to airdrop
- Providing staff supervision of technical training for personnel regarding the rigging and loading of supplies and equipment for airdrop and aerial re-supply

- In coordination with the TSC, integrating joint and commercial distribution capabilities
- Delivering supplies, materiel, equipment, and personnel over the theater ground distribution network from theater base to BCT's and forward distribution points as required
- Maintaining surveillance over the theater ground distribution network
- Operating forward distribution points to receive, store, issue, configure, and reconfigure materiel
- Conducting retrograde, redirection, frustrated cargo, and redistribution operations
- Establishing/maintaining total asset visibility/in-transit visibility TAV/ITV over commodities, equipment, personnel, units, and ground assets flowing in the distribution network to include what is inbound from the strategic base
- Integrating the battlefield distribution information network
- Executing the TSC's theater distribution plan
- Operating regional distribution hubs, to include CRSPs
- Synchronizing movements with the MEB through secured mobility corridors
- Leveraging the available distribution infrastructure and optimizing pipeline flow to meet requirements and priorities
- Projecting distribution pipeline volume, flow rates, contents, and associated node and port requirements Adjusting pipeline flow and responding to changing operational requirements
- Monitoring RSOI in order to integrate and prioritize unit moves and sustainment moves
- Monitoring distribution terminal operations and the flow of multi-consignee shipments
- Synchronizing reception of Army resources with theater movement control operations
- Providing advice and recommended changes to the distribution system to the TSC commander, JFC distribution managers, or HN/contracted providers
- Maintaining visibility of the physical, resource, communications, and automation networks within the assigned AO
- Identifying capacity problem areas and actions to take within the distribution system
- Managing and controlling the distribution pipeline flow through anticipatory support and the synchronization of materiel management and movement control
- Distribution planning
- Establish Convoy Support Centers (CSC) at the direction of the TSC Missions and Organizations

See following pages (pp. 2-24 to 2-25) for discussion of distribution and materiel management.

Distribution and Materiel Management

Ref: FMI 4-93.2, Sustainment Brigade (Feb '09), pp. 2-30 to 2-36. See also 1-42.

Distribution management is the process of planning and synchronizing the time definite delivery of materiel, equipment, units, personnel, and services to, within, and from the AO. Distribution management involves the fusion of information derived from a number of processes: commander's oversight, Army Battle Command Systems, physical distribution, and materiel management. The premise of the distribution operations in the current force is to reduce the time it takes the right supplies (both demand supported and bulk) to travel from the source of supply to the point of need. The transformed distribution management system will eliminate reliance on stockpiles and static inventories located forward at each echelon, which was a characteristic of the old Army of Excellence supply-based system. Distribution substitutes speed for mass, makes use of a COP providing situational understanding, and ensures efficiency of delivery systems, while ensuring visibility of assets in the pipeline. In essence, the distribution system becomes the "warehouse," representing "inventory in motion" reducing both the organizational and materiel footprint within the AO. Logisticians control the destination, speed, and volume of the distribution system. The key elements of distribution management (C2, physical distribution, and materiel management) are further amplified below. The commander (in concert with guidance from superior HQ) provides the priorities and mission plan for units. The commander's guidance, in conjunction with the readiness shortfalls for that unit, determines the priorities for materiel delivery and fulfillment. Included in physical distribution are not only the vehicles themselves, but also the management of movement, routing, ITV assets, facilities, and material handling equipment.

Command and Control of Distribution

The Army conducts distribution management at all levels from strategic to tactical. One of the key components of the modular force concept is to have centralized C2 of units at EAB. As such, the TSC is the central distribution manager for the theater of operations. The sustainment brigade is responsible for managing distribution within its assigned AO by balancing the existing capabilities of the distribution infrastructure with the day-to-day and projected operational requirements. The Sustainment brigade issues distribution directives to CSSB's, which in turn issues directives to transportation companies for execution. The Sustainment brigade and CSSB's command all sustainment forces inside their assigned areas. The BSB is not under the command of the TSC, or Sustainment brigade, but rather is their supported unit. The BSB SPO issues distribution directives to the Distribution Company to replenish the FSC's.

Effective distribution management applies the principles of managing distribution centrally, optimizing infrastructure, minimizing stockpiles, maximizing throughput, and maintaining a seamless pipeline. The TSC's role in distribution management is in the development of the theater's distribution. The TSC's DMC coordinates and monitors the strategic distribution flow with USTRANSCOM. The DMC collects, analyzes, and monitors ITV distribution flow and executes changes in the distribution priorities established by ASCC G-4. The ESC performs the same function except it is confined to its theater of operation.

The sustainment brigade collects and analyzes ITV distribution information to monitor routes and locations of its convoys. This assists in movement control for convoy protection through a unit's AO. The Sustainment brigade also uses the ITV to establish delivery schedules to its CSSB's in support of the TSC and/or its supported command's priority of supply and effort. The BSB focuses on delivering timely, dependable, accurate, and consistent support to the BCT through the FSC's. It monitors and tracks any inbound Sustainment brigade convoys to synchronize protection issues. The SPO synchronizes and establishes delivery schedules to the FSC's through the BSB's Distribution Company.

Physical Distribution

Physical distribution is defined as the facilities, installations, platforms, and packaging needed to physically store, maintain, move, and control the flow of military materiel, personnel, and equipment between the point of receipt into the military system and the point of issue to using activities and units; including retrograde activities. The Sustainment brigade manages distribution nodes in its assigned AO. The Sustainment brigade and CSSB track and maintain visibility of assets (ground and aerial platforms) that are available for distribution. The CSSB maintains visibility of its capacity to store commodities as another aspect of physical distribution. The level of physical distribution increases from the TSC to the BSB while the level of distribution management decreases. The TSC and ESC's roles in physical distribution are minimal. However, visibility is still maintained over theater distribution assets within the distribution network. The ESC maintains visibility of theater of operations distribution assets. The TSC and ESC can direct cross leveling of distribution resources to meet tactical requirements to optimize the distribution flow.

Materiel Management

Materiel management is the supervision and management of supplies and equipment throughout the strategic, operational, and tactical level areas of operation that includes cataloging, requirements determination, procurement, overhaul, and disposal of materiel. Materiel management is the monitoring and control of on-hand stocks, ensuring quality control, requirements determination, local purchase, retrograde, and distribution of materiel. It also provides visibility to the distribution management effort of assets in stationary stocks at all echelons. A portion of materiel management is maintenance management, which provides oversight of parts requirements and also projections of parts availability (return of re-parables, for instance).

Organizational Roles in material management include:

- **Army Sustainment Command (ASC)**. This is the single Army national materiel manager for units stationed in the CONUS. ASC is a subordinate unit of the Army Materiel Command. It provides continuous equipment and materiel readiness to CONUS forces through effective planning, resourcing, and materiel and distribution management in accordance with the Army Force Generation (ARFORGEN) process. It achieves this by synchronizing strategic with operational and tactical logistics and by integrating acquisition, logistics, and technology.
- **Theater Sustainment Command (TSC)**. The TSC manages materiel for all Army forces assigned or deployed within the assigned region and, as appropriate, for joint, multinational, and international forces. TSC managers are linked with the G–4s in their areas of operations for resource prioritization. The TSC also coordinates with the AMC Field Support Brigade Commander to support national-level system and materiel requirements.
- **Expeditionary Sustainment Command (ESC)**. The ESC synchronizes the AO distribution systems and provides distribution oversight. The ESC can assist in tracking where requests are in the supply system and coordinates distribution assets when appropriate to redirect essential items based on the priority of support and the division or corps commander's priorities.
- **Sustainment Brigades**. Sustainment brigades execute the materiel management and distribution guidance from the TSC or ESC (from ASC for those sustainment brigades stationed in CONUS when not deployed). When the ESC is deployed, the command relationship with the ESC enables the TSC to issue directives to redistribute and surge logistics capabilities across the theater of operations to fulfill requirements as needed. The sustainment brigade SPO interfaces with the TSC or ESC (or ASC) materiel managers for asset management, visibility, and distribution to support the division or any other assigned customer units.

Operational Distances

Ref: FMI 4-93.2, Sustainment Brigade (Feb '09), pp. 2-11 to 2-12.

As a general guideline, in order to prevent overreach of units in tactical environments, the recommended distance between a Sustainment Brigade and the BSB's it supports should be from 60 to no more than 175 KM. The 175 KM limit reflects one line haul trip a day (max 222 KM-20% = 177 KM), and is constrained by fuel consumption of the distribution platform(s). The lower distance of 60 KM reflects line haul in rough terrain (6 hr x 10 KM/hr). For both, the assumption is that the longest time a driver can continuously and safely drive in a shift is 6 hours (one way trip). There is also an assumption of two drivers per vehicle. Ideally, the BSB's should be from 30 KM to 45 KM from combat operations and the FSC's should be from 4 KM to 15 KM from combat.

Operational Distances

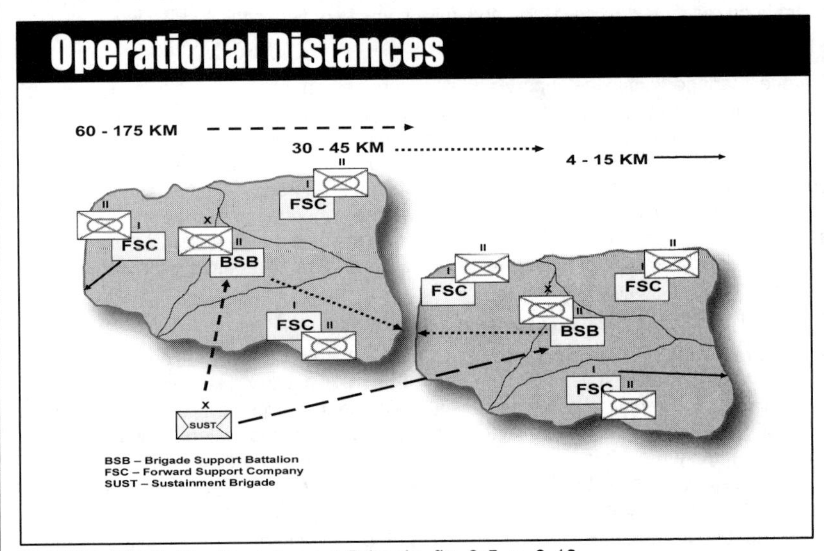

Ref: FM 4-93.2, The Sustainment Brigade, fig. 2-5, p. 2-12.

The following considerations should be used in determining operational distances:

- Sustainment Brigade's will locate near major transportation nodes (airfields, rail heads, inland, water ports)
- There are no CSC's inside a division's assigned area of operation
- Line haul convoys will not normally be refueled by BSB's
- Distances are constrained by the vehicle with the heaviest fuel consumption
- Distribution platforms should return with a 20 percent capacity fuel safety margin
- The longest time a driver can continuously and safely drive in a shift is 6 hours
- Critical items are distributed via throughput (normally by air)
- Sustainment Brigade's schedule of line haul replenishments to BSB's
- Poor roads reduce travel time by half from that of good roads
- Night driving reduces travel time by half from day driving

III. Organization and Support to the Warfighter

Ref: FMI 4-93.2, Sustainment Brigade (Feb '09), chap. 4.

In full spectrum operations, every unit—regardless of type—either generates or maintains combat power. Generating and maintaining combat power throughout an operation is essential to success. Commanders must have a thorough understanding of the sustainment warfighting function and the sustainment assets available in order to properly stage those assets in a way that preserves momentum.

This section discusses the sustainment warfighting function and sub-functions and describes the elements within the sustainment brigade that perform the functions. The intent is to provide and understanding of what must be done and what assets are available to provide support to the operating forces.

I. Sustainment Brigade Organizational Options

The sustainment brigade is designed to be a flexible organization that is task organized to meet mission requirements. The sustainment brigade has a command and staff structure capable of providing the full range of sustainment to the operational or the tactical level. It does not have the organic capacity to execute its assigned mission without the assignment of subordinate support units. The sustainment brigade is augmented by a number of different types of tailored organizations. The types and numbers of these organizations depend on the mission and the number, size, and type of organizations the sustainment brigade must support.

A Notional Sustainment Brigade

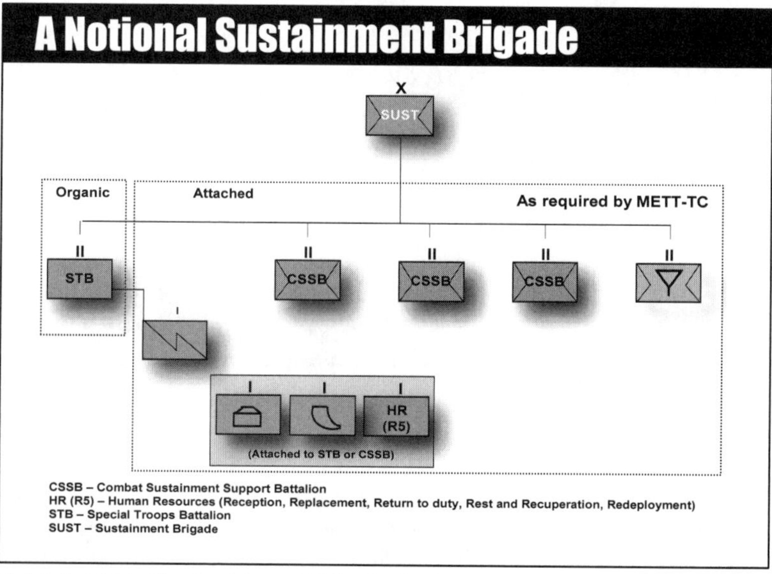

Ref: FM 4-93.2, The Sustainment Brigade, fig. 2-1, p. 2-2.

II. The Combat Sustainment Support Battalion (CSSB)

Ref: FMI 4-93.2, Sustainment Brigade (Feb '09), pp. 4-3 to 4-5.

The CSSB is a tailored, multifunctional logistics organization. It is a flexible and responsive team that executes logistics throughout the depth of their assigned AO. The CSSB subordinate elements may consist of functional companies providing supplies, ammunition, fuel, water, transportation, cargo transfer, MA, maintenance, field services, and HR management. This framework enables the employment of a tailored logistics unit capable of adapting quickly to changing tactical conditions. The CSSB works through the sustainment brigade in concert with the TSC for logistics operations to effectively support the maneuver commander.

Three to seven total subordinate battalions may be attached to a single sustainment brigade depending on the brigade's mission. The CSSB is under the C2 of the sustainment brigade commander. It is the base organization from which force packages are tailored for each operation. Through task organization, the CSSB is capable of providing support during all phases of operations. The CSSB is structured to optimize the use of resources through situational understanding and common operational picture (COP). The mission of the CSSB is to C2 organic and attached units; provide training and readiness oversight; and provide technical advice, equipment recovery, and mobilization assistance to supported units. The headquarters detachment provides unit administration and sustainment support to the battalion staff sections.

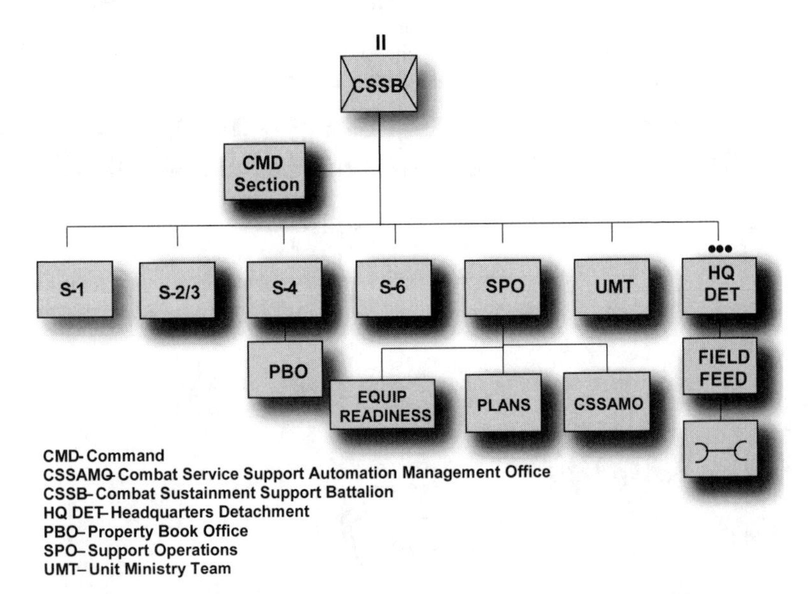

Supported units are reliant upon CSSB's to meet replenishment requirements beyond their internal capabilities. The CSSB is the entity of the sustainment brigade that provides the distribution link between theater base, APOD(s), SPOD's, and the supported units. The structure includes cargo transfer and movement control assets, fused with supply functions.

The CSSB also performs the function of transporting commodities to and from the BSB of the maneuver BCT and to and from theater repair or storage facilities. Its function is to maintain the flow of replenishment; to retrograde unserviceable components, end items, and supplies; to monitor the distribution of replenishment that is throughput directly from the theater base by assets of the sustainment brigade tasked with providing theater distribution; and to assist with coordination and delivery if necessary. The CSSB augments/supplements supported units during the normal replenishment cycle, delivering supplies that are not being throughput directly to units. A brief discussion of some of the CSSB's functions follows.

1. Ammunition

Ammunition elements assigned to the CSSB provide for the receipt, storage, issue, and reconfiguration of ammunition items. These elements provide flexibility and can be tailored in order to support the full spectrum of operations. Ammunition platoons can be attached as needed to meet surge requirements. Bulk Class V is received at the SPOD where ammunition elements of the TSC configure unit loads of Class V within its ASA. The ammunition is then throughput directly to using units, to CRSPs, or to the ASA, and ammunition transfer and holding points (ATHP) for distribution to supported units as required.

2. Transportation

Transportation assets of the CSSB provide mobility of personnel and all classes of supplies. When the CSSB is assigned to a sustainment brigade tasked to provide theater distribution, it will be heavily weighted with transportation assets. At the operational level, the CSSB's transportation assets will normally operate between the operational and the tactical levels. However, loads can be throughput from the strategic level direct to the tactical level (for example, BSB) based on METT-TC. At the tactical level, the CSSB's transportation assets will provide mobility from the CSSB base to the BSB and the FSC's within the BCT area.

3. Maintenance

Maintenance assets of the CSSB provide maintenance based on the two-level (field and sustainment) maintenance characteristics. The CSSB normally provides field maintenance support to the task force. Field maintenance is provided on an area basis. CSSB maintenance elements are designed with the capability to send slice elements forward to support a maintenance surge or to help clear maintenance backlogs at the BSB's and FSC's. Deployed CRC's and field repair activities remain under the C2 of the AFSB, but may collocate with a CSSB to provide sustainment maintenance support. Medical equipment maintenance, calibrations, and verification/certification in the sustainment brigade will be provided by medical equipment repairers from the supporting Medical Logistics Company (MLC).

4. Supply and Services

Supply and Services (S&S) assets of the CSSB provide all classes of supplies (less Class VIII) and quality of life operations for personnel operating in or transiting the AO. Supply involves acquiring, managing, receiving, storing, and issuing all classes of supply (less Class VIII). The CSSB provides field services such as laundry, shower, light textile repair, MA, and aerial delivery support. The CSSB may also provide billeting; food services, and sanitation when conducting base camp operations.

5. Human Resources and Financial Management

An organizational option is to attach the HR and FM companies to the CSSB instead of the STB of the sustainment brigade.

6. Command Group

The CSSB has a command group and staff that is the same as any other battalion and performs the same functions.

III. Support to the Warfighter - The Sustainment Warfighting Function

Ref: FMI 4-93.2, Sustainment Brigade (Feb '09), p. 4-1 to 4-2. See also pp. 1-2 to 1-3.

The sustainment warfighting function is comprised of the related tasks and systems that provide support and services to ensure freedom of action, extend operational reach, and prolong endurance. The endurance of Army forces is primarily a function of their sustainment. Sustainment determines the depth to which Army forces can conduct operations. It is essential to retaining and exploiting the initiative. Sustainment is the provision of the logistics, personnel services, and health service support necessary to maintain operations until mission accomplishment.

A. Logistics Sub-function

Logistics is the science of planning, preparing, executing, and assessing the movement and maintenance of forces. In its broadest sense, logistics includes the design, development, acquisition, fielding, and maintenance of equipment and systems. Logistics integrates strategic, operational, and tactical support of deployed forces while scheduling the mobilization and deployment of additional forces and materiel.

See pp. 2-31 to 2-62 for further discussion.

B. Personnel Services Sub-function

Personnel services are those sustainment functions related to Soldiers' welfare, readiness, and quality of life. Personnel services complement logistics by planning for and coordinating efforts that provide and sustain personnel.

See pp. 2-63 to 2-66 for further discussion.

C. Health Service Support Sub-function

Health service support consists of all support and services performed, provided, and arranged by the Army Medical Department. It promotes, improves, conserves, or restores the mental and physical well being of Soldiers and, as directed, other personnel.

See pp. 2-67 to 2-70 for further discussion.

Support Operations Under Centralized LOG C2

The design of the Theater Sustainment Command reduces command layers once present in the AOE logistics organizational hierarchy and integrates the other major sub-functions of the Sustainment Warfighting Function -- Personnel Services and Health Service Support. The realignment of support functions enables centralized control and decentralized execution of sustainment operations in accordance with the commander's priorities and intent. The TSC is the proponent for theater distribution and is responsible for theater RSO, movement, sustainment, and redeployment functions in support of Army forces. It is also responsible for establishing and synchronizing the intratheater segment of the distribution system.

The TSC, ESC, and sustainment brigade SPO translates the commander's operational priorities into priorities of support. The SPO prepares concept of support annexes to the OPLAN/OPORD and balances capabilities with requirements. The SPO coordinates, develops, and monitors the preparation and execution of plans, policies, procedures, and programs for external and area support. The TSC SPO monitors theater stocks, personnel, financial management, and the integration of aviation asset requirements into the overall support plan. The SPO is also responsible for coordinating movements and throughput of personnel, supplies, and equipment. The sustainment brigade SPO's materiel management effort is focused on the management of its supply support activities (SSA) in accordance with TSC plans, programs, policies, and directives.

IV. Logistics Support to the Warfighter

Ref: FMI 4-93.2, Sustainment Brigade (Feb '09), chap. 4.

Logistics is the science of planning, preparing, executing, and assessing the movement and maintenance of forces. In its broadest sense, logistics includes the design, development, acquisition, fielding, and maintenance of equipment and systems. Logistics integrates strategic, operational, and tactical support of deployed forces while scheduling the mobilization and deployment of additional forces and materiel.

See pp. 1-45 to 1-58 for discussion of logistics as a sub-function of the sustainment warfighting function.

Logisitics Support to the Warfighter

Supply

Field Services

Transportation

Maintenance

Distribution

Operational Contract Support

General Engineering Support

I. Maintenance

A. Replace Forward/Fix Rear

The overarching principle of replace forward/fix rear remains unchanged. Tailorable organizations execute the two-level maintenance system, composed of field maintenance and sustainment maintenance. Field maintenance involves platform tasks normally done by assets internal to the owning organization that return systems to a mission capable status. Sustainment maintenance involves platform tasks that are done primarily in support of the supply system (repair and return to supply), and will not normally be performed inside the brigade/BCT AO. There are no fixed repair time guidelines for performing field or sustainment repair. *See also pp. 1-55 to 1-56.*

The Army maintenance system employs tailorable field level maintenance units called support maintenance companies (SMCs). Sustainment level maintenance units called component repair companies (CRCs) are sent forward in the CSSB of the sustainment brigade. Each type of maintenance organization is built from a company HQ that can accept platoon and team level elements. These organizations are also capable of having 10 percent of their structure composed of contractors or DA/DOD civilians.

Support Maintenance Company (SMC)

The SMC provides area support to units in the sustainment brigade AO.

- Provides support field maintenance
- Supports theater opening packages (Army pre-positioned stocks)
- The SMC is capable of accepting modules (platoons/sections/teams) from Cry's and C&C's

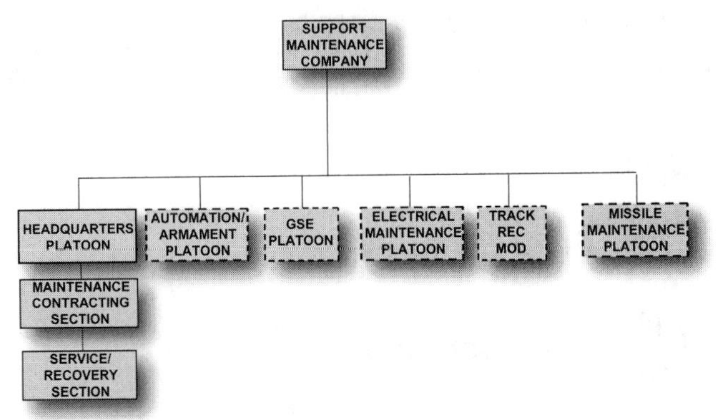

Component Repair Company (CRC)

The CRC provides repair and return to the theater supply system. CRY's are employed in the distribution system beginning at the national source of repair. These units can be pushed forward at the direction of the AFSB into the AO only as needed and will be work loaded by the AFSB.

These units:

- Provide repair and return to the supply system
- Operate in conjunction with a supply support activity (SSA)
- Can attach platoons/sections/teams to SMC or other sustainment units
- Integrated maintenance repair activities with AFSB

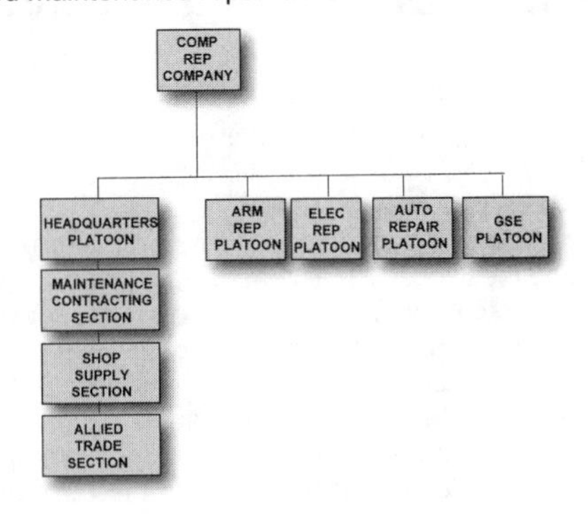

B. Aviation Maintenance

Ref: FM 3-04.500, Army Aviation Maintenance (Aug '06), chap. 2.

The aviation logistics transformation plan removed the multilayered pass-back maintenance concepts of the past. The new maintenance concept provides maintainers with the capability of replacing forward and repairing in the rear. In the contiguous and non-contiguous battle areas, the rear is generally defined as an area that provides higher security and capability. Aviation maintainers will continue to repair limited items forward within the ASB, returning repaired components to either the customer or the unit's PLL.

Sustainment Brigade Ops

Field Maintenance

Field maintenance is performed by combat aviation brigade personnel assigned to flight companies, aviation maintenance companies, and aviation support companies (ASCs). The aviation maneuver battalion's assigned flight companies perform authorized maintenance procedures within their capability. AMCs assigned to aviation maneuver battalions provide maintenance support to all flight companies.

Both the AMC and the ASC are authorized to perform field-level maintenance. The AMC is limited to performing unit maintenance according to the MAC. The ASC, in addition to performing unit maintenance and because of the additional SKOT that it is assigned, is authorized to perform intermediate maintenance.

Combat aviation brigade assets primarily perform field maintenance. Field maintenance includes both unit and intermediate maintenance. The brigade commander has complete operational command and control of all aviation maintenance operations at and below the division level. Field maintenance comprises aviation maintenance platoons (AMPs), AMCs, and ASCs. Aviation field maintenance is characterized by "on system maintenance," generally replacing components or performing component repair and return to the user. Aviation field maintenance capability varies based on SKOT, personnel assigned, and the authority directed by AR 750-1.

- **Field Maintenance Teams (FMTs).** Modularity allows combat aviation brigade units flexibility through the formation of field maintenance teams (FMTs). FMTs vary in composition depending on the support requirements, duration, and availability of personnel.
- **Flight Line/Company Operations.** Company maintenance activities primarily maintain Army aircraft by conducting scheduled maintenance. Unscheduled maintenance is conducted within the unit's capability. Allowing unit maintainers a degree of ownership in their assigned aircraft will generally enhance the quality and standards of maintenance performed, thus improving overall unit readiness.
- **Aviation Maintenance Company (AMC).** The AMC is organic to aviation maneuver battalions; for example, the general support aviation battalion (GSAB), attack reconnaissance battalion (ARB), and assault helicopter battalion (AHB) assigned to combat aviation brigades (CABs).

Sustainment Maintenance

According to FM 4-0, sustainment maintenance is the Army's strategic support. The strategic support base is the backbone of the National Maintenance Program and the sustainment maintenance system. At this level, maintenance supports the supply system by economically repairing or overhauling components. Maintenance management concentrates on identifying the needs of the Army supply system and developing programs to meet the supply system demands.

Sustainment maintenance support is divided and primarily performed by three separate entities:

- Original equipment manufacturers (OEMs) and their CFSRs
- Army depots, located at fixed bases in the continental United States (CONUS)
- National maintenance (NM) sources of repair (SORs)

C. Battle Damage Assessment & Repair (BDAR)

Ref: FM 4-30.31, Recovery and Battle Damage Assessment and Repair (Sep '06).

BDAR procedures apply to most operational levels (from the crew through field level maintenance) and depend on the extent of the damage, time allowances, available personnel with required skills, and accessible parts, tools, and materials. Maintenance personnel must act quickly to restore the vehicle to the combat-ready condition required to continue the mission or allow the vehicle to self-recover. *See also. p. 3-7.*

BDAR Principles

To be effective, BDAR should follow certain basic guiding principles:

- Ensure standard maintenance practice is always the first consideration
- Base decisions of using BDAR versus standard maintenance on the METT-TC
- Provide an accurate assessment
- Ensure economy of maintenance effort (use maintenance personnel only when necessary)
- Train multifunctional skills
- Repair only what is necessary to regain combat capability
- Remain flexible about repair priorities

Commanders should address using BDAR in the logistics section of their operation order (OPORD). This will provide the crews and maintainers with a clear understanding of when and at what risk level they can perform BDAR. In wartime, BDAR may have to be liberally applied at the discretion of the commander. In military operations other than war, local command policy will direct the degree of BDAR to apply and when to use standard maintenance. However, commanders at all levels must ensure that both crews and maintainers perform annual BDAR training.

Think Safety First

Personnel must be aware of live/loaded ammunition, damaged weapons or ammunition, fuel/oil spills, and electrical cables and wiring.

- Look for unexploded ordnance (UXO) in the area before performing the assessment
- Check the area for chemical contamination, to include depleted uranium (DU) when appropriate. Use a radiac meter to determine if DU is present.
- Avoid environmental contamination by spills of fuel and oil. All spills should be reported through the chain of command to the unit's logistical element/S4.

Beware of Booby Traps and IEDs

Booby traps and improvised explosive devices present a unique challenge when recovering abandoned vehicles. If equipment was abandoned or was unsupervised by friendly forces, the possibility of booby traps exists. To ensure the safety of BDAR/recovery operations, inspect equipment for tampering before attempting repairs.

Recognize Battle Damage Indicators (BDI)

Battle damage indicators (BDI) from an operator or crewman perspective include smoke, fire, unusual odor, unusual mechanical noise, leaking fluids, fault warming lights and alarms, and loss of mobility or system function.

Perform an Assessment

The senior man present decides when and if BDAR is performed during combat. This decision is based on METT-TC and the appropriate risk repair level. Do not attempt to operate systems or subsystems until the crew has performed an assessment to prevent further damage to equipment or personnel.

II. Supply

Providing the force with general supplies is the mission of the attached/assigned quartermaster elements of the CSSB and functional battalions. Supply operations must effectively support a brigade-based force. To accomplish this, Quartermaster units, tailored together with other sustainment units to form multi-functional CSSB's within Sustainment Brigades, provide logistics support at the operational level. The current force maneuver and support brigades have organic support capabilities that provide operational endurance. Command and control and management of supply operations begin at the TSC and its supporting multi-functional Sustainment Brigades, and is executed by CSSB's. Supply elements generally provide subsistence, general supplies, bulk fuel, heavy materiel, repair parts, and water. Personal demand items (Class VI) and medical supplies (Class VIII) are not typically provided by units under the C2 of the sustainment brigade, but must be considered during the planning process. *See also pp. 1-46 to 1-47.*

A. Class I, Food and Field Feeding

The current field-feeding standard is expected to remain in effect. This feeding standard dictates that a ration is three quality meals each day, with the capability to distribute, prepare, and serve at least one Unitized Group Ration–A (UGR-A) or UGR-Heat & Serve (UGR-H&S) meal per day (METT-TC dependent). The family of operational rations used to support this standard consists of individual meals (Meal-Ready-to-Eat; Meal, Religious, Kosher/Halal; Meal, Cold Weather/Food packet, Long Range Patrol; along with the emerging First Strike Ration and Compressed Meal) and unitized group meals (UGR-A and UGR-H&S, along with the emerging Unitized Group Ration–Express (UGR-E)), plus enhancements such as bread, cereal, fruit, and salad and the mandatory supplement of milk to ensure the nutritional adequacy of the group rations. The inclusion of a cook-prepared hot meal in the standard of three quality meals per day is based on units having the required personnel and equipment necessary for implementation. During extended deployments of 90 days and beyond, the feeding standard is expanded to include the UGR-A Short Order Supplemental Menus. This option provides easy to prepare breakfast and lunch/dinner short order menus and affords choices in menu selection for Soldiers. The feeding standard applied when troops are deployed beyond 180 days includes the incorporation of the DA 21-Day menu that can be prepared using organic field kitchens or contractor logistics support.

See following page (p. 2-36) for further discussion.

B. Water Production and Distribution

Water production and distribution operations in the current force can be characterized by a greater degree of self-reliance by maneuver units. This is due to a mobile storage capability, organic water purification, and improved distribution system within the brigade. This added water generation capability, along with the reduction in echelons and mobility improvements, will enhance the integration of sustainment into the operational battle rhythm. Modular Quartermaster Water Purification and Distribution Companies provide tailored water production package capabilities, storage, and bulk area distribution at the operational and tactical levels. Arrival of water units into the theater of operations is synchronized in such a way that the Water Distribution System will expand from commercial packaged water distribution early on, to water production and distribution as the theater of operations matures.

See following page (p. 2-37) for further discussion.

Class I Distribution

Ref: FMI 4-93.2, Sustainment Brigade, pp. 4-8 to 4-9. See also previous page (p. 2-35).

Class I items are pushed from the strategic level (vendors and/or depots) to the operational area based on supported unit strength reports. When logistic personnel, rations, transportation, and equipment are in place, a "Pull" or "Request" system will be implemented as UGR-H&S and UGR-A are introduced. The mix of perishable and semi-perishable rations depends on the Operational Commander's Feeding Policy and the availability of refrigerated storage and Subsistence Platoons. Class I stocks brought into the theater of operations will be moved to the Subsistence Platoon residing within Quartermaster Support Companies (QSC). The Commander's Stockage Policy will determine the number of days of Class I supplies to be maintained at various levels/locations. As an example, stockage levels set at 10 days of supply (DOS) at theater level are reduced to 3 to 5 DOS in the Subsistence Platoon of the sustainment brigade.

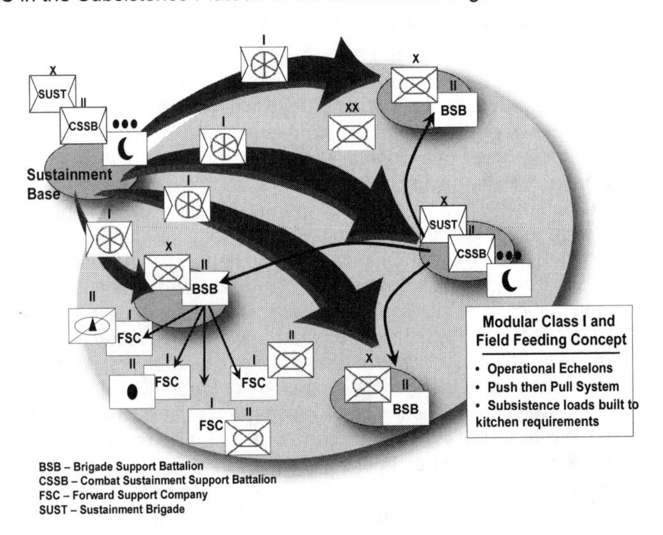

Class I bulk will be transported from the strategic level to the operational level in 20- or 40-foot refrigerated and non-refrigerated containers. The TSC will maintain the predominance of theater Class I stockage, which will be maintained by the Subsistence Platoon(s) of the QSC located in the operational level Sustainment Brigade or an equivalent civilian organization. At this location, the QSC's Subsistence Platoon(s) will also provide Class I area support to units located in the numbered Army level area. Rations are distributed in bulk from the operational level via internal single temperature refrigerated containers or leased refrigerated containers on semi-trailers, and via leased 20- or 40-foot ISO containers on semi-trailers for semi-perishable assets, to the Subsistence Platoon of the Sustainment Brigade CSSB supporting the higher tactical. It is envisioned that bulk Class I will be transported from the sustainment base to the Subsistence Platoon of the Sustainment Brigade CSSB Quartermaster Support Company, which will build support packages for the maneuver brigades. The rationale for the Subsistence Platoon shipping in bulk is: it more efficiently uses the limited transportation assets available; it most effectively supports large base-camp operations (as seen in Iraq); and it allows the QSC Subsistence Platoon to respond within the required 72 hour turnaround from the order placement to the order delivery for its customers. A 72-hour turnaround is required to respond to changes in operational pace and still support the Army's field feeding policy to provide the Soldiers with at least one hot meal a day, METT-TC.

Water Production and Purification

Ref: FMI 4-93.2, Sustainment Brigade, pp. 4-9 to 4-10. See also previous page (p. 2-35).

Water Purification

Quartermaster Water Purification and Distribution Companies and Augmentation Water Support Companies are assigned to CSSB's in Sustainment Brigades. The Water Purification and Distribution Company can provide up to 360,000 gallons of potable water per day from a fresh water source at up to eight water points. When using a salt water or brackish water source, production equals 240,000 gallons. This company can store 160,000 gallons; 80,000 per platoon. The Augmentation Water Support Company for arid environments can purify up to 720,000 gallons from a fresh water source or 480,000 from a brackish water source, and store 1.9 million gallons.

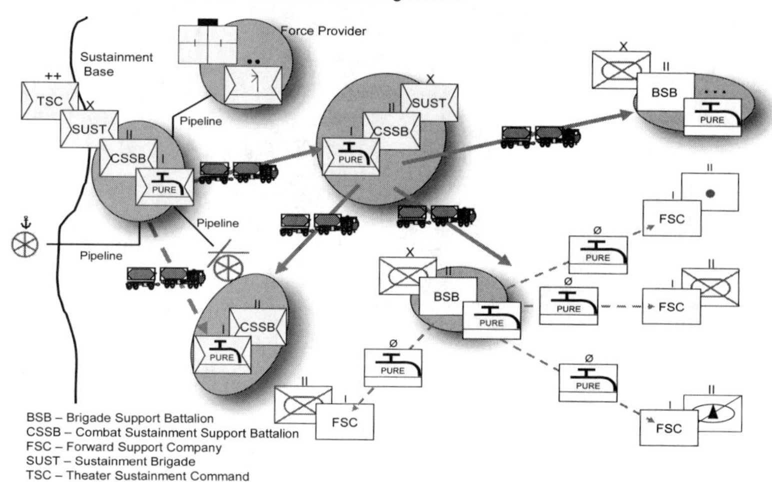

PURE = Purification Capability. At EAB; 3000 and 1500 GPH Reverse Osmosis Water Purification Units At the BSB; Tactical Water Purification System (TWPS) and Lightweight Water Purifier (LWP)

Water Support within Brigades

In temperate climates, water purification, packaging, storage, and distribution will take place in the BSB and JSOTF. The BSB has distribution assets to deliver limited package and bulk water forward to FSC's during replenishment operations. FSC's within BCTs do not have water support capability. Water will be issued using supply point distribution from the FSC's. Bulk water purification is able to be conducted forward of the BSB through the employment of high mobility multipurpose wheeled vehicle (HMMWV)-mounted light water purification systems. Water purification within the JSOTF may take place forward of the BSB.

Operations in Arid Environments

In an arid environment water sources are limited and widely dispersed, while requirements for potable water are increased. Therefore, operations in arid or other-than-temperate environments, place greater demand on purification, storage, and distribution capabilities than in other climatic conditions. A greater commitment of water assets at the operational level may be required due to limited raw water sources available and mobility requirements at tactical levels. Augmentation Water Support Companies may be employed to provide the additional capability required to meet the increased demand for water purification and storage. Line haul distribution will be accomplished using semi-trailer mounted fabric tanks transported by Transportation Medium Truck Companies.

C. General Supplies (Class II, III (P), IV)

Ref: FMI 4-93.2, Sustainment Brigade (Feb '09), pp. 4-11 to 4-12.

Supply distribution for general classes of supply is characterized by throughput of loads as far forward as the BSB and JSOTF (40-foot containers will not go forward of the sustainment brigade). Supply distribution is conducted by the QSC, assigned to the Sustainment Brigade CSSB's at the operational and higher tactical levels.

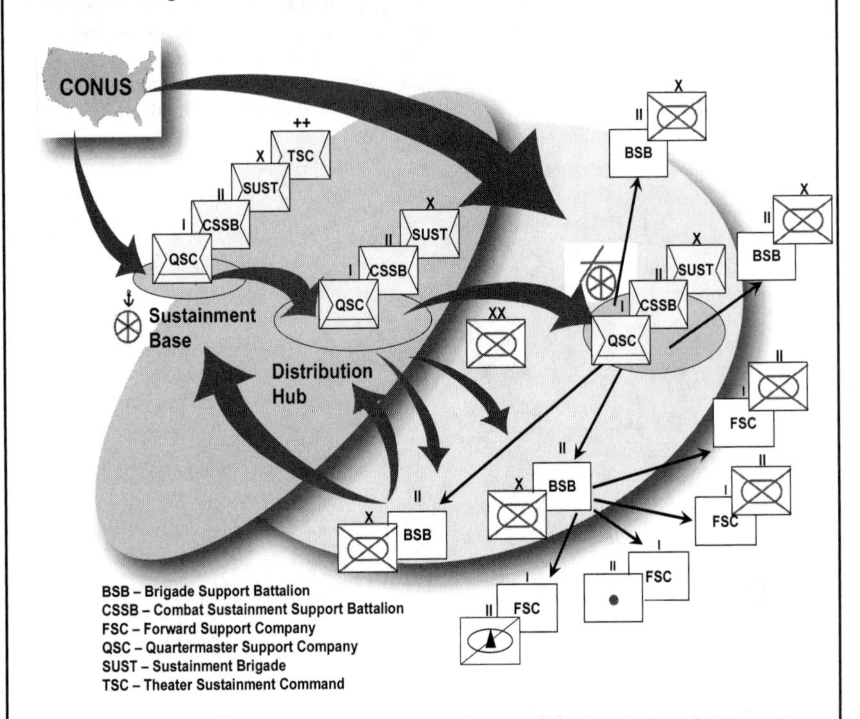

The QSC, as the Consolidated Supply Support Activity (CSSA), provides for the receipt, storage, and issue of 207.8 ST of Classes II, III (P), IV, VII, and IX (less aviation, missile repair parts, and communications security equipment) stocks per day. It packages/repackages supplies as required, to include retrograde and redistribution. The QSC is the center-post of the Distribution Hub at the sustainment base level, receiving all stocks as they enter the theater of operations that cannot be readily throughput to a unit destination. The QSC Area Support Platoons at the operational level configure pure pallet packages to satisfy unit requests or assemble support packages to resupply the QSC at higher tactical levels.

The QSC Area Support Platoons will build customized loads, combining pre-packaged loads received from designated distributors. These loads will be further combined with any required bulk supplies. Combined packages received from the strategic level will also be packaged for issue and forward movement to meet unit requests. These loads can be throughput as far forward as feasible.

Once supplies have been identified and configured for forward movement to the customer, the QSC will coordinate through its CSSB SPO for the most expeditious mode of transport. The TSC Support Operations/Distribution Management Center will establish the priority of effort for and coordination of the supply flow, maintenance and component repair activities, transportation, and distribution assets.

D. Class III (B)

Quartermaster petroleum units will provide fuel support for all US (and potentially coalition) land based forces. Sustainment brigades, specifically Quartermaster petroleum units and POL truck companies, will be involved in the reception and storage of POL from the refinery or terminal and the delivery to the BSB's distribution company. At theater level, POL Support Companies will be assigned to a POL Supply Battalion in a POL Group. At division/corps level, POL Support Companies will be assigned to a CSSB in a Sustainment Brigade. Theater opening or theater distribution sustainment brigades may be task organized with a Quartermaster pipeline and terminal operating company. Bulk petroleum will be distributed to the operational level for forward distribution to supported units. Fuel is throughput directly to Division locations and fuel support assets have been added to the maneuver BCTs and Support Brigades. Operational level fuel distribution operations use pipeline/hose line and large capacity long haul POL transportation. By moving POL capability to the Sustainment Brigades in the Corps/Division area, they can be more responsive to the BCT off-cycle demands and surge requirements. The TSC, sustainment brigades, and divisions can all maintain visibility of bulk fuel quantities on hand in the units and at the supply points via BCS3.

See following page (p. 2-41) for further discussion.

1. POL Supply Battalion

The POL supply battalion's mission is to provide C2, administrative, technical, and operational supervision over assigned or attached petroleum supply companies and petroleum truck companies. Its capabilities include C2 of two to five petroleum supply and truck companies; planning for the storage, distribution, and quality surveillance of bulk petroleum products; maintenance of theater petroleum reserves; and operation of a mobile petroleum products laboratory. POL supply battalions are typically assigned to either the TSC or to a petroleum group. However, these battalions may be attached to the sustainment brigade performing the theater-opening mission.

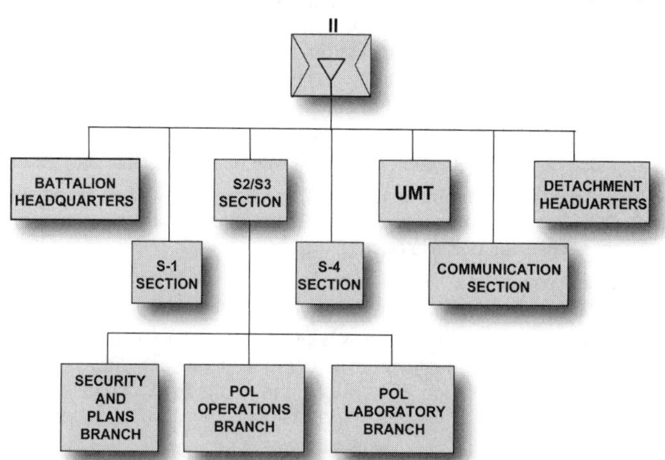

3. Quartermaster Petroleum and Terminal Operations Company (PPTO)

The Quartermaster petroleum and terminal operations company (PPTO) has as its mission to operate petroleum pipeline and terminal facilities for receipt, storage, issue, and distribution of bulk petroleum products. It is normally assigned to a petroleum pipeline and terminal operating battalion or a petroleum group. Its capabilities are to:

- Operate fixed terminal facilities for storage of up to 2,100,000 gallons of bulk petroleum. This normally consists of two tank farms, each with a capacity of up to 250,000 barrels or a tactical petroleum terminal (TPT) with a storage capability or up to 90,000 barrels.
- Operate up to 90 miles of pipeline for distribution of approximately 720,000 gallons per day
- Operate six pump stations, 24 hours per day, to deliver bulk product through 6- or 8-inch multi-product coupled pipeline
- Operate facilities for shipment of bulk product by coastal tanker, barge, rail, and tank trucks
- Maintain a prescribed reserve of bulk product for the theater of operations
- Operate a fuel system supply point for bulk issue operations

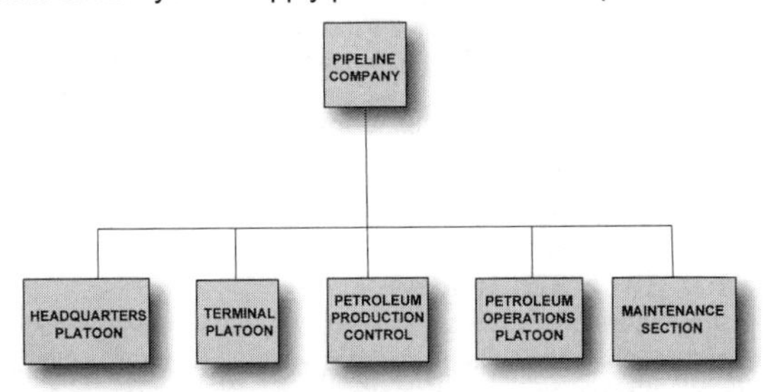

4. Quartermaster POL Support Company

There are two types of POL Support Companies: the POL Support Company (50K) and the POL Support Company (210K). The POL Support Companies receive, store, issue, and provide limited distribution of bulk petroleum products in support of division/corps or theater operations. Line haul distribution of bulk POL is performed by POL truck companies. The POL Support Company (210K) is normally located in the theater area and assigned to a POL supply battalion. The POL Support Company (50K) is normally located in the division area and assigned to a CSSB within a sustainment brigade. When task organized with three POL Support Platoons they have the following capabilities:

- Store up to 1,800,000 gallons when organized with three POL Support Platoons or 5,040,000 when organized with three POL Support Platoons (210K)
- Receive and issue up to 1,200,000 gallons per day when organized with three POL Support Platoons (50K) or 1,935,000 gallons per day when organized with three POL Support Platoons (210K)
- Establish and operate two hot refueling points using two Forward Area Refueling Equipment systems for transitory aircraft operating in their area

The area support section of each POL platoon can store up to 120,000 gallons of bulk petroleum at one location and 60,000 gallons at each of two locations. The distribution section of each POL platoon can distribute 48,750 gallons of fuel daily based on 75 percent availability of fuel dispensing vehicles at two trips per day.

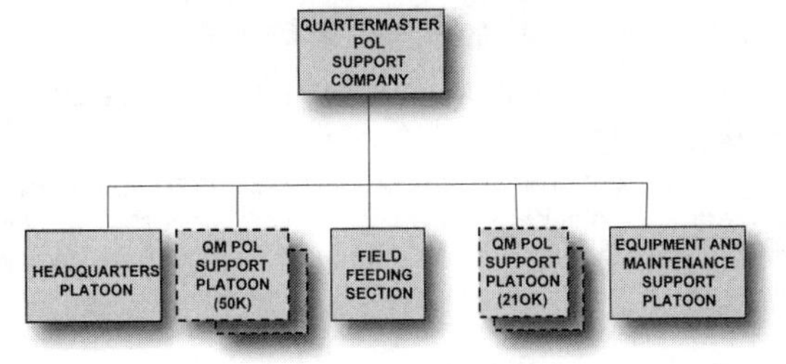

Theater Level Petroleum Operations

Ref: FMI 4-93.2, Sustainment Brigade (Feb '09), pp. 4-13 to 4-14.

In a developed theater of operations, the Fuel Distribution System includes: the Off-shore Petroleum Discharge System, inland tank farms Tactical Petroleum Terminals (TPT's), and pipeline systems. With a theater structure in place, Operational Petroleum Pipeline and Terminal Operating (PPTO) Companies in a TSC POL Group establish the theater petroleum support base for products received from ocean tankers at marine petroleum terminals. The petroleum support base serves as a hub for receiving, temporarily storing, and moving fuels to Petroleum Support Companies (PSC) at the operational and higher tactical levels. Petroleum Supply Battalions at the operational level provide theater stocks and deliver fuel to the Sustainment Brigade PSC's at the higher tactical level. PSC's support the area distribution mission and POL truck companies deliver fuel to the BSB's in the BCTs.

Pipelines/hose lines (the most efficient mode) will be used to deliver fuel products as much as possible during initial operations (usually to the Sustainment Brigade supporting the division). Large-scale combat operations may justify the construction of coupled pipelines/hose lines using the Inland Petroleum Distribution System or a future more rapidly emplaced pipeline system to move bulk petroleum from theater storage locations forward. Air bases and tactical airfields are serviced by pipeline when feasible. When available, pipeline distribution is supplemented by tank type vehicles, railcars, and barges.

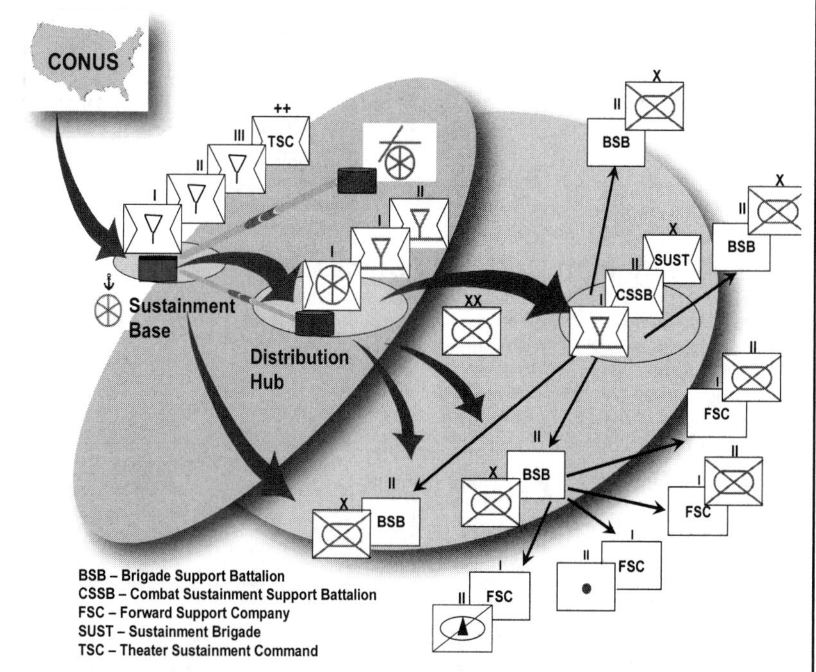

POL supply battalions at the operational level provide theater stocks and deliver fuel forward to the petroleum support companies (PSC) attached to sustainment brigades at the higher tactical level. PSC's in turn support the area distribution mission, while POL truck companies deliver fuel to the BSB's in the BCTs.

E. Class V, Ammunition Support

The sustainment brigade provides ammunition support with various organizations and from various locations based upon the maturity of the theater of operations and the mission of the supported units. Sustainment brigades operate Theater Storage Areas (TSA) at sustainment bases and ammunition storage points (ASP) at the distribution hub and other forward locations. Sustainment brigades operate Ammunition Support Activities (ASA) that provide the capability to receive, store, & issue Class V.

See following pages (p. 2-44 to 2-45) for further discussion.

1. Ammunition Battalion

The ammunition battalion's mission is to C2 ammunition companies and other attached units. It also provides supervision of ammunition support operations. The ammunition battalion is usually assigned to an operational-level sustainment brigade. Its core capabilities are:

- C2 and staff planning for two to five subordinate units
- Technical direction over ammunition support missions of subordinate units, except inventory management functions for which the TSC distribution management center (DMC) is responsible
- Maintaining a consolidated property book for assigned units

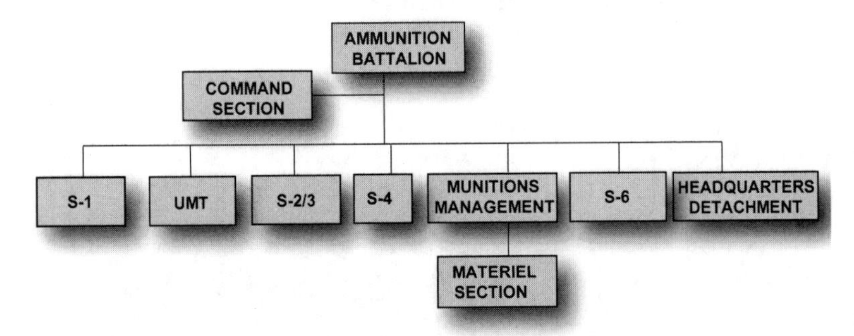

2. Ammunition Ordnance Company

The mission of the ammunition ordnance company is to provide command, control, and administrative, planning, and logistical support for ammunition platoons. It is normally assigned to an ammunition battalion or a CSSB.

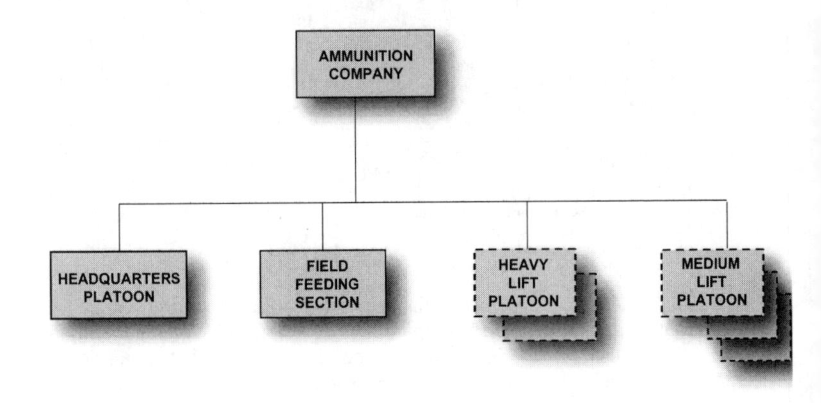

Class V Control Procedures

Ref: FM 4-30.13, Ammunition Handbook: TTP for Munitions Handlers (Mar '01).

Ammunition Supply Rates

The procedures used to control ammunition consumption are the required supply rate (RSR) and controlled supply rate (CSR). The Standard Army Ammunition System—Modernization (SAAS-MOD) is the management information system used to support these control procedures.

1. Required Supply Rate (RSR)

The RSR is the amount of ammunition a maneuver commander needs to sustain tactical operations, without restrictions, over a specified time period or for a specific mission. The RSR is expressed as rounds per weapon per day or, for selected items such as mines or demolition materials, as a bulk allotment per day or per mission. As the threat or mission changes, RSRs should change to reflect revised ammunition expenditure estimates. Maneuver commanders develop RSRs and submit them to the next higher HQ through operations channels. Each HQ reviews, adjusts, and consolidates RSRs and forwards them through operations channels. At the HQ that has ammunition management responsibilities, normally at TA/ASCC level, the total ammunition requirements are compared against total ammunition resupply capabilities for that period. If there is a shortfall in capability, a CSR will be established.

2. Controlled Supply Rate (CSR)

The CSR is that amount of ammunition that can be allocated based on the availability of ammunition types or quantities, Class V storage facilities, and transportation assets over a specific time period. The CSR is expressed in the same terms as the RSR. Commanders should use CSRs to allocate or prioritize the ammunition flow to units engaged in combat and to units held in reserve. They could also withhold some ammunition, especially high-lethality, low-density ammunition, to meet unforeseen requirements.

Ammunition Basic Loads (ABLs)

ABLs originate with a tactical force's planned deployment. An ABL is that quantity of ammunition either allocated to or issued to a unit [depending on the MACOM's policy] to sustain its operations in combat until it can be resupplied.

Basic load requirements are based on unit weapon density and mission requirements and are designed to meet a unit's anticipated initial combat needs. Units must be able to transport ABLs in one lift on organic weapon systems, equipment, and unit personnel. An ABL is normally expressed in rounds per weapon but may be expressed IAW MACOM policy as a number of required combat loads (example: battalion loads for artillery systems). The following factors influence ABLs:

- Nature of the enemy threat
- Type of mission
- Intensity of engagement
- Resupply transport availability
- Ammunition availability
- Number and types of weapons in unit

Lift Capability

Ammunition units' capabilities are measured in lift. A lift uses MHE to pick up ammunition and set it down, with each pickup and set down constituting one lift. A lift is measured in short tons (STONs) (2,000 pounds). Ammunition units' expressed lift capabilities are limited by personnel and MHE availability.

Ammunition Support

Ref: FMI 4-93.2, Sustainment Brigade (Feb '09), pp. 4-17 to 4-18. See also previous pages (p. 2-42 to 2-43).

The ammunition logistics system provides to the force the right type and quantity of ammunition in any contingency. The challenge is to move required amounts of ammunition into a theater of operations from CONUS and other pre-positioned sources in a timely manner to support an operation. The system must be flexible enough to meet changing ammunition requirements in simultaneous operations around the world. The objective of the system is to provide configured Class V support forward to the force as economically and responsively as possible with minimized handling or reconfiguring and quickly adapt to changes in user requirements. The unique characteristics of ammunition complicate the system. These factors include its size, weight, and hazardous nature. It requires special shipping and handling, storage, accountability, surveillance, and security. Munitions are managed using different methods depending on the level of command.

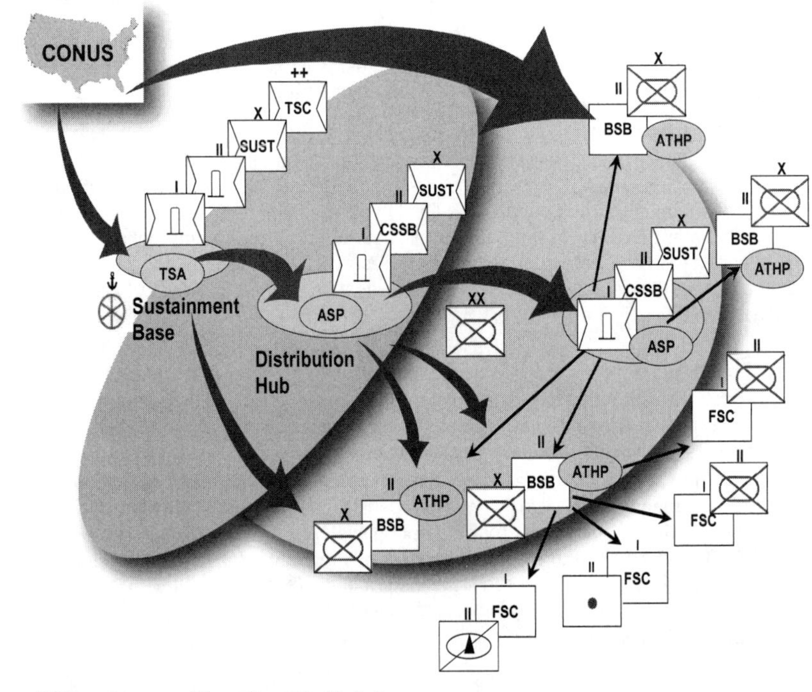

ASP – Ammunition Supply Point
ATHP – Ammunition Transfer and Holding Point
BSB – Brigade Support Battalion
CSSB – Combat Sustainment Support Battalion
FSC – Forward Support Company
SUST – Sustainment Brigade
TSA – Theater Storage Area
TSC – Theater Sustainment Command

Ammunition Flow

Ammunition issued to users is replaced by ammunition moved up from theater storage areas. In turn, ammunition stockage levels at the theater storage areas are maintained by shipments from CONUS or out of other theater locations. The quantity of ammunition shipped forward is determined by the amount on hand, current and projected expenditures, and the controlled supply rate (CSR).

Each battalion S-4 transmits a request for re-supply of ammunition for units through the brigade S-4 to the BAO. The BAO coordinates and controls the use of Class V supplies for the brigade, consolidates the brigade requests, and submits them to the sustainment brigade, ESC/TSC. The TSC, in coordination with the numbered Army G-4, reviews all requests and balances them against the CSR issued by the theater storage activity. The numbered Army issues the CSR to support the units. Some ammunition requirements are prioritized due to scarcity and some may not be issued due to unavailability. The BAO through TSC coordinate for the shipment of ammunition to the ATHP.

Captured enemy ammunition (CEA) must be kept separate from US munitions; however, it must be accounted for, stored, and guarded using the same criteria that applies to US munitions. When an enemy ammunition cache is found or captured, the commander must assess the combat situation. He/She must decide whether to destroy the CEA because of the situation or to secure it and request explosive ordnance disposal support. During retrograde operations, leaders must ensure safety policies and procedures are carefully observed as these operations can be particularly hazardous and serious injury has occurred in the handling of CEA. Close control of CEA is required. Positively identified and serviceable CEA may be compatible for use in US or allied forces weapon systems. These munitions can potentially ease the burden on the ammunition supply system. CEA can also be used as a substitute for bulk explosives during demolition operations.

1. Theater Storage Area (TSA)

The TSA encompasses the storage facilities located at the operational level. This is where the bulk of the theater reserve ammunition stocks are located. Ammunition companies, with a mixture of heavy- and medium-lift platoons, operate and maintain TSA's. The primary mission of the TSA is to receive munitions from the national level, conduct the bulk of operational level reconfiguration, and distribute munitions to forward ASA locations and BCT ATHP's. The TSA will build those configured loads that cannot be shipped into a theater of operations due to explosive compatibility conflicts for international shipment. Ammunition will be managed by either an ammunition battalion or CSSB based upon METT-TC

The sustainment brigade must keep the TSC DMC informed of storage or handling limitations or shortages in each TSA. When mission analysis indicates more than one TSA or port facility is required, the GCC should plan for early deployment of an ordnance ammunition battalion to provide mission C2 of munitions distribution at the TSC level.

2. Ammunition Supply Points (ASP)

Ammunition supply points (ASPs) provide the capability to receive, store, issue, and perform limited inspections and field level munitions maintenance support. The sustainment brigade gains such capability when it is assigned one or more ammunition ordnance platoons. The CSSB's attached to the sustainment brigade will contain ammunition ordnance companies and ASP. The number of companies and ASPs varies based upon the role of the sustainment brigade to which they are attached and the size and mission of the supported organizations.

ASPs receive, store, issue, and maintain ammunition based on the capabilities of assigned ammunition platoons. ASP stockage levels are based on tactical plans, availability of ammunition, and the threat to the re-supply operation. Additionally ASPs are the primary source of re-supply of ATHP located in BCTs.

Sustainment Brigade Ops

F. Class VI

Soldiers usually deploy with a 60-day supply of health and comfort items. Health and comfort packs (Class VI) can be supplied through supply channels. Tactical field exchanges provide Class VI supply support beyond the health and comfort packs. Class VI support can be limited to basic health and hygiene needs or expanded to include food, beverages, and other items based upon the requirements outlined by the theater commander. The availability of health and comfort packs and Class VI items can greatly enhance morale.

G. Class VII (Major End Items)

Due to their cost and critical importance to combat readiness, major end items (Class VII), are intensely managed and controlled through command channels. They will be distributed to the brigade support battalion (BSB) distribution company from the quartermaster support company (QSC) in the sustainment brigade or from the strategic level.

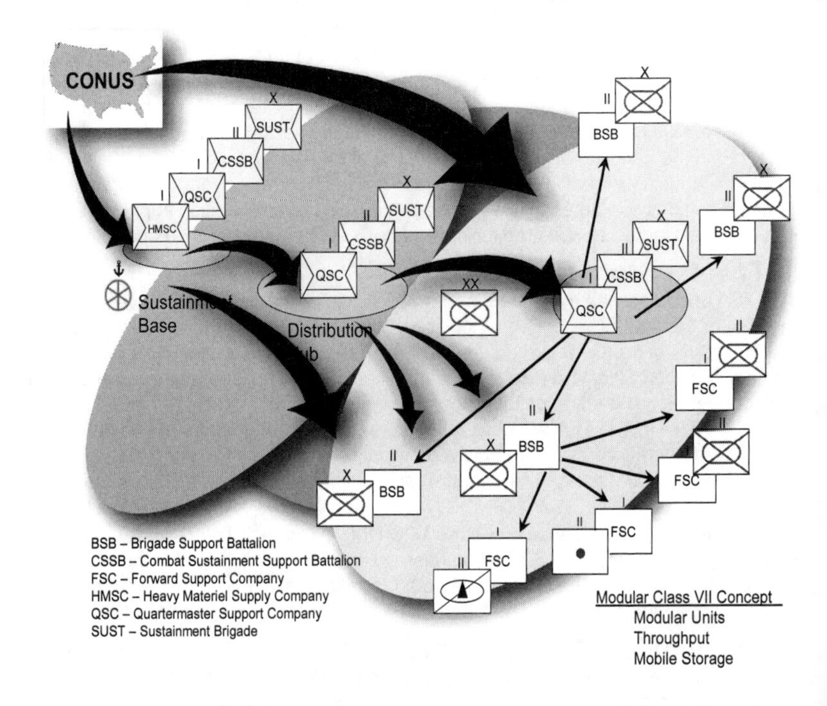

H. Class VIII

Ref: FMI 4-93.2, Sustainment Brigade (Feb '09), pp. 4-20.

Theater-level inventory management of Class VIII will be accomplished by a team from the medical logistics management center (MLMC). The medical community performs all supply functions for Class VIII, but relies on the sustainment brigade and subordinates to transport the supplies unless they are transported by ambulance. The MLMC support team collocates with the DMC of the TSC/ESC providing the medical command with visibility and control of all Class VIII inventory. When an ESC is deployed, an element from the MLMC will also collocate in their DMC. The medical logistics company (MLC) in the MMB will serve as the consolidated forward distribution point for Class VIII.

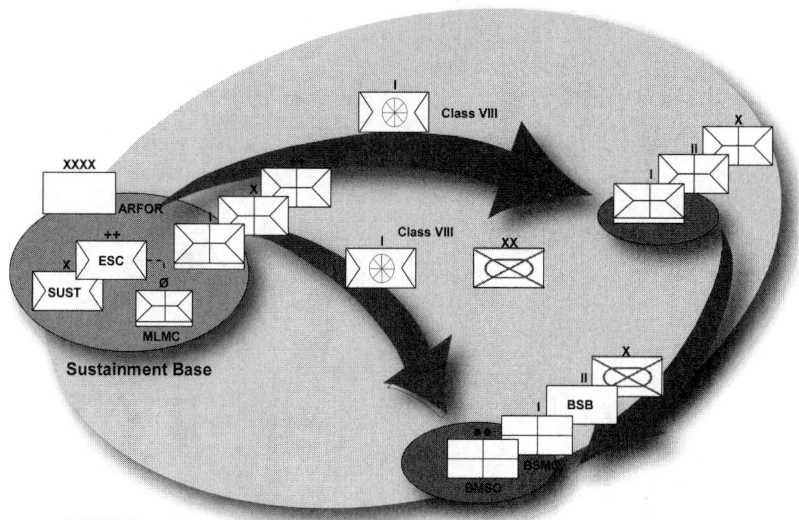

ARFOR – Army Forces
BMSO -
BSB – Brigade Support Battalion
BSMC – Brigade Support Medical Company
ESC – Expeditionary Sustainment Command
MLMC – Medical Logistics Management Center
SUST – Sustainment Brigade

Once supplies are identified and configured for forward movement to the customer, the MLC will submit transportation movement requests to the sustainment brigade and the MCB for appropriate transportation assets for forward movement.

H. Class IX - Repair Parts

Ref: FMI 4-93.2, Sustainment Brigade (Feb '09), pp. 4-21.

Class IX may be throughput from the theater sustainment base to the BSB depending on if the item(s) are critical in the fight. Other Class IX may be throughput to the sustainment brigade.

CSSB's receive, store, and issue Class IX items and configure/reconfigure loads, as required, in support of sustainment operations, to include retrograde and redistribution. The area support platoons assigned to the QSC, can receive, store, and issue Class IX items at either one or two locations, supporting 8,000 Soldiers at one location or 4,000 Soldiers at each of two locations. The area support platoons provides field and sustainment support as part of the QSC mission. These functions may be accomplished separately. The number of QSC's required in the division is METT-TC tailored.

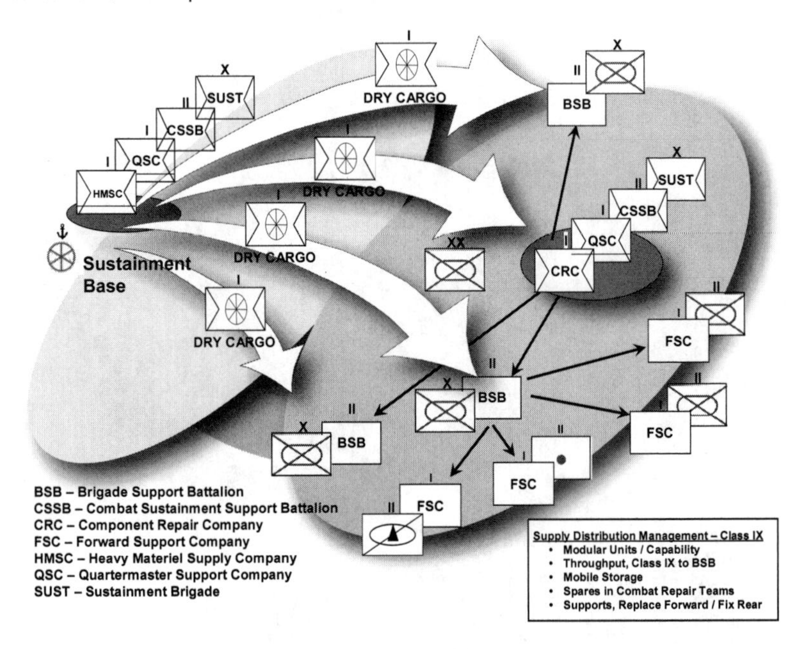

The QSC in the CSSB of the sustainment brigade receives the Class IX repair parts and configures loads as required, to include retrograde and redistribution responsibilities. The QSC establishes the CSSA that will receive, store, and issue the theater stockage levels established by the GCC. The QSC also provides support to units within the theater base. Since the repair of Class IX reparable items will be conducted at the sustainment maintenance facilities at the theater base, repaired items will be re-introduced into the supply system at the QSC CSSA.

Repair Parts Supply Operations

Ref: FM 4-30.3, Maintenance Operations and Procedures (Jul '04), chap. 5.

The supply system includes the wholesale level, retail level, and unit level. Wholesale supplies are managed at the strategic management level, retail supplies are managed at the operational and tactical levels, and unit level supplies are managed at the unit level.

Prescribed Load List (PLL)

The PLL is a list of the authorized quantities of supplies required by a unit to do its daily unit maintenance. Units that are authorized personnel, tools, and equipment to perform maintenance maintain a prescribed load of repair parts. Units that regularly support other units without maintenance capabilities include the supported unit's equipment in their PLL computations. A PLL consists of repair parts and other stocks. The unit PLL consists of unit-level maintenance repair parts that are demand-supported (15 DOS), non-demand-supported, and specified initial stockage for newly introduced equipment.

Authorized Stockage List (ASL)

The ASL consists of those parts stocked in DS repair parts supply units for issue to user units and to support DS-level maintenance operations. The MMC, based on priorities established by the commander, will establish the guidelines for issue, ASL design, or distribution. The MMC is the common exit point for requisitions and other supply documents for the division.

Mandatory Parts Lists (MPLs)

MPLs, which are published as DA pamphlets, are used to standardize the combat PLLs. The MPL is the mandatory portion of the standardized combat PLL. Parts on the MPL must be on-hand or on order at all times.

Weapon System Replacement (WSRO)

Weapon System Replacement Operations (WSRO) is a management tool used to supply the Combat Commander with fully operational major weapon systems, including both the required equipment and trained crews. Procedures for issue of weapon systems differ from those for other Class VII items. Weapon systems replacement is managed at each level of command. Two terms often used to describe WSRO are ready-for-issue and ready-to-fight.

Operational Readiness Float (ORF)

ORF is a quantity of selected end items or major components of equipment authorized for stockage at CONUS installations and overseas support maintenance activities, which extends their capability to respond to materiel readiness requirements of supported activities. It is accomplished by providing supported units with serviceable replacements from ORF assets when their like items of equipment cannot be repaired or modified in time to meet operational requirements.

Controlled Exchange

Controlled exchange is the removal of serviceable parts, components, or assemblies from unserviceable, but economically reparable equipment and their immediate reuse in restoring a like item of equipment to combat operable or serviceable condition. *See also p. 3-7.*

Cannibalization

Cannibalization is the authorized removal of parts, components, or assemblies from economically non-repairable or disposable end items. Cannibalization supplements and supports the supply operation by providing assets not readily available through the normal supply system. *See also p. 3-7.*

III. Field Services

A. Mortuary Affairs (MA)

The MA program is designed to support war and stability operations including mass-fatality situations. For non-contingency situations each Service provides or arranges support for its deceased personnel. Roles and responsibilities for MA are outlined in the Department of Defense Directive 1300-22, which designates the Secretary of the Army as the executive agent for MA.

The Theater Sustainment Command (TSC) is the Army organization responsible for providing CUL within the theater of operations, to include MA support. The TSC commander, or a designated representative, determines the responsibility and placement of theater MA assets within the JOA. Normally, this is accomplished through coordination with the JMAO. The sustainment brigades exercise tactical/operational control over MA assets while the TSC exerts technical control over the theater-wide MA mission and executes any required adjustments in mission support in coordination with the sustainment brigades. If neither the TSC nor the ESC deploys to an AO and the sustainment brigade is the senior sustainment HQ in the AO, then the sustainment brigade will exert technical control over the MA mission and would perform all functions/actions listed in the following discussion as being performed by the TSC.

The Army MA Program is divided into three distinct programs: Current Death Program, Concurrent Return Program, and Graves Registration Program. Theater MA support during major operations is provided through the Concurrent Return and Graves Registration Programs. This support includes the initial search and recovery, tentative identification, coordinated evacuation, and decontamination (if necessary) of remains and personal effects (PE). Each MA program addresses the federal responsibility for recovery and evacuation of US military personnel, government employees, and US citizens not subject to military law. Per the Geneva and Hague conventions, commands also have to address the recovery and evacuation of Multinational, Enemy Prisoners of War, Enemy, and host nation dead. Mortuary affairs operations for Army forces rests within the GCC's theater organization. The Joint Mortuary Affairs Office (JMAO) has responsibility for the Joint Mortuary Affairs Points (JMAP) within a JOA.

During multinational operations, MA staff planners must ensure that joint doctrine takes precedence. Specifically, it is imperative that MA planners and MA personnel process all remains in the same manner in accordance with JP 4-06 and all applicable international and cooperative agreements in the specific theater of operations.

See following page (p. 2-51) for further discussion.

B. Shower and Laundry Services

The current force conducts field hygiene operations through the utilization of field sanitation teams located at the company level. The Sustainment Brigade, Combat Sustainment Support Battalion's Quartermaster Field Service Company supports the field sanitation teams through shower, laundry, and clothing repair (SLCR) sections dispatched from echelons above brigade. Shower, laundry, and clothing renovation (SLCR) capabilities resident within the Quartermaster field services company are provided from the sustainment brigades with projection as far forward as possible. The mission is to provide Soldiers a minimum of a weekly shower and up to 15 pounds of laundered clothing each week (comprising two uniform sets, undergarments, socks, and two towels). The Quartermaster Field Service Company will provide SLCR for supported units. The Quartermaster Field Service Company can be moved forward to provide field services for the BCT. Each SLCR Platoon has 3 SLCR Teams which can each support 500 Soldiers per day/3,500 Soldiers per week.

Mortuary Affairs (MA) Concept of Operations

Ref: FMI 4-93.2, Sustainment Brigade (Feb '09), pp. 4-23 to 4-24.

The MA program starts at the unit level with limited search and recovery operations and continues until remains are returned to the person authorized to direct disposition and all personal effects (PE) are returned to the person eligible to receive effects. Unit commanders are responsible for the initial search and recovery operations within their AO. Every unit is responsible for designating a search and recovery team to conduct search and recovery operations during the combat phase of operations. Instilled with the MA motto that all remains are to be treated with the utmost "dignity, reverence, and respect," these unit teams must be briefed on local customs and courtesies on dealing with remains. Upon recovering remains, a unit search and recovery team will evacuate those remains to the closest MA collection point (MACP). MA personnel set up MACP's, theater mortuary evacuation points (TMEP's), and PE depots throughout the theater of operations. Remains recovered by unit initial search and recovery operations are evacuated to the nearest MACP. From the receiving MACP, MA personnel evacuate the remains to the TMEP. The preferred method of evacuation is directly from the initial MACP to the TMEP; however, the tactical situation may dictate that remains be evacuated through several MAP's before reaching the TMEP.

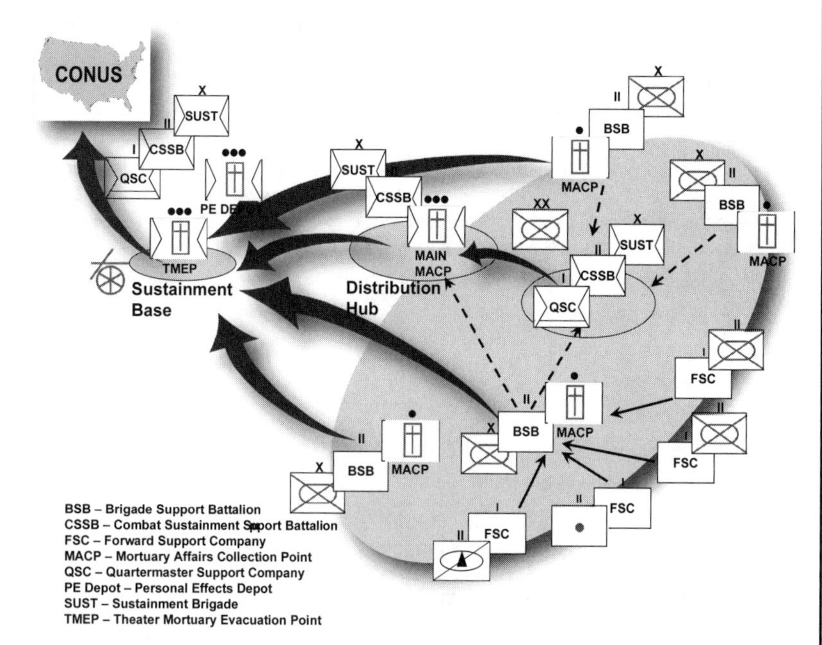

MA assets are managed as theater assets. MA units are deployed as augmentation elements and do not have the personnel, equipment, or supplies required for self-support. MA augmentation elements are sustained by the units, which they support. A MA Company Support Operations Section is deployed as an Early Entry Module (EEM) to support operational planning. Upon arrival, it will report to the Theater Mortuary Affairs Officer (TMAO) or a Sustainment Brigade Mortuary Officer, to plan and coordinate RSOI of follow-on MA elements.

IV. Transportation

The mobility branch of the TSC SPO provides staff supervision of all transportation and coordinates directly with the MCB. The MCB is attached to the TSC/ESC and manages movement control across the entire theater of operations. The MCB would be attached to a sustainment brigade when the MCB is deployed to an AO in which the sustainment brigade is the highest sustainment C2 HQ. The MCB and its MCT's coordinate all movement in the JAO/AO to include all divisional areas.

Transportation expertise in the sustainment brigade is provided by the transportation branch of the distribution operations division. This section works with the other elements of the brigade SPO integrating movements with materiel management. The theater-opening element, when assigned, augments the staff of a sustainment brigade and provides the technical expertise to open a theater of operations and establish the initial theater distribution network. The motor transportation battalion SPO receives technical supervision and guidance from the brigade SPO to support the theater distribution mission. Transportation assets of the CSSB and functional transportation battalions provide distribution from the sustainment brigades forward and retrograde of damaged or surplus items.

A. Movement Control

Movement control is the planning, routing, scheduling, controlling, coordination, and ITV of personnel, units, equipment, and supplies moving over multiple lines of communication. It involves synchronizing and integrating logistics efforts with other elements that span the spectrum of military operations.

See following pages (pp. 2-54 to 2-55) for further discussion.

B. Motor Transport Planning

Motor transport planning, particularly in its early stages, must be based upon a set of broad planning factors and assumptions. These factors should be used only in the absence of specific data relating to the current situation. Because of the different services performed, loads carried, and terrain crossed, caution should be exercised when analyzing the following factors:

- Task vehicle and driver availability rate
- Vehicle payload capacity
- Operational hours per shift
- Operational day
- Daily round trips
- Operational distance per shift
- Rate of march in the hour
- Delay times
- Threats and protection requirements

C. Container Management

Recent history has shown that inter-modal operations are critically affected by the manner in which container management policies are enforced and container management is subsequently executed. Container management is a command function, not just a sustainment function. Due to the nature of container use, commanders at all echelons must be involved in container proper control. Therefore, a container management program must be established at a theater-level echelon that permits centralized management. For Army forces, the ASCC will provide further direction and control measures in order to maintain greatest discipline at the least cost. Sustainment brigade commanders must implement the theater management program for subordinate units. *See also p. 1-54.*

Our industry partners will continue to use this method of packaging and distribution within the global environment for the foreseeable future; therefore, it is vital to maintain the control and flow of containers. Operations must fully integrate container management into the distribution system. Full spectrum inter-modal distribution management capabilities include systems to maintain visibility, manage disposition at destination, and enable a rapid return to the distribution system to ensure adequate numbers are available to maintain deployment, employment, sustainment, and retrograde operational pace.

Successful container management ensures the following:

- Expeditious movement of throughput and high-priority container shipments
- Minimal time for holding and/or consolidating cargo
- 100 percent in-transit visibility of containers and contents
- Economical movements via container use
- Movement of containers as fast as mode operators and consignees can handle them
- Integration of military and commercial container management systems
- Consolidation of single consignee shipments
- Detention and demurrage reduction
- Management of container availability to support retrograde movements

1. Flatrack Management Operations

Flatracks and containers offer tactical efficiencies that serve to increase the pace of sustainment operations. The key to these efficiencies and maintaining this pace is congruent flat rack/container management procedures at each stage or level of support (FSC, BSB, sustainment brigade, ESC, and TSC). An increased operational depth and the reduction of redundant logistics force structure challenge flatrack/container management and, ultimately, the sustainment of combat power. Flatrack/container employment, management, and retrograde operations are the responsibility of distribution managers. In the sustainment brigade, the support operations officer must track flat racks and containers dispersed throughout the distribution system within its operational span of control.

2. Logistics Support Area Flatrack Management

Sustainment brigades operating a logistics support area (LSA) face increased flat rack management challenges especially when transportation assets to move supplies and retrograde flat racks are limited. Management responsibilities within the LSA rest with the support operations officer and the distribution division. Their responsibilities include:

- Identifying a flatrack collection point upon occupation of the LSA
- Managing all common user flatracks on an area basis
- Ensuring flatrack exchange (providing a back hauled flat rack for every one received) procedures are adhered to as a matter of priority
- Maximizing the use of PLS/HEMMT-LHS platforms for retrograding flat racks from the flat rack collection point back into the distribution system
- Reporting flatrack on-hand quantity by location, status, and condition to the flatrack control office established by the senior manager in theater of operations (in accordance with AR 710-2, Inventory Management Supply Policy Below the National Level)
- Coordinating with the servicing MCB for supplemental transportation support when retrograding flat racks from the LSA flatrack collection point (FRCP)

Movement Control

Ref: FMI 4-93.2, Sustainment Brigade (Feb '09), pp. 4-27 to 4-29.

Movement control is the planning, routing, scheduling, controlling, coordination, and ITV of personnel, units, equipment, and supplies moving over multiple lines of communication. It involves synchronizing and integrating logistics efforts with other elements that span the spectrum of military operations.

See chap. 6, Deployment & Redeployment Operations, for further discussion.

1. Movement Control Battalion (MCB)

The MCB will also coordinate with host nation authorities for cargo transfer locations, road clearances, border clearances, escort support, and transportation support. The MCB will have as many subordinate MCT's as needed to operate in its area of operations, based on the number of customers, air terminals, rail terminals, seaports, and MSR's it must support. The MCB will provide logistics support to the MCT's under its C2. However, MCT's operating away from their HQ will require logistics support from other units. *See also p. 2-58.*

The MCB will—

- Continue to provide command, control, and technical guidance to 4 to 10 MCT's
- Provide asset visibility and maintain ITV of tactical and non-tactical moves within its assigned geographical area (including unit moves and convoys)
- Assist in planning and executing plans and operations
- Apply and meet movement priorities provided by the TSC and sustainment brigade
- Support the entire spectrum of distribution

2. Movement Control Team (MCT)

The MCT is the basic and most critical level in the movement control process. MCT's are the common point of contact for mode operators and users of transportation. The MCT is a 21-Soldier team created with the capability to perform every type of movement control mission. It is designed to provide maximum flexibility in its employment. Each team has a headquarters section and four identical subunits (or sections). The MCT can operate as a single team or separately at up to four different locations. For example, a single MCT can be deployed initially to provide movement control functions at an airfield while simultaneously providing cargo documentation. As the mission expands, the team can deploy a section onto the MSR's to conduct movement control operations. As the operation matures, that same MCT can operate at a second airfield or seaport. The operational use of the MCT can be specifically tailored to the mission and operational environment. The standardization of MCT's increases the number of teams available for deployment, since each unit is tailorable in the truest sense of that term. If METT–TC factors dictate that the MCT needs to be split into four sections in different locations, each section can be properly equipped with the vehicles, communications equipment, STAMIS, and generators it needs to operate independently.

MCT's process movement requests and arrange transport for moving personnel, equipment, and sustainment supplies. They process convoy clearance requests and special hauling permits. MCT's coordinate with the MCB for the optimal mode (air, rail, inland waterway, or highway) for unprogrammed moves and commit the mode operators from the sustainment brigade, LOGCAP, multinational elements, and the host nation. They also assist in carrying out the movement program.

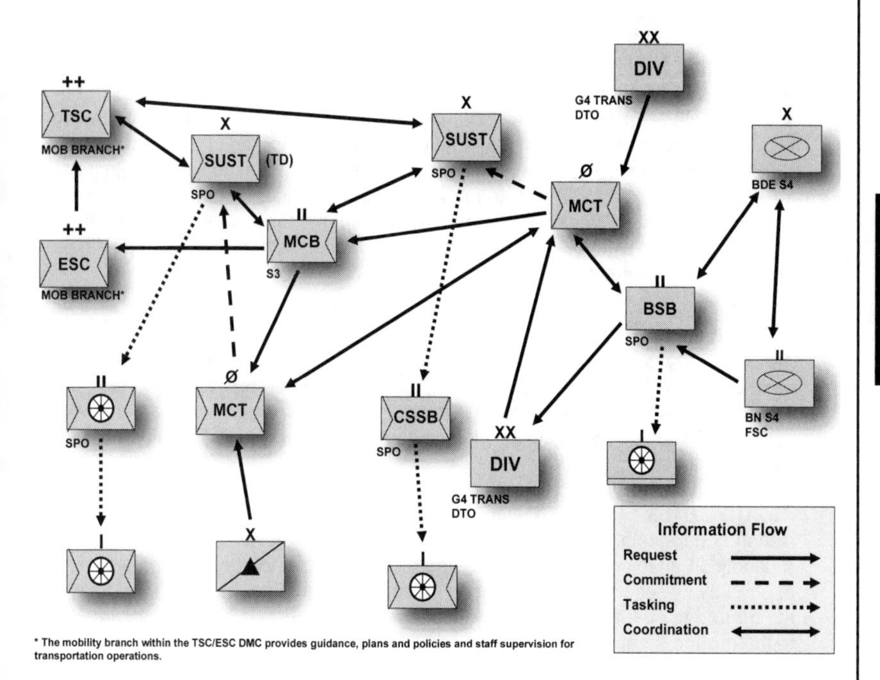

* The mobility branch within the TSC/ESC DMC provides guidance, plans and policies and staff supervision for transportation operations.

MCT's are arrayed on the battlefield at various nodes to best support the TSC concept of support. In general, a MCT collocates with a Sustainment Brigade to provide movement control support on an area basis to EAB units or units not assigned/attached to a Division. An additional MCT can work directly for the Division G-4 Transportation Officer (DTO) to assist in the planning and execution of movement control in the division area of operations. MCT's can operate at an APOD to facilitate the movement of ground cargo (primarily palletized), assist in passenger manifesting, and provide air load planning capability. MCT's operating at SPOD's facilitate the ground movement of containerized cargo and assist in the marshalling/staging of equipment for onward movement. Elements of a MCT can locate at other critical nodes (for example Central Receiving and Shipping Point (CRSP), Convoy Support Center (CSC), TDC, and container yards, as needed) to provide cargo documentation capability or for the added visibility and management of distribution operations. MCT's are also used to enforce the highway regulation plan established by the TSC and will deploy its sections along Main Supply Routes (MSR) to augment convoy tracking operations and provide on the ground traffic de-confliction.

The sustainment brigade SPO distribution integration branch coordinates with the MCT, MCB, HN transportation agencies, transportation mode operators, and customers. This office assists in planning and executing plans for the reception, staging, onward movement, and retrograde of personnel, equipment, and sustainment supplies. This includes actions associated with marshaling and staging areas. When serving in a sustainment brigade that has the port mission, the distribution integration branch will have the responsibility to monitor, manage, and execute the movement and port clearance plans and programs. The branch also monitors the use of trailers, containers, and flat racks located in sustainment brigade AO and coordinates with users to expedite return of these assets to the transportation system.

Flatrack exchange is the preferred method for retrograding flat racks. FRCP's outside of the LSA are designated for flat rack consolidation purposes when required and these proposed locations are reported to the distribution division.

3. Flatrack Reporting Procedures

Accurate daily reporting of flatracks in a unit's area of operations by location, status, and condition is critical to efficient management of these assets within the distribution system. Separate reports may not be required for reporting flatrack status, since distribution managers may roll flat rack status into existing reports. Requests for supplemental transportation to retrograde flat racks in the operational area are submitted as routine transportation requests through support operations channels. Flatrack procedures should be reinforced and clarified with unit tactical standing operating procedures (TACSOP's).

D. Aerial Delivery

Aerial delivery consists of Service fixed-wing, rotary-wing and, in some cases Unmanned Aircraft Systems (UAS's), conducting air-land, airdrop, and sling-load cargo operations. It supports the sustainment requirements of combat units on a noncontiguous AO with extended LOC's, reducing the amount of traffic on major ground supply routes.

1. Air-Land Operations

Air-land operations provide for greatest cargo tonnage movement and are conducted routinely on an inter-theater basis. Using this method, air carriers normally terminate at a relatively secure APOD, physically descending, landing, and spending time on the ground to off-load. In doing so, supplies and equipment are normally introduced to the ground for onward movement and distribution. Continued, intratheater, air-land movement to forward austere airfields commonly restricts the use of fixed-wing aircraft due to threat capabilities, necessary airfield preparation, and off-load capabilities. When continued intratheater, air-land operations are conducted, rotary-wing assets conducting internal cargo carrying operations are normally employed. However, use of rotary-wing aerial delivery platforms can present range limitations.

2. Airdrop Operations

Airdrop permits throughput of supplies from as far rearward as the National level, directly to the using unit, and reduces the need for forward airfields, landing zones, and materiel handling equipment, effectively reducing the forward battlefield footprint, as well as mitigating the enemy threat to traditional surface methods of distribution. Airdrop provides the capability to supply the force, even when landlines of communication have been disrupted, adding flexibility to the distribution system. New aerial delivery platforms increase the flexibility and responsiveness to move supplies quickly and precisely to a BSB or throughput to a maneuver unit from a sustainment base. Certain contingencies may require airdrop re-supply support from the beginning of hostilities. However, the requisite airdrop support structure may not be in a theater of operations due to deployment priorities. In such cases, the strategic-level commander should consider having a portion of the supporting airdrop supply company deploy to designated APOE's responding to the deployment and where supply support to the contingency area is more abundant. This might include airdrop capability at Intermediate Staging Bases, Depots, and DLA locations. When in theater, airdrop supply companies, functioning as part of the Sustainment Brigade, should be located at major transportation, supply and distribution hubs with capable airfields. Such hubs should also include depot and DLA sites, where supplies may be rigged or pre-rigged and readily available for tasked aircraft. Using the range and speed of air carriers, forces then fly intra-theater airdrop missions, supplying directly to the airdrop location and then returning to a transportation hub.

E. Transportation Units

Ref: FMI 4-93.2, Sustainment Brigade (Feb '09), pp. 4-44 to 4-46.

Motor Transportation Battalion

The mission of the motor transportation battalion is to command, control, and supervise units conducting motor transport operations and terminal operations (less seaport). Motor transportation battalions are typically assigned to the TSC upon arrival into a theater of operations, and are further attached to a Sustainment Brigade. Its core capabilities are providing C2 and technical supervision for three to seven motor transport or cargo transfer companies. The battalion plans and schedules requirements to conform to the overall movement program. The staff—

- Translates transportation requirements from higher HQs into specific vehicle or unit requirements
- Evaluates highway traffic plans affecting road movement, to include terrain, road conditions, and security
- Supervises the operation of truck terminals, trailer transfer points, and/or a trailer relay system
- Coordinates for host nation support as available

Transportation Terminal Battalion

The mission of the Transportation Terminal Battalion is to command, control, and supervise units conducting terminal operations. Transportation Terminal Battalions are typically assigned to the TSC upon arrival into a theater of operations and are further attached to a Sustainment Brigade. Terminal operations include truck, rail, air, as well as marine terminals and ports. Its core capabilities are providing C2 and technical supervision for three to seven terminal operations and/or watercraft companies. The battalion plans and schedules requirements to coincide with the strategic and operational distribution and movement programs. The objective is to balance the flow of materiel and personnel from strategic transportation modes with the ability of the operational providers to clear the terminal or port. Achieving this is paramount to keeping the ports of debarkation un-congested and units, personnel, and supplies flowing smoothing into and out of a theater of operations. The staff—

- Translates transportation requirements from higher HQs into specific vessel discharge and terminal/port clearance schedules
- Evaluates terminal and port facilities for usability
- Evaluates ingress and egress routes and highway traffic plans affecting road movement, to include terrain, road conditions, and security
- Supervises the operation of truck, rail, air and marine terminals
- Coordinates for host nation support as available

Transportation Companies

Transportation companies provide lift capability for both the operational and tactical sustainment mission of the sustainment brigade. Truck companies move personnel and materiel throughout the distribution system, while cargo transfer and terminal companies provide capabilities essential to the theater opening, port operations, and hub operations. Transportation companies are typically assigned to the CSSB or functional transportation battalions that are attached to sustainment brigades. This section of the chapter identifies various transportation companies that might be attached to a sustainment brigade and provides information on general capabilities of each company.

Convoy Support Centers (CSC)

Ref: FMI 4-93.2, Sustainment Brigade (Feb '09), pp. 4-46 to 4-49.

CSC provide mess, maintenance, crew rest facilities, and other personnel and equipment in support of convoys moving along Main Supply Routes/Alternate Supply Routes (MSR/ASR). CSC's can also serve as Life Support Areas and, in some instances, a site for supply point distribution. They are a cross between a trailer/cargo transfer point and an intermediate truck terminal with the mission to support, enhance, and otherwise facilitate direct haul convoy operations. CSC's are located along the MSR/ASR, generally every 200 miles or as required by METT-TC. When practical, they will be collocated with existing organizations such as the TDC/hub, an inter-modal terminal, or a Central Receiving and Shipping Point (CRSP). When operating as a stand-alone facility, they may require additional capability for protection above Level 1. Services provided by all CSC's are refueling, quick-fix maintenance, and asset recovery. They are established at the discretion of the Combatant Commander, and are normally operated by a CSSB. Three levels of CSC support are described below.

1. Truck Stop

A truck stop CSC is a full service CSC that operates as part of a larger support area, supporting convoys moving through the area en route to another hub or final destination. The sustainment brigade designates a CSSB to provide C2 of a CSC providing full support. Support units required to man the truck stop type of CSC are: POL support platoon (50K) with bag farm, MCT, maintenance contact team, medical treatment team, and an infantry platoon for security (METT-TC). The truck stop CSC is capable of providing bulk and retail Class III, bulk and bottled water, Class I (MKTUGR/MRE), Class IX (ASL and prescribed load list), Class V for protection, and replenishing combat lifesaver (CLS) bags. The truck stop medical treatment team is authorized to carry limited CLS bag re-supply stocks to support contingency re-supply operations for convoys. Limited stocks consist of common high use CLS restock items minus controlled substances (narcotics). A truck stop CSC provides life support for all convoy personnel and might include one or more 150-person modules of the Force Provider set (tents and cots), a shower, laundry, and clothing repair (SLCR) team from a field services company and Army and Air Force Exchange Service support.

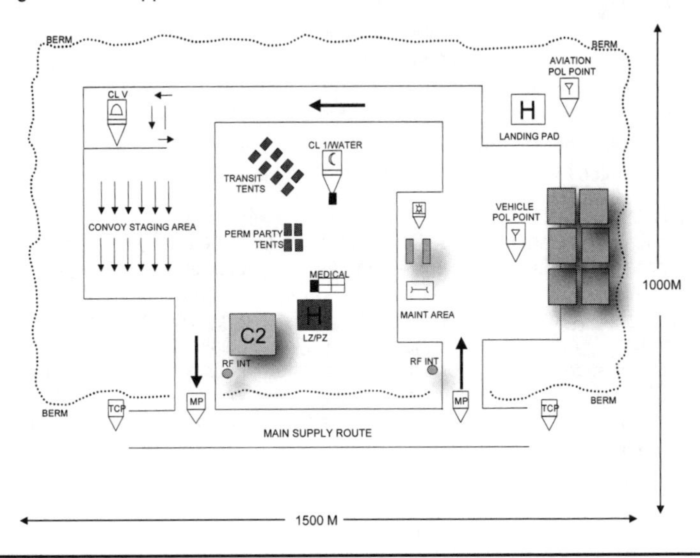

2. Mini-Mart

Mini-Mart CSC is a medium service CSC that operates as part of a smaller support area supporting convoys moving through the area en route to another hub or final destination. The CSSB provides C2 for the Mini-Mart CSC. Support units required to man the truck stop type of CSC are:

- POL support platoon (50K) with bag farm, MCT, maintenance contact team, medical treatment team, and an infantry platoon for security (METT-TC). The Mini-Mart CSC is capable of providing bulk and retail
- Class III, bulk and bottled water, Class I (MKT-UGR/MRE), and replenishing combat lifesaver (CLS) bags using procedures. There is no life support capability except for CSC personnel.

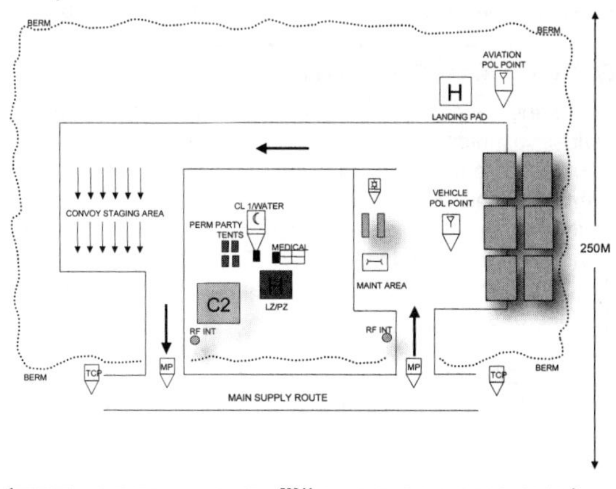

3. Pit Stop

Pit stop CSC provides minimal service. It is normally limited to Class III and Class I (MRE/water) and possibly minimal maintenance support. The CSSB provides C2 for the Pit Stop CSC. Support units required to man the truck stop type of CSC are: POL support platoon (truck to truck), MCT, maintenance contact team, CLS capability, and an infantry squad for security (METT-TC). The truck stop CSC is capable of providing bulk and retail Class III, bulk and bottled water, and Class I (MKTUGR/MRE). There is no LOG automation. There is no life support capability except for CSC personnel.

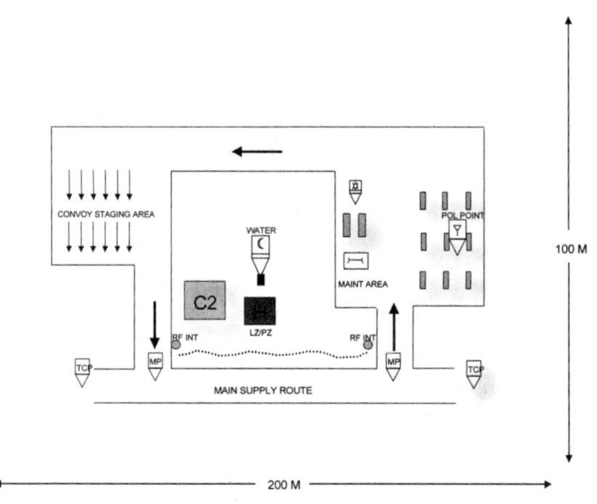

Airdrop operations provide for the ability to supply and distribute cargo, using parachutes and platforms to release supplies and equipment from an aircraft while in flight. When conducted in range, this method of aerial delivery requires no descent or landing deep in a combat area or remote area.

3. Rotorcraft Sling Loading

Rotorcraft sling loading provides for rapid movement of heavy, outsized cargo directly to the user, bypassing surface obstacles. This provides greater responsiveness and flexibility to the ground commander by reducing planning cycle time. It also allows the use of multiple flight routes and landing sites, which enhance survivability of the aircraft and crew. Rotorcraft are highly flexible, forward positioned, aerial delivery platforms. Operationally, they are able to perform missions surrounding all three methods of aerial delivery in support of Full Spectrum Operations.

F. Movement Control Battalion (MCB)

The MCB commands and controls between four and ten movement control teams (MCTs), provides technical supervision, and coordinates the use of common user transportation assets theater-wide. The MCB also provides visibility of unit moves, convoy movements, and operational moves. The MCB is assigned to the TSC and accountable for the execution of the movement program and performance of the theater transportation system. In the current force, an MCB will be under the TACON or administrative control (ADCON) of a sustainment brigade when the sustainment brigade is the senior sustainment HQ in an AO.

Movement Control Teams (MCT)

The MCT is the lynch pin of the Movement Control System. At least one will normally be collocated with the sustainment brigade to support brigade operations. Elements of an MCT may also collocate with operational area under the command of a sustainment brigade such as ports or convoy support centers. An MCT is a 21-person team with the capability to perform every type of movement control mission on a 24-hour basis. It is designed to provide maximum flexibility in its employment. Each team has a headquarters section and four identical sub-units (or sections). The MCT is capable of operating as a single team or separately at up to four different locations. The MCT is capable of conducting the following missions:

- Coordinate transportation support, highway clearance, and inbound clearance for moving units, personnel, and cargo
- Coordinate transportation movements, diversions, re-consignments, and transfers of units, cargo, and personnel
- Provide technical expertise to transportation users within its assigned area of responsibility
- Provide ITV of unit equipment and sustainment cargo movements in an assigned area of responsibility
- Observe, assess, and report on the progress of tactical and nontactical transportation movements along MSRs or alternate supply routes and through critical nodes
- Adjust movement schedules as necessary to coordinate the movement of authorized traffic
- Provide first destination reporting points
- Provide as many as four sub-units to four separate locations, each performing a different aspect of movement control
- Commit transportation assets

V. Distribution

Distribution is defined as the operational process of synchronizing all elements of the logistics system to deliver the right things to the right place and right time to support the CCDR. It is a diverse process incorporating distribution management and asset visibility.

See pp.1-42 to 1-43 and 2-22 to 2-25 for further discussion of distribution.

Theater Distribution Center (TDC)

Unlike the CRSP, the theater distribution center (TDC) holds and stores supplies in addition to performing the functions of a hub. As well as tracking en route cargo, the TDC will perform many of the same receiving and issuing functions of an SSA and would cross-level excess materiel to cover shortages among the units it supports. A TDC must perform the following:

- Supply management functions found in a quartermaster support company
- Cargo tracking and convoy movement control functions found in movement control teams
- Cargo handling functions found in cargo transfer companies

C2 would be provided by a sustainment brigade (TD). The TDC would require a large area for operations to include covered climate-controlled storage for cargo, open storage for containers and MILVAN's, secure storage for high value cargo, and possible capability for refrigerated storage if contractor support is unavailable for food acquisition and storage. To prevent pilferage, the entire area should remain secured with controlled access. Unless the center establishes operations in a prepared location, engineering assets may be necessary to prepare the site. The distribution center (DC) should be able to handle all classes of supply except ammunition. POL storage and issuance is also a possibility.

The DC must be located near major road networks, airfields, and railheads to receive and distribute supplies through a variety of means. The platoon would require a variety of equipment to perform its mission, especially materials handling equipment (MHE) to include all terrain capable, organic bulk transport equipment, long haul transportation assets, light sets, generators, computers, RF interrogators, and radios. A TDC must have the equipment and personnel capable of organizing convoys, handling materiel inside the warehouse, and handling containers and 463L pallets. Personnel should have the skill to stuff and un-stuff containers and to properly build 463L pallets. TDC's would provide for receipt, storage, issue, and distribution of supplies. It would configure LOGPAC loads for forward distribution and supports line haul and local haul motor transport operations. A TDC serves as a functional base for subsistence distribution. The maintenance activity provides emergency refueling and repair of vehicles transiting the DC as well as complete support of owned equipment.

VI. Operational Contract Support

Operational contract support is the process of planning for and obtaining supplies, services, and construction from commercial sources in support of operations along with the associated contractor management functions.

See p. 1-37 and 1-56 for further discussion of operational contract support.

VIII. General Engineering

General engineering activities modify, maintain, or protect the physical environment (see FM 3-34.400).

See pp. 1-56 to 1-58 for further discussion.

Centralized Receiving and Shipping Point

Ref: FMI 4-93.2, Sustainment Brigade (Feb '09), pp. 4-44 to 4-46.

A Centralized Receiving and Shipping Point (CRSP) is a dock-to-dock distribution center using the hub and spoke method of efficiently delivering cargo. Generally, cargo is not warehoused at a CRSP, with the common holding period being 24 hours or less. The objective is to move cargo as quickly and efficiently as possible, the exception being frustrated cargo, cargo destined to low volume consignees, or battle damaged equipment, which might require inspection and processing. Although traditional use of CRSPs was for container handling only, the mission was expanded to great advantage during Operation Iraqi Freedom. Use of CRSP operations reduced transit times and provided greater security, as the convoy operators were more familiar with assigned terrain and threats than in-theater personnel who had previously been performing the convoy operations all the way to the tactical level. Under the CRSP concept, theater convoys deliver to a CRSP(s) with CSSB's operating convoys delivering to the consignee, forward operating base (FOB), or other CRSP. Each CRSP would arrange for backhaul both from the FOB's to the CRSP and from the CRSP to the theater-level supply units. Convoys should pick up all retrograde cargo from a FOB, regardless of whether or not the consignee is part of the CRSP's network as it is easier for the CRSP to use the CRSP network to trans-ship the retrograde to the appropriate CRSP. AMC could, as arranged, come to the CRSP to pick up equipment for turn-in and remove the equipment from the units' property books.

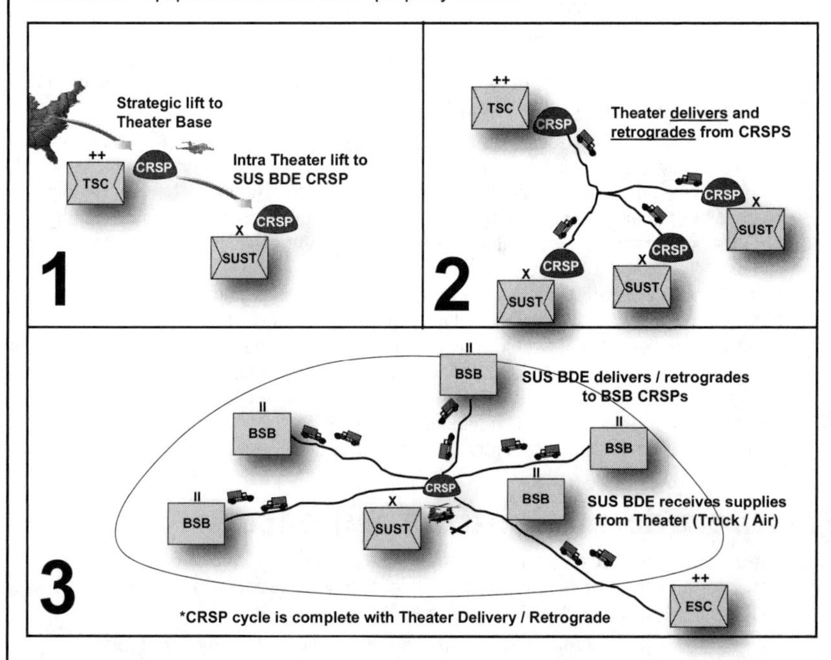

CRSPs should be used for all classes of supply except ammunition. Any mode of distribution should be accessible. Personnel should be capable of properly packing containers, loading helicopters, or building 463L pallets for fixed wing aircraft. A CRSP may be collocated with an MCT, A/DACG, or USAF Aerial Port.

V. Personnel Services Support to the Warfighter

Ref: FMI 4-93.2, Sustainment Brigade (Feb '09), chap. 4.

Personnel services are those sustainment functions related to Soldiers' welfare, readiness, and quality of life. Personnel services complement logistics by planning for and coordinating efforts that provide and sustain personnel.

See pp. 1-59 to 1-68 for discussion of personnel services as a sub-function of the sustainment warfighting function.

Personnel Services Sub-Function

Human Resources (HR) Support

Financial Management (FM) Operations

Legal Support

Religious Services

Band Support

Ref: FM 4-0, Sustainment, chap. 5.

I. Human Resources (HR) Support

The objective of HR support is to maximize operational effectiveness and to facilitate enhanced support to the total force (Soldiers, DOD civilians, Joint service personnel, and others that deploy with the force). Reliable, responsive, and timely HR support in the operational area is critical to supporting the operational commander and the force. It relies on secure, non-secure, robust, and survivable communications and digital information systems. These provide a common operational picture, asset visibility, predictive modeling, and exception reporting—important to making accurate and timely manning decisions. HR support as described in FM 1-0, Human Resources Support, consists of 10 core functions and capabilities. These include:

- Personnel readiness management (PRM)
- Personnel accountability and strength reporting (PASR)
- Personnel information management (PIM)
- Reception, replacement, redeployment, rest and recuperation, and return-to-duty (R5)

- Casualty operations
- Essential personnel services (EPS)
- Postal operations
- MWR
- Band operations
- HR planning and operations

At the unit level, the focus of HR support encompasses all core functions and are conducted by brigade and battalion S-1 sections for assigned or attached personnel. HR units attached to the sustainment brigade provide HR support on an area basis.

The commander of sustainment organizations is responsible for the training, readiness oversight, and mission execution of assigned/attached HR organization. FMI 1-0.02, Theater Level HR Operations, provide additional detail on these area support functions. The commander has an overall responsibility for:

- **Postal**. The commander has the overarching mission to ensure adequate postal forces and activities are in place to ensure the timely delivery of mail to/from Soldiers in the brigade footprint and for ensuring compliance with USPS directives. The commander must also ensure postal force is trained and mission capable and that adequate facilities are established and sustained.
- **Casualty**. The commander has a critical role to ensure casualty elements (platoons and teams) are resourced and distributed in compliance with theater policy that ensures timely and accurate reporting. The commander must be prepared to provide or coordinate augmentation in the event of mass casualty. Through the assigned HR Company, the commander ensures casualty personnel are trained and capable to execute this mission.
- **R5**. The commander has the responsibility to ensure R5 operations are resourced and adequate to account for all personnel entering or exiting at all APOD/E and SPOD/E when an R5 Team is assigned. The critical HR task accomplished in R5 operations is personnel accountability, which updates the theater database (DTAS). The commander ensures all life support, transportation, and coordination requirements are provided in a timely manner to minimize any delays, particularly for personnel joining their units.

The HR Branch within the SPO is the commander's hub for HR integration, synchronization, planning, coordination, sustainment, and operations. This is a valuable asset that provides the commander a well-trained and capable staff to plan and coordinate sufficient HR organizations needed to provide HR support based on the commander's intent. It is critical that the HR Branch is included in the planning process of the staff (SPO). The HR Branch provides technical guidance and resources to the SRC12 organizations (supporting units) ensuring they have the capability to provide the required postal, R5, and casualty support directed in the HR concept of support. Guided by supported/supporting relationships, the G-1/S-1s request support and resources for postal and R5 operations through the HR Branch. The HR Branch processes the request, prioritizes the request based on the available HR resources and scope of requested support to determine supportability. If the HR Branch is unable to support the request with HR assets internal to its sustainment brigade, the HR Branch forwards the request to the ESC/TSC's SPO for resolution.

Human Resources Company

The HR company provides theater-directed HR support for the postal, R5, and casualty core competencies. The company is tailorable and scalable, allowing Sustainment commanders to tailor the support to meet the requirements of the supported population. The HR Company has specific capabilities and associated tasks and is not intended to augment S-1 sections for the delivery of HR support outside the postal, R5, and casualty core competencies.

The MMT team provides tailorable HR support to the theater of operations by establishing, coordinating, and executing military mail terminal operations under the control of the sustainment brigade with the augmentation of an HR company. The MMT mission is a specific task required to ensure the overall effectiveness of the theater postal system.

The TG R5 team provides tailorable HR support to establish the TG R5 center at the inter-theater APOD to establish and maintain the personnel accountability of entering, departing, and transiting personnel and to coordinate the execution of sustainment tasks within the Sustainment brigade and joint elements operating in and around the APOD. The TG R5 center mission is a critical element of the early theater opening sustainment mission and has an enduring requirement during theater distribution operations, as it ensures the establishment and maintenance of an accurate theater deployed accountability system required by Congress.

II. Financial Management (FM) Operations

The finance operations mission is to sustain Army, joint, and multinational operations by providing timely commercial vendor and contractual payments, providing various pay and disbursing services, maintaining battle flexibility for combat units to engage enemy forces and following policies and guidelines established by the National FM providers. Financial management units provide support during all operational phases. The level of support varies according to the nature of the operation and depending on METT-TC factors.

Financial Management SPO (FM SPO)

The Sustainment Brigade FM Support Operations Team (FM SPO) monitors and tracks financial management operations throughout the area of operations (AO). The team integrates all FM operations; plans the employment of FM units; coordinates FM requirements; synchronizes the FM network; and manages the FM systems. It also coordinates for additional operational and strategic FM support when needed. All functions of the FM SPO are closely coordinated with the FM CO commander and either the Division or ARFOR G-8. Some of these functions include:

- **Integrate all FM operations within the AO**. The FM SPO evaluates the adequacy of support throughout the AO by analyzing workload data, supporting population size, scheduling support at forward operating bases, and making recommendations to maximize the efficiency of limited FM resources.
- **Plan employment of FM units**. The FM SPO works with the Financial Management Center (FMC), G-8, and FM COs to adapt financial management support to changes in the operational area.
- **Synchronize division FM network**. The FM SPO coordinates financial management efforts for the sustainment brigade across all supporting FM COs, contracting offices, and civil affairs units.
- **Manage all FM systems**. In coordination with the FMC, G-8, and FM CO, monitors the employment and update of financial management systems used for disbursing, military pay, travel pay, vendor pay, and resource management. Planning would include coordinating communications support for expanding electronic banking initiatives.
- **Coordinate FM requirements**. In coordination with the FMC, G-8, and FM CO, coordinates cash requirements for special funding sources such as Commander's Emergency Response Program, Soldier support (check cashing, casual pays, and vendor payments).
- **Coordinate operational FM support**. The FM SPO establishes mutually supporting flow of information among the FMC, G-8, and the FM CO, enabling responsive FM support in the operational area.

FM Company (FM CO)

The FM CO is assigned to either the sustainment brigade STB or a CSSB. The FM CO analyzes the supported commander's tasks and priorities to identify the financial resource requirements that will enable mission accomplishment. The FM CO performs the following functions:

- Ensures regulatory guidelines, directives, and procedures are adhered to by all operational elements of the FM CO and subordinate FM detachments
- Establishes Disbursing Station Symbol Numbers and Limited Depository accounts
- Provides limited US and non-US pay support
- Funds subordinate FM detachments and determines the need for currency (US and foreign) and its replenishment
- Maintains regulatory accountable records
- Provides EPW, civilian internee, and Local National pay support
- Establishes banking relationships and procedures
- Makes payments on prepared and certified vouchers
- Receives collections
- Receives and controls all currencies and precious metals
- Cashes negotiable instruments
- Converts foreign currency
- Protects funds from fraud, waste, and abuse
- Establishes a management internal control process providing reasonable assurance that government assets are protected and safeguarded
- Ensures funds and other assets are protected and revenues and expenditures are properly accounted for in accordance with congressionally mandated accounting and reporting requirements

III. Legal Support

Members of The Judge Advocate General's Corps (JAGC) provide proactive legal support on all issues affecting the Army and the Joint Force and deliver quality legal services to Soldiers, retirees, and their families. Legal support centers on six core disciplines across full-spectrum operations. The six core disciplines are: military justice, international and operational law, contracts and fiscal law, administrative and civil law, claims, and legal assistance. *See pp. 1-64 to 1-66 for further discussion.*

IV. Religious Support

Religious support facilitates the free exercise of religion, provides religious activities, and advises commands on matters of morals and morale. The First Amendment of the U.S. Constitution and Army Regulation (AR) 165-1 guarantees every American the right to the free exercise of religion. Commanders are responsible for fostering religious freedoms. Chaplains and chaplain assistants functioning as Unit Ministry Teams (UMT) perform and provide RS in the Army to ensure the free exercise of religion (see FM 1-05). *See pp. 1-66 to 1-67 for further discussion.*

V. Band Support

Army bands provide critical support to the force by tailoring music support throughout military operations. Music instills in Soldiers the will to fight and win, foster the support of our citizens, and promote America's interests at home and abroad. (see FM 1-0 and FM 1-19). *See p. 1-68 for further discussion.*

VI. (HSS) Health Service Support to the Warfighter

Ref: FMI 4-93.2, Sustainment Brigade (Feb '09), chap. 4.

Health service support consists of all support and services performed, provided, and arranged by the Army Medical Department. It promotes, improves, conserves, or restores the mental and physical well being of Soldiers and, as directed, other personnel.

See pp. 1-69 to 1-72 for discussion of health service support as a sub-function of the sustainment warfighting function. See also pp. 4-31 to 4-36 for HSS considerations in support of joint and multinational operations.

HSS Sub-Function

HSS includes casualty care, which involves all Army Medical Department functions, to include:

- **Organic and area medical support**
- **Hospitalization**
- **Dental care**
- **Behavioral health**
- **Clinical laboratory services**
- **Medical evaluation**
- **Medical logistics**

Ref: FM 4-0, Sustainment, chap. 5.

Army Health System (AHS) Support

AHS supports a subcomponent of the military health system capabilities to deliver HSS and FHP in support of full spectrum operation. AHS support involves the delineation of support responsibilities by capabilities (roles of care) and geographical area (area support). The AHS that executes the HSS/FHP initiatives is a single, seamless, and integrated system. It is a continuum from the point of injury or wounding through successive roles of care to the CONUS-support base. The AHS encompasses the promotion of wellness and preventive, curative, and rehabilitative medical services. It is designed to maintain a healthy and fit force and to conserve the fighting strength of deployed forces.

A. The Sustainment Brigade Surgeon

Ref: FMI 4-93.2, Sustainment Brigade (Feb '09), pp. 2-20 to 2-23.

The sustainment brigade surgeon ensures that all AHS support functions are considered and included in operation plans and operation orders. He/She coordinates for AHS support for both Health Service Support (HSS) and force health protection (FHP). The sustainment brigade command surgeon coordinates AHS support operation with both the division surgeon and the medical brigade (MEDBDE) commander and helps establish medical guidelines for the division and the sustainment brigade.

Health Service Support (HSS)

The sustainment brigade surgeon's duties and responsibilities for AHS may include:

- Advising the commander on the health of the sustainment brigade units
- Plans and coordinating for HSS for sustainment brigade units (including but not limited to, medical treatment, medical logistics, medical evacuation, hospitalization, dental support, preventive medicine [PVNTMED], behavioral health, and clinical medical laboratory support)
- Developing and coordinating the HSS portion of AHS operation plans to support the sustainment commander's decisions, planning guidance, and intent in support of full spectrum operations
- Determining the medical workload requirements (patient estimates)
- Advises the sustainment brigade commander on policy regarding the eligibility of care for non-US military personnel
- Maintaining situational understanding by coordinating for current HSS information with surgeons of the next higher, adjacent, and subordinate HQ
- Recommending task organization of medical units/elements in support to sustainment brigade units to satisfy all HSS mission requirements
- Recommending policies concerning medical support of stability operations (that include civil military operations)
- Monitoring troop strength of medical personnel and their utilization
- Coordinating and synchronizing health consultation services
- Evaluating and interpreting medical statistical data
- Monitoring medical logistics and blood management operations in the theater.
- Monitoring medical regulating and patient tracking operations for sustainment brigade personnel
- Determining sustainment brigade training requirements for first aid and for maintaining wellness of the command
- Ensuring field medical records are maintained on each Soldier assigned to the TSC at their primary care medical treatment facility
- Establishing, in coordination with the chain of command, and promulgating a plan to ensure individual informed consent is established before administering investigational new drugs as described in Executive Order 13139
- Recommending disposition instructions for captured enemy medical supplies and equipment
- Submitting to higher HQ those recommendations on medical problems/conditions that require research and development

Coordinates and synchronizes:

- Health education and combat lifesaver training for the brigade
- Mass casualty plan developed by the S-3
- Medical care of enemy prisoners of war (EPW), detainees, and civilians within the brigade AO
- Treatment of sick, injured, or wounded Soldiers
- Medical evacuation, including use of both the Army's dedicated medical evacuation (MEDEVAC) platforms (air and ground)
- Medical logistics including Class VIII re-supply, blood management, and medical maintenance
- Health-related reports and battlefield statistics
- Collection and analyses of operational data for on-the-spot adjustments in the medical support structure and for use in post operations combat and materiel development studies
- Army Health System support for stability and civil support operations

Force Health Protection (FHP)

The sustainment brigade surgeon's duties and responsibilities for FHP may include:

- Identifying potential medical-related commander's critical information requirements (priority intelligence requirements and friendly force information requirements) as they pertain to the health threat, ensuring they are incorporated into the command's intelligence requirements
- Coordinating for veterinary support for food safety, animal care, and veterinary preventive medicine to include zoonotic diseases transmissible to man
- Planning for and implementing FHP operations to counter health threats

Force health protection operations may include:

- Planning for and accomplishing redeployment and post deployment health assessments
- Establishing and executing a medical surveillance program
- Establishing and executing an occupational and environmental health surveillance program
- Recommending combat and operational stress control, behavioral health, and substance abuse control programs
- Ensuring the general threat, health threat, and medical intelligence considerations are integrated into AHS support operation plans and orders
- Advising commanders on FHP chemical, biological, radiological, and nuclear (CBRN) defensive actions, such as immunizations, use of chemoprophylaxis, pretreatments, and barrier creams
- Identifying health threats and medical-related commander's critical information requirements
- Maintaining situational understanding by coordinating for current FHP information with surgeon staffs of the next higher, adjacent, and subordinate HQs

Coordinates and synchronizes:

- Combat and operational stress control program with the division surgeon section (DSS) and supporting medical brigade
- Veterinary food inspection, military working dogs and other animal care, and veterinary preventive medicine activities of the command, as required
- Preventive medicine services to include identification of health threats
- Preventive dentistry support program for the prevention of cavities and gum disease
- Area medical laboratory support to include the identification of biological and chemical warfare agents, as required

Brigade Surgeon Section (BSS)

The BSS monitors and tracks operations with medical communications for Medical Communications for Combat Casualty Care (MC4) System and provides updated information to the surgeon and the SPO chief for building capabilities to meet the sustainment brigade's medical requirements identified by the surgeon. The BSS consist of two cells (a plans and operations cell and a medical logistics (MEDLOG) and sustainment cell). Also, under the technical control of the surgeon is the medical treatment team and evacuation squad.

- **Medical Plans and Operations Cell.** The medical plans and operations cell is normally staffed with medical operations officers, a medical operations NCO. The primary function of this cell is medical planning to ensure that adequate AHS support is available and to provide, in a timely and efficient manner, for the sustainment brigade and its attached units. This cell coordinates with the DSS and, as authorized, with medical brigade for the placement and support requirements of medical units and elements located in the sustainment brigade AO.
- **Medical Logistics and Sustainment Cell.** The medical logistics and sustainment cell is normally staffed with a MEDLOG officer (Major, 04, AOC 70K00). This cell receives daily updates on the status of Class VIII within the brigade and from attached medical units/elements. This cell may update priorities with the supporting MEDLOG activity to correct deficiencies in the delivery system. The supporting MEDLOG company or SSA will forward information to the MEDLOG and sustainment cell on items filled and shipped and on those requisitions that were not filled. This cell provides daily updates to the sustainment brigade surgeon and SPO chief.

B. Medical Brigade (MED BDE)

The MED BDE may be OPCON to a sustainment brigade when the sustainment brigade is in a command relationship with the senior tactical headquarters. The MED BDE provides a scalable expeditionary medical C2 capability for assigned and attached medical functional plugs task-organized for support of deployed forces. The MED BDE brings all requisite medical C2 and planning capabilities to provide responsive and effective AHS throughout the AO. Some MED BDE subordinate elements will collocate with sustainment units in LSA's and FOB's, because MTF's require essential non-medical supplies and services and the LSA's and FOB's will require AHS support since most sustainment units do not contain organic medical assets.

C. Medical Reporting

The MC4 and Theater Medical Information Program support the information management requirements for the brigade surgeon section and BCT medical units. The brigade surgeon section uses BCS3, FBCB2, and MC4-TMIP to support mission planning, coordination of orders and subordinate tasks, and to monitor/ensure execution throughout the mission.

The MC4-TMIP is an automated system, which links health care providers and medical support providers, at all levels of care, with integrated medical information. The MC4-TMIP receives, stores, processes, transmits, and reports medical C2, medical surveillance, casualty movement/ tracking, medical treatment, medical situational awareness, and medical logistics data across all levels of care.

See pp. 1-69 to 1-72 for discussion of health service support as a sub-function of the sustainment warfighting function. See also pp. 4-31 to 4-36 for HSS considerations in support of joint and multinational operations.

VII. Theater/Expeditionary Sustainment Commands

Ref: FMI 4-93.4, Theater Support Command (Apr '03) and FM 4-0, Sustainment (Aug '09), chap. 2.

The Theater Sustainment Command (TSC) serves as the senior Army sustainment HQ for the Theater Army. The TSC provides C2 of units assigned, attached, or under its OPCON. The mission of the TSC is to provide theater sustainment (less medical).

TSC-ESC-Sustainment Brigade

++ TSC	■ Assigned to the ASCC ■ AOR focused ■ C2 logistics across a GCC AOR ■ Performs theater-wide materiel and distribution management
+ ESC	■ Forward presence of the TSC - focuses on a particular area of operations or JOA ■ Performs specific materiel and distribution management in assigned area of operations or JOA
X SUST	■ TSC's face to Brigades / Divisions / Corps ■ Multifunctional capability supporting on an area basis ■ Executes multi-nodal and multi-modal transportation ■ Standard HQs design: theater opening, theater distribution and theater sustainment mission sets ■ Theater opening and distribution missions enabled with TTOE and TDAE ■ Performs some materiel management at the local level

The TSC is capable of planning, preparing, executing, and assessing logistics and human resource support for Army forces in theater or JFC. As the distribution coordinator in theater, the TSC leverages strategic partnerships and joint capabilities to establish an integrated theater-level distribution system that is responsive to Theater Army requirements. It employs Sust Bdes to execute theater opening (TO), theater sustainment, and theater distribution operations.

The TSC includes units capable of providing multifunctional logistics: supply, maintenance, transportation, petroleum, port, and terminal operations. Other specialized capabilities, such as MA, aerial delivery, human resources, sustainment to I/R operations, and FM, are available from the force pool.

This section deals with the TSC & ESC as they relate to Sustainment Brigade operations. For discussion of additional operational-level sustainment organizations and activities, see pp. 1-18 to 1-19.

I. Theater Support Command (TSC)

The TSC is a multifunctional support headquarters that works at the operational level with links to strategic- and tactical-level support organizations and agencies. The ASCC commander supervises the TSC's peacetime contingency planning. When the TSC, or any part of it, deploys to an AO, it reports to the commander, Army forces. The ARFOR commander may be the ASCC commander or a lower-level commander depending on the scale of operations. During peacetime planning the ASCC commander provides guidance for the types of combat support (CS) and CSS capabilities that may be attached to the TSC for a given contingency. This is done in accordance with the Joint Operations Planning and Execution System (JOPES). (See Chairman of the Joint Chiefs of Staff Manual [CJCSM] 3122.03.)

The TSC has some permanently assigned major subordinate units. The ASCC commander may attach other units to the TSC for specific operations. Support requirements at the operational-level vary considerably depending on the type of operations and the scale of the deployment. The ASCC commander has the flexibility to tailor the support presence in the AO appropriately.

TSC Mission

The mission of the TSC is to maximize throughput and follow-on sustainment of Army forces and other supported elements regardless of the scale of operations. Throughput in this sense means that the TSC ensures that unit personnel, unit equipment, and commodities move to their point of employment with a minimum number of intervening stops and transfers. For this reason the TSC establishes command of support operations and control of the distribution system before those elements arrive in the AO. The TSC provides area support to the operational-level units in the AOs and overall sustainment support to Army forces. This support may include interim tactical-level support to early deploying corps and divisional elements. The TSC also executes those lead service CUL support requirements that the ASCC commander assigns.

The TSC commander has a vital interest in the security and terrain management of the rear area. Depending on the joint force commander (JFC) and ARFOR commander decisions, the TSC responsibility may range from the inherent responsibility for the internal security of TSC elements to being formally designated as the joint rear area coordinator (JRAC).

Area Support Groups (ASGs)

Area support groups (ASGs) are subordinate units assigned to the TSC. They are responsible for area support in the AO and may provide support to corps or other forces. The mission of the ASG is to provide DS logistic support to designated units and elements within its AO. This support typically includes DS supply (less ammunition, classified map supply, and medical supply and support), DS maintenance, and field services, as well as other support directed by the ARFOR commander through the TSC. ASGs can also provide GS supply and sustainment maintenance support to TSC and CZ DS supply organizations and sustainment maintenance to support the theater mission. If an operational-level ammunition group is not established, specialized battalions assigned to the ASG provide ammunition support. ASGs can support ISBs and RSOI operations. EEMs of specialized units may be attached to an ASG headquarters EEM during the initial stages of an operation.

ASGs provide a variety of support to units stationed in or passing through their areas. An ASG area depends on the density of military units and materiel to support and on political boundaries and identifiable terrain features. One ASG is assigned to a TSC for every 15,000 to 30,000 troops supported in the AO. ASGs are located along the LOC to take advantage of the transportation network and provide responsive support to the units they support.

FM 4-93.40 (FM 54-40) contains additional details on ASGs.

A. TSC Command Relationships

Ref: FM 4-93.4, Theater Support Command (Apr '03), pp. 2-4 to 2-10. See also p.2-4.

In organizing the TSC, the ASCC commander may elect to reduce his span of C2 over these specialized commands. If he does, he has available to him the three command relationships spelled out in FM 5-0 (FM 101-5) and defined in FM 1-02 (FM 101-5-1)-attachment, OPCON, and TACON. In brief, these relationships are defined as follows:

1. Attachment

Attachment is the placement of units in an organization where such placement is relatively temporary. Subject to limitations imposed by the attachment order, the commander of the organization receiving the attachment has the responsibility to provide the attached units with sustainment support beyond their organic capabilities.

2. Operational Control (OPCON)

OPCON is the authority to perform those functions of command over subordinate forces involving organizing and employing commands and forces, assigning tasks, designating objectives, and giving authoritative direction of military operations and joint training necessary to accomplish missions assigned to the command. In the Army, a unit OPCON to a command/unit continues to receive logistics support from its parent unit.

3. Tactical Control (TACON)

TACON in the Army allows commanders to apply force and direct the tactical use of logistics assets but does not provide authority to change organizational structure or direct administrative and logistical support. As with OPCON, the parent unit retains responsibility for logistics support to a unit under the TACON of another unit.

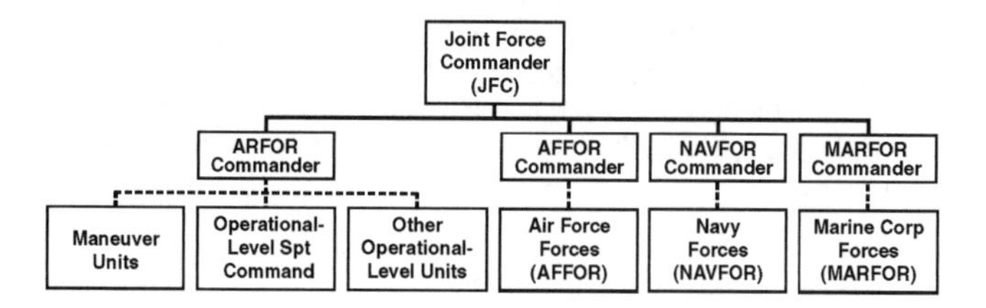

Service Component Alignment

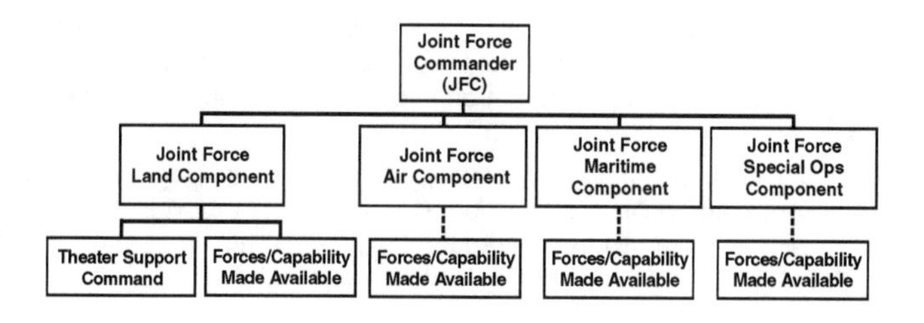

Functional Component Alignment

Regardless of the option the ASCC commander chooses for the specialized commands, they retain technical linkages with their respective national provider-level command and ASCC staff principals in order to execute their ASCC special staff functions.

B. TSC Operational-Level Support

Ref: FM 4-93.4, Theater Support Command (Apr '03), pp. 2-11 to 2-14.

JP 4-0 and FM 4-0 identify three levels of sustainment-strategic, operational, and tactical. Each level provides critical yet different types of support. Strategic- and operational-level sustainment supports wars, contingencies, campaigns, and major operations. Tactical sustainment supports battles and engagements. This section discusses the role of the TSC in force projection operations.

Given the range of responsibilities confronting the ARFOR commander, it is often necessary to consolidate and delegate selected responsibilities to subordinate commands. The TSC is uniquely designed to execute many of the sustainment responsibilities for the ARFOR commander. During theater opening, the TSC focuses primarily on supporting RSO&I. As the AO matures, the TSC shifts focus to sustaining operations.

See chap. 6, Deployment & Redeployment Operations, for further discussion.

Theater Force Opening Package (TFOP)

Deploying U.S. forces requires an in-theater support infrastructure capable of executing RSO&I operations and sustaining and redeploying the force. Recent operations in Somalia, Haiti, and Bosnia demonstrate a need for establishing early adequate support infrastructures in places where they did not previously exist. The theater force opening package (TFOP) is the Army mechanism to do this.

The TFOP is a modularly configured, multifunctional support task force comprised of specialized CSS and related CS modules. A typical TFOP needed during the initial stages of deployment includes transportation, engineer, supply, contracting, maintenance, and medical modules. The JFC may also elect to include strategic CSS cells from the USAMC, USAMMA, DLA, MTMC, and DESC.

The composition of the TFOP varies throughout the stages of a force projection operation until it becomes a TSC. The TFOP's mission remains identical to that for the TSC: to maximize throughput and follow-on sustainment of Army forces and other designated supported elements.The only difference is scale. As the TSC (forward), the TFOP is also uniquely responsible for building the theater infrastructure from a combination of existing and deploying assets.

1. Deployment

The ARFOR commander receives deploying forces, stages them, moves them forward, and integrates them into the theater structure. RSO&I is critical to successful force projection. RSO&I is complete when deploying units are determined combat effective by the operational commander. The ARFOR commander, based on guidance from the geographic combatant commander/JFC, determines the necessary level of combat effectiveness and the indices for determining this level. The TSC and other units track the build-up of the force by providing appropriate reports. The operational commander retains responsibility to track and report through operational channels the build-up of combat capabilities.

To meet requirements in this process, the Army developed a modular concept for opening theaters in which the TSC is a critical component. Modularity involves incrementally deploying only the minimum capabilities required to an AO. The TSC early entry module (EEM) provides C2 for many of the elements initially conducting RSO&I as directed by the ARFOR commander.

2. Employment

Though the line between entry and decisive operations may not be clear-cut, once the ARFOR commander has sufficient forces integrated into the total force to meet the commander's requirements, the emphasis for the TSC shifts from RSO&I support to sustaining the force.

The ARFOR commander plans and conducts force sustainment operations throughout the AO. The TSC conducts operational-level force sustainment to support the ARFOR, and tactical-level sustainment to forces operating in or passing through the TSC's AO. The TSC may also support other services, multinational partners, and NGOs and/or OGAs in accordance with the ARFOR commander's lead service responsibilities. As the AO develops, the EEM matures into a TSC, with all required capabilities and other required commands (as determined by the ARFOR commander).

The TSC distribution system can provide DS and GS to all designated forces operating within the rear/sustainment area and to any forces requiring related sustainment support as they transit the TSC AO. The primary customers of tactical-level support in the rear and sustainment area are the elements of the TSC and any specialized CS and CSS commands in theater. However, the TSC through the distribution systems may also be involved in some direct support to tactical forces.

The TSC is also involved in reconstitution either as part of sustaining decisive operations or as preparing for redeployment. The ARFOR commander plans and directs reconstitution operations. The TSC, typically through an area support group/area support battalion (ASG/ASB) is even more involved in regeneration. It usually establishes the regeneration site and provides most of the CSS elements of the regeneration task force. (FM 4-100.9 provides details on reconstitution).

3. Redeployment

The TSC is actively involved in redeployment in a number of ways. It may help redeploying units move to assembly areas, and plays a major role in reconstitution. It also controls the movement of units to the port of embarkation (POE) and provides life support at all nodes in the TSC AO.

Redeployment starts for forward units when they close into assembly areas (AAs) and continues at redeployment assembly areas (RAAs) activated and supported by the TSC. CSS activities are paramount during this period. Logistics functions include: identifying, separating, and reporting excess supplies and equipment to the appropriate materiel managers for disposing or redistributing as appropriate; initiating detailed equipment maintenance and cleaning; and canceling requisitions. Accounting for personnel and processing awards are two of the critical personnel activities under the responsibility of the PERSCOM. Combat health support (CHS) is an important factor throughout the redeployment process under the responsibility of the MEDCOM. Before redeployment, medical screening for clinical signs of disease and injuries and medical surveillance is required to ensure a fit and healthy force. If the ARFOR commander assigns the mission to the TSC commander, the TSC may oversee these administrative activities.

In all anticipated cases, the TSC receives, identifies, and determines disposition; maintains accountability; and stores, prepares for shipment, and arranges for movement of Class I, II, III (packaged), IV, V, VI, VII, and IX items to the port or designated storage location. Carrying out these functions may require augmentation from other military elements and/or contractor personnel. Contracted support may be the preferred solution to support Army forces leaving the mission area by operating seaports and aerial ports of debarkation. This includes operating wash racks and providing life support for redeploying units. The USAMC LSE or contractors may also repair items in the theater or send them to designated forward stations or CONUS GS or depot maintenance activities. USAMC's LSE also has major responsibilities for retrograde of Army pre-positioned stocks (APS) in the theater.

The TSC staff plans to transfer its responsibilities to another organization as the theater draws down. This may be an organization of another service or multinational partner, the USAMC LSE, an host nation support (HNS) organization, or an international agency.

See chap. 6, Deployment & Redeployment Operations, for further discussion.

C. TSC Support Operations

TSC support operations focus on establishing and maintaining the Army portion of the theater distribution system and sustaining the force in the AO consistent with the ARFOR commander and JFC strategic support priorities.

The TSC support structure responsible for support operations consists of three components —the support operations elements of the headquarters (in conjunction with the specialized commands), the control centers, and the operating units and organizations. The support operations staff with specialized commands and planning and coordination cells plan support operations, ensure plans are executed IAW the commander's intent, provide staff supervision over operating units, work to resolve support issues, and synchronize the operations of all TSC elements. The specialized control centers manage supply, transportation, and maintenance operations.

Support Operations Section

The support operations section supervises the provision of all TSC external mission support. It supervises supply, maintenance, field services, and movement control units and activities involved with external support. It also integrates transportation, aviation, medical, personnel, finance, and engineer mission support requirements into the overall support plan. In order to do this, the TSC support operations section requires planning and LNO cells from the specialized commands co-located with it.

D. TSC Role in Force Protection

The TSC's role in force protection comes within the context of sustaining operations as discussed in the decisive-shaping-sustaining operations framework in FM 3-0.

The Joint Rear Area (JRA)

The JFC is responsible to the geographic combatant commander for force protection at the operational level in a designated JOA. The ARFOR commander has responsibility for force protection of ARFOR, and receives resources from the ASCC to fulfill this mission. The JRA is a specified land area within a JFC's operational area that the JFC designates to facilitate protection and operation of installations and forces supporting the joint force. In the context of the overall battlespace, the JRA is a and area near, or contiguous with, the CZ, where land component forces are conducting tactical operations. The JRA often shares borders with both sea and air battlespace areas. The JFC may assign responsibility for the JRA to a service component commander, such as the ARFOR commander. The ARFOR commander may then serve as the JRAC, or appoint a subordinate commander or staff officer to serve as the JRAC. Designating the TSC as the JRAC implies a significant increase in the TSC's mission that requires staff augmentation and additional unit capabilities.

The TSC supports ARFOR operations primarily from within the JRA and executes Army lead service responsibilities to the joint force as assigned by the ARFOR commander. The TSC is normally the ARFOR commander's largest subordinate element in the JRA. Some TSC assets may be positioned in a sanctuary location, such as an ISB. However, the TSC's most critical facilities and base areas are in the JRA. These include SPODs; APODs; road, rail, and water networks; petroleum storage and distribution facilities; maintenance sites, and other critical facilities. The TSC commander, therefore, is a key player in security, terrain management, and movement control within the JRA. The TSC commander interacts closely with the JRAC, and TSC subordinate units interact closely with other services' security and support forces.

Joint Rear Area Coordinator (JRAC)

The joint rear area coordinator (JRAC) integrates the rear area security and intelligence efforts of all functional and service component commands. The JRAC also interfaces with the AADC and the NCWC functional elements that control security for the battlespace adjacent to the JRA.

E. TSC Role in Distribution Management

Ref: FM 4-93.4, Theater Support Command (Apr '03), chap. 5.

Distribution is a critical component of support operations. Though many elements of the TSC are involved in distribution, and the entire support operations section—along with the specialized commands and modules—play key roles, this chapter focuses on the TSC DMC, MCA, MMC, and MLMC. TSC elements involved in the distribution mission operate in a joint and often multinational environment.

Doctrine for the Army's role in theater distribution is explained in FM 4-01.4 (FM 100-10-1). JP 4-01.4 discusses joint distribution, JP 4-0 addresses the joint boards and centers, and JP 4-08 covers multinational considerations.

Distribution Management Center (DMC)

The distribution management center (DMC) acts as the distribution management support element for the DCSO. It provides staff supervision to the TSC MMC and MCA, and coordinates with the MLMC. It synchronizes operations within the distribution system to maximize throughput and follow-on sustainment, and executes priorities in accordance with ARFOR commander directives. Specialized commands and organizations provide liaison personnel to integrate distribution aspects of other CSS functions (such as postal or replacement operations, Class IV and V support to engineer operations, Class VIII and medical materiel operations, and contracting activities) into the overall distribution operation.

Distribution Planning

Detailed planning for distribution operations is a key part of the environment of the distribution manager. The distribution plan is closely related to the Logistics Preparation of the Theater (LPT) and is a part of the service support plan with its associated annexes and appendices. *See pp. 2-24 to 2-25 for additional discussion.*

Force Tracking

Force tracking is the process of gathering and maintaining information on the location, status, and predicted movement of each element of a unit while in transit to the specified operational area. These elements include the unit's command element, personnel, and unit-related supplies and equipment The ARFOR G3 tracks readiness and location of all ARFOR. The TSC support operations sections support the ARFOR force tracking by monitoring the logistical readiness of ARFOR and responding to shifting support priorities IAW ARFOR commander intent.

Maneuver and Mobility Support (MMS)

Maneuver and mobility support (MMS), formerly known as battlefield circulation and control (BCC), refers to functions of MP forces to support movement control operations. MP forces support TSC operations in a variety of ways, from law and order to forming tactical response forces, as required, to meet JRA threats. However, MMS is perhaps the most direct MP contribution to the TSC's movement control role as well as its core distribution management function. The highway traffic division (HTD) of the responsible road network controlling authority determines routes classification. In the TSC AO, the HTD is in the TSC MCA.

Movement Control Agency (MCA)

The Army executes movement control at the operational-level through a movement control agency (MCA). The MCA operates under the C2 of the TSC. The MCA helps develop and execute the Army portion of the joint movement program developed by the JMC. The MCA synchronizes its operations with those of the JMC, US-TRANSCOM, and lower echelon movement control organizations, and follows the priorities established by the ARFOR commander. The Army's theater MCA is pending a reorganization that incorporates its functions within transportation command elements (TCEs) that are subordinate to the Army theater TRANSCOM.

II. Expeditionary Sustainment Command (ESC)

Expeditionary Sustainment Commands (ESC) are force pooled assets that, under a command and control relationship with the TSC, command and control sustainment operations in designated areas of a theater. The ESC plans, prepares, executes, and assesses sustainment, distribution, theater opening, and reception, staging, and onward movement operations for Army forces in theater. It provides operational reach and span of control. It may serve as a basis for an expeditionary joint command when directed by the GCC or designated multinational or joint task force commander. It normally deploys to provide command and control when multiple Sustainment Bdes are employed or when the TSC determines that a forward command presence is required. This capability provides the TSC commander with the regional focus necessary to provide effective operational-level support to Army or JTF missions.

See pp. 2-3 for further discussion of the relationships between the TSC, ESCs and sustainment brigades.

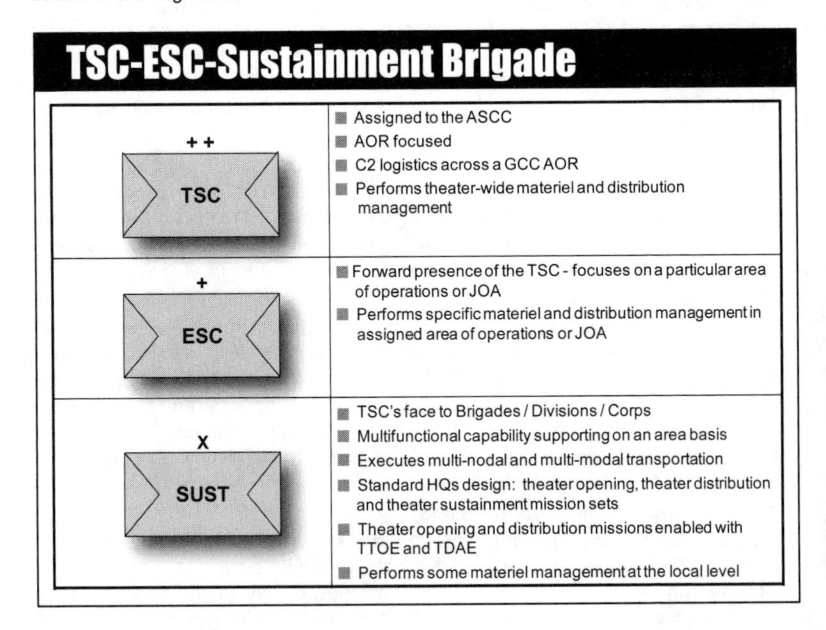

VIII. Protection Considerations

Ref: FMI 4-93.2, Sustainment Brigade (Feb '09), app. B.

Protection consists of those actions taken to prevent or mitigate hostile actions against DOD personnel (to include family members), resources, facilities, and critical information. Additionally, counter proliferation and consequence management actions associated with chemical, biological, radiological, nuclear, and high yield explosive weapons, which includes toxic industrial material and improvised explosive devices (IED) should be addressed.

I. Responsibilities of the Sustainment Brigade

The sustainment brigade is designed to plan and conduct base and base cluster self-defense. It has the capability to defend against level I threats, assist in destruction of level II threats, and escape or evade against level III threats. When faced with a threat beyond its capabilities, the sustainment brigade relies on resources within supported maneuver units to assist in defeating the threat. Conducting an active defense against any level of threat will degrade the sustainment brigade's capability to conduct its primary mission to a greater or lesser degree dependent upon the level of the threat. All duties and functions, which are required of a sustainment brigade, might also be required of a CSSB, especially base commander and base defenses functions.

Threat Levels

Defense planning includes these levels of threat:

Level I

Threats that base or base cluster self-defense measures can defeat.

Level II

Threats that initial response forces, but not base or base cluster self-defense measures, can defeat. Bases and base clusters can delay level II threats until response forces arrive.

Level III

Threats targeting several friendly elements as part of a larger, coordinated effort, rather than individual, separate entities. They require a tactical combat force (TCF) to defeat them.

A. The Threat

Hostile action against US forces may occur at any time, any place, and under any conditions. Recognizable, armed combatants or persons who are or appear to be civilians may commit hostile actions. Sustainment personnel must exercise vigilance against such attacks because the enemy seeks soft targets and assumes that these can be found in the areas and facilities that sustain US forces. Therefore, sustainment personnel should approach their tasks with the same warrior spirit that their combined arms counterparts exercise in their duties. This is especially true on the evolving battlefield with increased lethality, larger AO's, and more noncontiguous operations. Either the commander has to use more assets to secure sustainment activities or accept greater risk and reduced levels of sustainment activities when sustainment personnel substitute protection measures for sustainment operations. In any case, sustainment personnel understand, train for, and plan security operations within the context of their support activities.

Sustainment elements are prime targets for the asymmetrical threat forces on the noncontiguous operational area. Terrorists, saboteurs, opposition special operations forces, and others all pose a threat to sustainment personnel, installations, and convoys. By effectively cutting LOC's between sustainment and maneuver elements or by damaging and disrupting C2, C2 information systems, sustainment automation, and/or facilities these opposing forces hope to have significant negative impact on US maneuver forces with minimum risk to their own.

B. Combat Action

The combat threat in the sustainment brigade AO may include individual acts of sabotage; inserting large, organized forces; snipers, and artillery, mortar, air, and missile attacks. Large-scale enemy attacks may require committing US reserve forces, combat units from forward areas, HN resources, or multinational resources.

An understanding of the threat to the sustainment brigade and detailed IPB and LPB products help to protect the support structure. Threat forces conduct operations in the sustainment brigade areas and bases to seize and maintain the initiative, facilitate strategic and operational level penetrations, and degrade or destroy forces' ability to conduct support operations. To achieve these aims, enemy activities target:

- Command and control nodes
- Air defense artillery sites
- Critical support facilities and units such as:
  - Ammunition and weapon storage sites and delivery systems
  - SPOD's
  - APOD's
  - POL terminals and facilities
  - Maintenance, supply, and services activities
- Regeneration sites
- Key choke points along LOC's

II. Fire Support Considerations

The sustainment brigade, and all subordinate commanders and staffs, must have a thorough understanding of fire support procedures, organizations, and assets that are available within their AO. Normally, the ESC will have a dedicated fires battalion assigned for fire support that a sustainment brigade may call upon. However, this unit might not be available until the theater of operations is mature.

The sustainment brigade S-3 must include fire support considerations into their planning and the fire support information must be disseminated to all subordinate elements. All fire support assets need to be assessed (artillery, mortar, UAS, CAS, and naval) to determine availability and coordination requirements. Planning should focus on close coordination with the fires units and detailed procedures to request fires support. It should be routine for all personnel to receive training on call for fire procedures and to rehearse the procedures. The sustainment brigade S-3 should be prepared to provide the fires unit recommended targets and target indicators within the sustainment brigade AO.

III. Protective Measures

Ref: FM 4-93.2, Sustainment Brigade (Feb '09), pp. B-8 to B-10.

Sustainment brigade units take several measures to reduce their vulnerability to enemy operations. These measures include dispersion, cover, concealment, camouflage, intelligence gathering, obstacles, and air and missile defense.

A. Dispersion

Sustainment brigade organizations disperse as much as possible throughout the assigned AO. Dispersion as a protective measure is balanced against the potential deficits to support operations and the base cluster defense system. Dispersion helps avoid catastrophic damage from air and mass destruction weapons. Even if a sustainment brigade unit is not the primary target, it may be attacked as a target of opportunity. The dispersion required depends on the following:

- Type of threat
- Terrain
- Defensibility

B. Engineer Support

Engineer survivability support will be important for the elements of the sustainment brigade. When available, it may be used for a variety of protection hardening measures in support of the sustainment brigade to include survivability (see FM 5-103, Survivability Operations) support and support to camouflage, concealment, and decoys. Engineers may also provide geospatial support to the sustainment brigade's protection efforts.

C. Cover, Concealment, and Camouflage

The enemy cannot target sustainment brigade resources that it cannot detect. Cover, concealment, and camouflage remain critical to protecting sustainment units, facilities, and supplies from enemy detection and attack. Cover includes natural and artificial protection from enemy observation and fire. When selecting sites, advance parties consider the type of cover available. Concealment includes natural or artificial protection from enemy detection. Sustainment brigade units use concealed ingress and egress points and halt locations within support locations. Camouflage consists of using natural or artificial objects or tactical positions to confuse, mislead, or evade the enemy.

D. Obstacles

Obstacles slow, impede, or channel enemy movement and incursion. They buy time until reaction forces can deploy or a response force can arrive. Effective use of obstacles involves sound counter-mobility planning and early warning. Obstacles in urban environments are as important as in the field. Strategically placed obstacles provide protection against terrorist access to buildings.

E. Air and Missile Defense

Air defense artillery (ADA) forces cannot provide dedicated air and missile defense (AMD) for all sustainment brigade assets in the AO. The commander positions brigade organizations to take advantage of coverage that available AMD forces provide. Using base clusters makes it possible for AMD units to cover more Brigade assets than if units disperse throughout the AO, but reduces the benefits of dispersion. Brigade assets identified AMD priorities that do not receive dedicated support are positioned to take advantage of the coverage provided by AMD units protecting higher-priority assets.

Passive air defense operations include the means a unit uses to avoid enemy detection, along with measures to minimize damage when attacked. Sustainment brigade units use OPSEC to conceal their location from enemy visual and electronic surveillance. Elements within base clusters disperse as much as possible. Dispersal along with field fortifications and obstacles significantly reduce casualties and damage from air and missile attack.

IV. Bases and Base Clusters

Ref: FM 4-93.2, Sustainment Brigade (Feb '09), pp. B-14 to B-16.

The sustainment brigade must integrate its HQ and subordinate elements FP plans into the base and base cluster defense plans. These plans are integrated with the defense plans of the MEB or BCT having the overall responsibility for the AO, and are copied to the higher sustainment commander. This defense method protects elements from level I threats in their assigned areas. Commanders ensure all bases and base clusters in their AO's train and prepare for their roles. Cooperation and coordination elements are critical.

Bases and base clusters form the basic building block for planning, coordinating, and executing base defense operations. The sustainment brigade S-3, with input from the sustainment brigade SPO, organizes units occupying the support HQ AO into base clusters. The sustainment brigade SPO recommends appointments of base or base cluster commanders from units in the cluster to the sustainment brigade S-3. The base cluster commander is usually the senior commander in the base cluster. The base cluster commander forms a base cluster defense operations center (BCOC) from the staff and available base assets.

A base may be a single-service or a joint-service base. The base cluster commander appoints the base commanders. Base commanders form base defense operating centers (BDOC's).

Within the base cluster, three commanders have distinct responsibilities. These three — the individual unit commander, the base commander, and the base cluster commander — are discussed below. Non-sustainment units residing within a base which a sustainment brigade commands will have a command relationship and responsibilities established by order to the sustainment brigade for protection.

Individual Unit Commanders

The commanders of units in a base are responsible for the following:

- Participating in base defense planning
- Providing, staffing, and operating base defense facilities in accordance with base defense plans
- Conducting individual and unit training to ensure their forces' readiness to perform their assigned tasks in defense of the base
- Providing appropriate facilities and essential personnel for the BDOC and the base commander
- Providing liaison personnel to advise the base commander on matters peculiar to their units
- Providing internal security of the base
- Providing communications systems, including common-user communications, within the command

Sustainment brigade units use observation posts, listening posts, or unattended sensors on likely avenues of approach to collect intelligence on threat activity. In areas where the populace is friendly, local law enforcement or government agencies can provide information on threats in the area. BCOC's implement an integrated warning plan within their cluster and with adjacent bases or base clusters.

Base Commander

The base commander is responsible for base security and defense. All forces assigned to the base are under OPCON for base defense purposes. The base commander's responsibilities for base defense include:

- Establishing a BDOC from available base assets to serve as the base's tactical operations center (TOC) and focal point for security and defense. The BDOC assists with planning, directing, coordinating, integrating, and controlling base defense efforts.

- Establishing an alternate BDOC from base resources or, if base assets are not available, designating a HQ element from units dedicated to the base for its local defense
- Planning for including transient units by ensuring that base defense plans include provisions for augmenting the regularly assigned base defense forces with units present at the base during periods of threat

Base Cluster Commander

The base cluster commander is responsible for securing the base, coordinating the defense of bases within the base cluster, and integrating base defense plans into a base cluster defense plan. Specific responsibilities include:

- Establishing a BCOC from the staff and available base or base cluster assets to serve as the base cluster's TOC and focal point for planning, directing, coordinating, integrating, and controlling base cluster defense activities
- Providing appropriate facilities, housing, and services for necessary liaison personnel from bases from within the cluster

Base and Base Cluster Defense Plan

Base and base cluster commanders develop and implement comprehensive defense plans to protect their support capability. The defense plan includes measures to detect, minimize, or defeat level I and defend level II threats. To maximize mutual support and prevent fratricide, the base and base cluster commanders coordinate defense plans with adjacent base and base clusters and joint, multinational, and HN forces. The sustainment brigade S-3 ensures that all plans conform to the overall TSC or BCT security plans.

Area Damage Control (ADC)

Commanders of bases and installations within the JSA coordinate requirements for area damage control (ADC) with the JSC through their respective chains of command. Commanders establish priorities for ADC missions as part of their planning process at the base or installation level. All units are responsible for providing ADC within their base or installation to the extent of their capabilities.

The sustainment brigade takes ADC measures before, during, and after hostile action or natural disasters to reduce the probability of damage, to minimize its effects, and to reestablish normal operations. Necessary repair begins after the damage is contained.

Other forces and assets that contribute to the ADC mission include—

- Ordnance
- MP
- Chemical
- CA
- Maintenance
- Medical
- Signal
- Supply
- Transportation
- Transiting units

HNS can be a vital resource for ADC in the AO. Early HNS identification and coordination are essential to supplement ADC efforts. Responsibilities and support from HN assets are negotiated at theater level and are part of the status-of-forces agreements and treaties.

V. Convoy Security

The most serious threat faced by the deployed sustainment brigade elements occurs when moving. The sustainment brigade must coordinate with maneuver, military police, and medical units operating in the AO for additional security capability, medical support, and/or route security assessments.

Movement control always includes convoy defense considerations. Supply routes are assumed not to be secure on a high threat area. Therefore, convoy movements between supported unit areas are combat operations. The TSC and supported unit commanders and staffs work together to integrate defensive capabilities into convoys. Adequate convoy security depends on two critical components. These are thorough staff planning to counter enemy plans and capabilities and individual Soldier training to counteract enemy action. All convoy personnel must be familiar with published rules of engagement and local tactics, techniques, and procedures developed for the current situation on the convoy route.

The convoy commander ensures that troops are trained in convoy defense techniques. The damage a convoy prevents or incurs when attacked often depends on the adequacy of convoy defense training. It also depends on the route and timing of the convoy in relation to the enemy situation and the adequacy of the intelligence and information convoy leaders receive in advance of the operation. The following paragraphs discuss in the broadest of terms the considerations of convoy protection.

There is no such thing as an administrative move on the noncontiguous operational area. Once outside the base perimeter you have a tactical convoy. A tactical convoy is a deliberately planned combat operation. Its mission is to move personnel and/or cargo via a group of ground transportation assets in a secure manner to or from a target destination. Tactical convoys operate under the control of a single commander in a permissive, uncertain, or hostile environment. Tactical convoys should always have access to the COP and be characterized by an aggressive posture, agility, and unpredictability. There should never be less than two personnel in the cab, one to drive and one for protection.

Each tactical convoy must be prepared to take appropriate action in the face of ambush and once contact is made based on rules of engagement and TTP's. Training, experience, and unit SOPs will accelerate tactical convoy preparation and prepare unit personnel to take appropriate actions on contact.

Standard troop leading procedures (modified to reflect convoy operations) are included in Multi-Service Tactics, Techniques and Procedures for tactical convoy operations and should be used to ensure all planning elements are considered when preparing to conduct convoy operations.

A. Movement Corridors

The Maneuver Enhancement Brigade (MEB) is a multifunctional headquarters, task organized according to METT-TC that provides security and protection within its assigned areas. The sustainment brigade is likely to operate within a MEB AO and it will rely on the protection MEB elements provide throughout designated movement corridors.

A movement corridor (MC) is part of a layered and integrated security approach to LOC security. Layered security constitutes concentric rings that increase in survivability and response measures. The first ring (the center ring) being the ability of every convoy to defeat a Level I threat and to delay a Level II threat. The next ring (middle ring) provides the increased security/protection capability in support of the center ring activities, capable of defeating Level I and Level II threats, and supports the defeat of Level III threats. The middle ring is also capable of integrating fires, CAS, MEDEVAC, safe havens, vehicle removal/recovery operations, and so forth in support of the center ring and central effort that is the MC concept. The final ring (the

B. Main Supply Routes (MSRs)/ Alternate Supply Routes (ASRs)

Ref: FM 4-93.2, Sustainment Brigade (Feb '09), pp. B12 to B-13.

MSR's are routes designated within the MEB's or higher headquarters AO upon which the bulk of sustainment traffic flows in support of operations. An MSR is selected based on the terrain, friendly disposition, enemy situation, and scheme of maneuver. Supply routes are selected by the MEB S-4 in coordination with the BSB support operations officer and MEB S-3. They also plan ASRs for use if a MSR is interdicted by the enemy or becomes too congested. In the event of CBRN contamination, either the primary or alternate MSR may be designated as the dirty MSR to handle contaminated traffic. All ASRs must meet the same criteria as the MSR. MPs assist with regulating traffic and engineer units maintain routes. Security of supply routes in a noncontiguous AO may require the MEB commander to commit non-logistics resources.

Some route considerations are:

- Location and planned scheme of maneuver for subordinate forces to include combined arms forces, artillery units, and other forces moving through the MEB's AO
- Route characteristics such as route classification, width, obstructions, steep slopes, sharp curves, and type roadway surface
- Two-way, all-weather trafficability
- Weight classification of bridges and culverts
- Requirements for traffic control such as at choke points, congested areas, confusing intersections, or along built-up areas
- Number and locations of crossover routes from the MSR to ASRs
- Requirements for repair, upgrade, or maintenance of the route, fording sites, and bridges
- Route vulnerabilities that must be protected. This may include bridges, fords, built-up areas, and choke points.
- Enemy threats such as air attack, conventional and unconventional tactics, explosive hazards, ambushes, and chemical strikes
- Known or likely locations of enemy penetrations, attacks, chemical strikes, or obstacles
- Known or potential civilian/refugee movements that must be controlled or monitored

See also p. 3-17 for discussion of supply route considerations.

outer most ring) is the final ring of protection and brings with it the ability to defeat all level of threats through the integration of all joint capabilities.

An MC is a protected LOC that connects two support areas. Within the MC are main and alternate roads, railways, and/or inland waterway supply routes used to support operations. Within an AO, there is an MC network that consists of multiple MCs that connect inter-theater APOD and SPOD; intra-theater Aerial Port of Embarkation and APOD's; operational level support, distribution, and storage areas; and brigade support areas. The MEB is responsible for tactical LOC operations and security within its AO and, as assigned, within its higher headquarters AO. The width and depth of an MC will be dependent on METT-TC factors and the commanders' guidance.

The establishment of an MC network is the result of applying multiple functions and establishing required command, control, and support relationships. The collective integration and synchronization of units, capabilities, and facilities will provide a comprehensive three-dimensional protection capability for the designated LOC's, the unit and convoy movements on the LOC's, and the units supporting LOC and movement operations. A fully developed MC will consist of military police units providing route regulation and enforcement, straggler and dislocated civilian control, area and route security, convoy escort, response force operations, and logistical units conducting and managing movement control. Supporting functions include units and capabilities for vehicle recovery and storage, cargo transfer, refueling, road maintenance and repair, MC safe haven support facility construction and repair, CBRN detection and response, aerial reconnaissance, and medical treatment and evacuation.

The sustainment brigade S-3 coordinates with the MEB when planning tactical convoys through movement corridors. This may be done through movement control elements collocated with the MEB or the division transportation officer.

C. Danger Areas

Intersections, ramps, traffic circles, over and underpasses, rest halts, or halts to recover disabled vehicles can all be danger areas for convoys. See FM 4-01.45 for TTP's in dealing with these areas.

D. Battle Drills

Battle drills are the pre-planned, rehearsed responses to contact. FM 4-01.45 provides a simple decision matrix for reaction to contact and examples of how convoy participants might respond. Battle drills on how to react to convoy ambushes must be developed and rehearsed.

E. Improvised Explosive Devices (IED's) and Vehicle Borne IED's (VBIED's)

IED's are one of the greatest threats to convoys and are often used to initiate an ambush. Convoy personnel should always expect an ambush immediately following an IED detonation. Convoy commanders should brief convoy personnel on the latest IED threat: what types of IED's are being used and where they have previously been emplaced along the route.

IED's and VBIED's represent an attractive attack weapon for asymmetrical threat forces. They can inflict substantial damage to convoys (equipment and personnel) while providing the threat with very limited exposure counter-attack by US forces.

IED's are rigged from any explosive materiel available to threat forces and detonated by timing device, trip wire or pressure trigger, or are command detonated. IED's are most easily emplaced during periods of limited visibility making morning periods exceptionally dangerous for exposure to IED's. IED's are often used as the opening weapon in an ambush. Convoy participants must be prepared to counter an assault by direct fire from hidden roadside positions immediately following an IED encounter.

Convoy Operations

Ref: FM 4-93.2, Sustainment Brigade (Feb '09), p. B-13.

Layered Convoy Protection

The tactical convoy's physical security elements are composed of three layers:

- Organic security elements
- Convoy escorts
- Corridor security

The organic security element is the responsibility of the convoy commander. These assets (personnel, weapons, and vehicles) are taken from the organic assets of the unit conducting the convoy operation. Vehicles selected for this mission should be fast, maneuverable, and hardened to the extent possible (these vehicles are known as "gun trucks"). Crew served weapons, automatic weapons, and grenade launchers mounted on these gun trucks provide suppressive fire as the convoy initially responds to an ambush.

Convoy escorts may be provided by any organization tasked to provide convoy security. These organizations may include MPs, CAS, and/or security detachments from a maneuver element. The mission of the convoy escort is to provide protection from direct fire and complex ambushes. Convoy escort elements may be used as reconnaissance ahead of the convoy and/or as a trail security element. Either armor or mechanized units provide greater firepower for the escort mission. See FM 4-01.45 for suggested employment of ground and CAS convoy escort assets. Corridor security is the responsibility of the BCT or MEB whose AO includes the convoy route.

Convoy Communication

Radio communication is vital to the support the convoy. A convoy commander needs to consider and plan for the following:

- MEDEVAC operations
- Internal convoy vehicle-to-vehicle communications
- Communications with security vehicles (gun trucks and/or convoy escorts)
- External communications to higher headquarters, quick reaction force, artillery support

Alternate means of communication should be planned for the following:

- Internal to vehicle
- Between vehicles
- When dismounted due to ambush or IED
- Radios capable of secure communications are used in convoys. Three types of communications must be considered and planned for. Alternate means of communications (vehicle signal system, hand and arm signals, pyrotechnics, and so on) should also be planned.

Mounted Tactics

The four principles of mounted tactics for convoys are:

- 360-degree security – situational awareness, interlocking fires, mutual support
- Deterrence – aggressive posture, display a willingness to engage
- Agility – Ability to adapt to environment and conditions
- Unpredictability – No observable routine

Every participant in the convoy must understand the areas in which they are responsible for observing and into which they must fire in the event of enemy contact. Interlocking fires and mutually supporting fires are ensured to the extent that fratricide considerations allow.

VI. CBRN Defense

The ability to sustain combat operations with an appropriate level of support is vital to operational success. Operations in CBRN environments place significant burdens on the sustainment system. Sustainment operations and facilities are at particular risk to CBRN attack to the degree that they rely on fixed sites (ports, airfields, and so on) or must remain in particular locations for extended periods of time. The need to operate in CBRN environments will add to the physical and psychological demands of military operations, with degrading effects on the performance of individuals and units. The sustainment brigade S-3 is responsible for CBRN planning.

A. Protective Equipment

Sufficient equipment must be available to protect not only the uniformed force but also the essential supporting US and civilian work forces. This is of particular concern for sustainment operations where potentially significant numbers of contract and HN personnel support operations. Individual and unit training for proper sizing, use of, and care for this individual and crew-served equipment is required to take full advantage of its capabilities.

B. CBRN Defense Principles

Three principles that specifically address the hazards created by CBRN weapons include: avoidance of CBRN hazards, particularly contamination; protection of individuals and units from unavoidable CBRN hazards; and decontamination in order to restore operational capability.

1. Avoidance

Successful contamination avoidance prevents disruption to operations and organizations by eliminating unnecessary time in cumbersome protective postures and minimizing decontamination requirements. Avoiding contamination requires the ability to recognize the presence or absence of CBRN hazards in the air, on water, land, personnel, equipment, and facilities. Surveillance and detection capabilities enable units to recognize CBRN hazards. The fusion of these capabilities with information from other sources yields an overall surveillance picture supporting decisions for specific avoidance, protection, and decontamination actions. These surveillance and detection results also establish requirements for other avoidance measures such as sounding alarms, marking hazards, and warning forces. Leaders at all levels must implement measures designed to avoid or limit exposure consistent with mission requirements. These measures should include increased use of shelters during CBRN employment windows and providing key information for movement before, during, and after CBRN attacks. In planning for contamination avoidance, leaders must include an assessment of the capabilities of available detection systems. Particular challenges include the unanticipated use of biological agents and the capabilities and limitations of current remote and standoff detection systems.

2. Protection

CBRN protection requires the planning, preparation, training, and execution of physical defenses to negate the effects of CBRN weapons and hazards to personnel and materiel. As staffs analyze their mission requirements and conditions, the planning process will yield specific actions required before, during, and after CBRN attacks. As commanders anticipate and identify CBRN risks, these actions should be clearly communicated and rehearsed from command to individual levels. CBRN protection conserves the force by providing individual and collective protection postures and capabilities.

Commanders adopt a mission oriented protective posture (MOPP) to establish flexible force readiness levels for individual CBRN protection. MOPP analysis (the process of determining a recommended MOPP) integrates CBRN protection requirements (derived from CBRN threat assessments) with mission requirements in light of the performance degradation caused by wearing protective equipment. MOPP analysis relies on accurate IPB and CBRN hazard prediction as well as a clear understanding of the force's ability to quickly increase its CBRN protection. To facilitate adapting to varying mission demands across a combatant command's AO, MOPP decisions should be delegated to the lowest level possible and retained at higher levels only in exceptional cases. The JFC has overall responsibility for providing guidance for levels of protection and ensuring timely warning of CBRN risks. Force components may require variations of configurations, such as "mask only" for identified situations, but should standardize configurations where possible. Tactics techniques and procedures (TTP's) that address specific techniques and procedures for MOPP analysis and donning protective equipment can be found in the Multi-service Tactics, Techniques and Procedures (MTTP) for chemical, biological, radiological, and nuclear (CBRN) protection.

Sustaining operations in CBRN environments may require collective protection equipment, which provides a toxic free area (TFA) for conducting operations and performing life support functions such as rest, relief, and medical treatment. Contamination transfer into the TFA compromises the health and safety of all occupants and jeopardizes their ability to support the mission. Therefore, training must include procedures for TFA entry and exit. When collective protection is not available and mission requirements permit, plans must be developed, exercised, and evaluated to move personnel to alternative TFA's that are well away from the contaminated areas. If evacuation is not possible, building occupants may be able to gain limited protection by closing all windows and doors, turning off ventilation systems, and moving to closed, inner rooms. If there is some advance warning, occupants may be able to increase protection by sealing windows, doors, and openings, while recognizing that the building or space may quickly become uninhabitable without cooling or ventilation.

3. Decontamination

Decontamination supports the post-attack restoration of forces and operations to a near-normal capability. Decontamination is intended to minimize the time required to return personnel and mission-essential equipment to a mission capable state. Because decontamination may be labor intensive and assets are limited, commanders must prioritize requirements and decontaminate only what is necessary. Commanders may choose to defer decontamination of some items and, depending on agent type and weather conditions, opt to either defer use of equipment or allow natural weathering effects (temperature, wind, and sunlight) to reduce hazards. Decontamination is organized into three categories that reflect operational urgency: immediate, operational, and thorough. Decontamination also entails special considerations for patients, sensitive equipment, aircraft, fixed sites, and the retrograde of equipment. The extent and time required for decontamination depends on the situation, mission, degree of contamination, and decontamination assets available. TTP's provide details for the technical aspects of decontamination and can be found in FM 4-11.5, MTTP for CBRN Decontamination.

Retrograde cargo may require extensive decontamination measures; specialized, highly sensitive monitoring equipment; extended weathering, or destruction. Retrograde of previously contaminated equipment may be delayed until after conflict termination.

Fixed site decontamination techniques focus on fixed facilities and mission support areas such as C2 information systems, supply depots, aerial and seaports, medical facilities, and maintenance sites.

C. Sustainment Operational Considerations - CBRN

Ref: FM 4-93.2, Sustainment Brigade (Feb '09), pp. B-5 to B-8.

Sustainment elements will operate throughout the theater of operations. Sustainment elements directly supporting engaged forces will be small, mobile units. Sustainment elements providing area support may involve larger, more complex transportation, maintenance and supply activities conducted at fixed or semi-fixed sites.

Mobile units seek to avoid CBRN contamination to the maximum extent possible. If contaminated, units identify clean areas, and on order, move along designated routes from contaminated areas. Units decontaminate equipment and conduct MOPP gear exchange during the move to clean sites.

Sustainment units at the operational level, in contrast, may be required in emergency situations to conduct operations from contaminated fixed sites until they can relocate to clean areas.

In most circumstances, the sustainment infrastructure in a theater of operations operates with a substantial complement of nonmilitary personnel. In a typical theater, sustained operations will rely heavily on military personnel, DOD civilians, HNS personnel, other nation support personnel, and contractor provided sustainment support personnel. During the early phases of deployment, the sustainment infrastructure may rely on HNS personnel for port operations and transportation requirements. Protection of all types of personnel in the AO is required and must be included in the CBRN planning and preparation.

1. Warning Systems

In order for individuals and units to take necessary self-protection measures, timely warning of CBRN attacks and subsequent spread of contamination is essential. The JFC has the responsibility, in coordination with the HN, to establish an effective and timely warning system, and to exercise this system on a recurring basis. Sustainment commanders tied to a fixed site should monitor CBRN warning systems continuously and should be capable of passing warnings to workers and units throughout their sites. Because of the variety of delivery methods for CBRN weapons and the limitations of detection capabilities, personnel and units may not receive warning before exposure occurs. Warning systems should be designed to alert workers promptly upon initial detection of an attack. Since workers may be widely dispersed throughout the area, a site-wide alarm system, capable of being activated immediately upon receipt of warning, must be available, maintained, and exercised regularly. At many sites, military throughput will rely on civilian labor. Sustainment planners must consider the vulnerability of HN and other civilian workers to attack and plan accordingly.

2. Materials Handling Equipment (MHE)

Typically, even during high capacity operations, much of the MHE at a facility is not in use. Commanders should protect idle MHE from exposure to chemical or biological agents in the event of attack. Housing and covering MHE with plastic, or otherwise protecting it from exposure, can ensure that it will be readily available to resume operations after the attack.

3. Aerial Ports

Regardless of an aerial port's CBRN preparedness, some aircraft will not be able to land at or depart from contaminated airfields. Of particular importance are limitations in CBRN environments on the employment of the civil reserve air fleet, civilian, and other aircraft under contract to support military operations. Sustainment plans must provide for replacing these aircraft with other airlift assets or conducting trans-load operations from bases outside the immediate threat area. These replacement aircraft would have to operate from trans-load airbases to shuttle the affected cargo and passengers to the theater of operations. If that is not feasible, alternate means (sea, rail, wheeled transport, and so on) must be made available to accomplish the mission.

4. Sea Ports

In large-scale operations, US equipment and materiel normally enter the theater of operations on strategic sealift ships and off-load at SPOD. The vital importance of these seaports to US power projection capability makes them an attractive target for CBRN attack. However, conducting successful attacks against SPOD's presents significant challenges to the adversary. If port managers and operators are properly prepared to survive the attack and sustain operations, CBRN attacks may not cause significant long-term degradation of throughput capacity. This is especially true at large ports where many piers, storage areas, and much of the MHE may escape contamination. Operations in these cases may be limited more by the effects of the attacks on the local workforce and nearby civilian population. In some cases, it will be possible to continue operations at a contaminated port.

5. Reception Staging and Onward Movement (RSO)

The permanency of sites for RSO of arriving forces can vary widely between theaters of operations. Theaters of operations with large forward-deployed forces rely on fixed sites for a wide variety of activities, such as pre-positioned stock maintenance and control, supply and maintenance, materiel and transportation management, and sustainment network operations. Theaters of operations with limited forward presence normally rely more heavily on temporarily fixed sites (facilities that are transportable or mobile but, due to ongoing operational constraints, may not be rapidly moved).

Staging areas for personnel or equipment near APOD's and SPOD's may be attractive targets for CBRN attack. Sustainment planners must assess the relative value of the convenience provided by establishing large centralized facilities, which are more easily targeted, and the enhanced security that results from having more smaller dispersed facilities that are more difficult to C2 but less vulnerable to CBRN attack. While the anticipated threat will influence the staging area selection process, adequate facility and area space availability may be the determining considerations. Planning must consider equipment-marshalling areas and rail yards (which may not be in close proximity of APOD and SPOD facility complexes); sustainment hubs and bases (which may be fixed facilities with large forward deployed forces); and force integration assembly areas (where deploying units complete deployment recovery, equipment receipt, and processing and preparation for movement to TAA's).

6. Main Supply Routes (MSR's)

The vulnerability of MSR's to CBRN attack may vary widely among theaters of operations. In those that rely on a few major MSR's and have limited alternative routes and off-road capability, CBRN attacks may have a greater impact on operations than in those with more extensive supply routes and where obstacles can more easily be traversed.

7. Contractor and Host Nation Support (HNS) Considerations

The three basic categories of external support for US military operations are wartime HNS, contingency contracts, and current contract agreements. These usually exist in conjunction with one another and collectively provide a full sustainment capability to the theater of operations. Contracts or agreements will clearly specify services to be provided during periods of crisis or war. Sustainment commanders should not expect unprotected or untrained individuals to continue to provide essential services under the threat of CBRN attack or during operations in CBRN environments. CBRN protection includes individual and collective survival skills as well as operational training. Survival skills refer to the capability to take required, immediate action upon CBRN attack, to include masking, proper wear and care of protective clothing and equipment, personal decontamination, and buddy aid.

VII. Risk Management

Ref: FM 4-93.2, Sustainment Brigade (Feb '09), pp. B-1 to B-2.

Risk management is a process that assists decision makers in reducing or offsetting risk by systematically identifying, assessing, and controlling risk arising from operational factors and making decisions that weigh risks against mission benefits. Risk is an expression of a possible loss or negative mission impact stated in terms of probability and severity. The risk management process provides leaders and individuals a method to assist in identifying the optimum course of action. Risk management must be fully integrated into planning, preparation, and execution. Commanders are responsible for the application of risk management in all military operations. Risk management facilitates the mitigation of the risks of threats to the force.

Military operations are inherently complex, dynamic, and dangerous, and by nature, involve the acceptance of risk. Because risk is often related to gain, leaders weigh risk against the benefits to be gained from an operation. The commander's judgment balances the requirement for mission success with the inherent risks of military operations.

The fundamental goal of risk management is to enhance operational capabilities and mission accomplishment with minimal acceptable loss. The basic principles that provide a framework for implementing the risk management process include:

- **Accept no unnecessary risk**. An unnecessary risk is any risk that, if taken, will not contribute meaningfully to mission accomplishment or will needlessly endanger lives or resources. No one intentionally accepts unnecessary risks. The most logical choices for accomplishing a mission are those that meet all mission requirements while exposing personnel and resources to the lowest acceptable risk. All military operations and off-duty activities involve some risk. The risk management process identifies threats that might otherwise go unidentified and provides tools to reduce risk. The corollary to this axiom is "accept necessary risk" required to successfully complete the mission.
- **Make risk decisions at the appropriate level**. Anyone can make a risk decision. However, the appropriate level for risk decisions is the one that can make decisions to eliminate or minimize the threat, implement controls to reduce the risk, or accept the risk. Commanders at all levels must ensure that subordinates know how much risk they can accept and when to elevate the decision to a higher level. Ensuring that risk decisions are made at the appropriate level will establish clear accountability. The risk management process must include those accountable for the mission. After the commander, leader, or individual responsible for executing the mission or task determines that controls available to them will not reduce risk to an acceptable level, they must elevate decisions to the next level in the chain of command.
- **Accept risk when benefits outweigh the cost**. The process of weighing risks against benefits helps to maximize mission success. Balancing costs and benefits is a subjective process and must remain a leader's decision.
- **Anticipate and manage risk by planning**. Integrate risk management into planning at all levels. Commanders must dedicate time and resources to apply risk management effectively in the planning process and where risks can be more readily assessed and managed. Integrating risk management into planning, as early as possible, provides leaders the greatest opportunity to make well-informed decisions and implement effective risk controls. During the execution phase of operations, the risk management process must be applied to address previously unidentified risks while continuing to evaluate the effectiveness of existing risk control measures and modify them as required.

Chap 3

I. Brigade Combat Team Sustainment Operations

Ref: Adapted from FM 4-90.7, Stryker Brigade Combat Team Logistics (Sept '07) and related sources. When published, FM 4-90.1, Brigade Support Battalion, will describe how sustainment operations take place in heavy and infantry BCT area of operations.

I. Brigade Combat Team (BCT) Sustainment

BCTs are organized with the self-sustainment capability for up to 72 hours of combat. Beyond 72 hours, sustainment organizations at the division and corps levels are required to conduct replenishment of the BCT's combat loads. That replenishment is a function of the higher headquarters' sustainment brigade(s).

Bde Cbt Tm Sustainment

The Brigade Support Battalion (BSB)

The brigade support battalion (BSB) is the core of sustainment to the brigade combat team (BCT). The BSB is organic to the BCT and consists of functional and multifunctional companies assigned to provide support to the BCT. Field Manual (FM) 4-90.1, Brigade Support Battalion, when published, will describe how sustainment operations take place in the heavy and infantry BCT area of operations.

See pp. 3-3 to 3-10 for further discussion.

Sustainment Brigades

All logistics requirements (less medical) beyond the BSB's ability are either furnished by or coordinated through the supporting sustainment brigade. The sustainment brigade supports the BCT on an area basis. When properly task organized, the sustainment brigade is capable of supporting BCT requirements for all classes of supplies (less Class VIII), maintenance, field services, contracting, and other logistics requirements. Through its distribution capability, the sustainment brigade normally provides distribution of supplies to the BCT BSB in support packages. The sustainment brigade operates ATHPs for the distribution of Class V. The sustainment brigade SPO is the POC for BCT logistics requirements above the capacity of the BCT BSB.

See chap. 2, Sustainment Brigade Operations, for further discussion.

Combat Sustainment Support Battalions (CSSB)

The combat sustainment support battalions of the sustainment brigade are the base organization from which sustainment units are task organized for various operations. The combat sustainment support battalion subordinate elements consist of functional companies that provide supplies and services, ammunition, fuel, transportation, and maintenance. Additionally, personnel and finance units can either be assigned to or administratively controlled by the combat sustainment support battalions to perform essential human resources and finance functions. The combat sustainment support battalions provide the distribution link between theater aerial/sea ports of debarkation and the BCT's BSB. The structure includes cargo transfer and movement control assets, performing the function of transporting commodities to and from the BCT BSB, and to/from repairing or storage facilities at the theater base. Its function is to ensure and maintain the flow of replenishment using expeditionary support packages, to include retrograde of unserviceable components, end items and supplies.

See pp. 2-28 to 2-29 for further discussion.

II. Sustainment Reach Operations

Adapted from FM 4-90.7, Stryker Brigade Combat Team Logistics (Sept '07), pp. 1-9 to 1-11 and related sources. See also p. 1-40.

The BSB is organized to be reinforced by echelons above brigade (EAB) sustainment reach operations. Sustainment reach operations are using and positioning all-available sustainment assets and capabilities from the national sustainment base through the Soldier in the field to support full spectrum operations. The goal of sustainment reach is to reduce the amount of supplies and equipment in the AO to sustain combat power more quickly and fully exploit all available sources of support. Reach operations include, but are not limited to, external sources of information and intelligence, sustainment planning and analysis conducted outside the AO, telemedicine, and other temporarily required capabilities. The BSB exploits regionally available resources through joint, multinational, host nation (HN), or contract sources for certain bulk supplies and services. Army field support brigades are part of Army Sustainment Command, a major subordinate command of the United States Army Materiel Command.

The **support operations officer** is the principal staff officer for coordinating sustainment reach operations for supported forces. The support operations section is the key interface between the supported units and the source of support. The support operations officer advises the commander on support requirements versus support assets available. Sustainment reach operations involve risk analysis, and the commander ultimately decides which support capabilities must be located within the AO and which must be provided by a reach capability. This ratio is based on METT-TC factors and command judgment. The support operations officer also determines which reach resources can be directly coordinated and which must be passed to the next higher support level for coordination.

The **support operations officer** continually updates sustainment reach requirements based on the sustainment plan. Planning is the process of gathering data against pertinent battlefield components, analyzing their impact on the sustainment estimate, and integrating them into tactical planning so that support actions are synchronized with maneuver. It is a conscious effort to identify and assess those factors that facilitate, inhibit, or deny support to combat forces.

Host nation support (HNS) is one of the more commonly used sources of sustainment reach support. HNS is provided to Army forces and organizations located in or transiting through HN territory and includes both civil and military assistance. This support can include assistance in almost every aspect required to support military operations within the AO. Commanders must consider support requirements generated by using HNS. The TSC support operations section includes a HNS directorate.

When planning sustainment reach operations, commanders must conduct a thorough risk analysis of the mission. Reach operations are vulnerable and highly susceptible to many factors including: US Army commanders' distrust of non-US support, changes to the political situation, direct or indirect terrorist activity, local labor union activities, language differences, quality assurance/quality control challenges, compatibility issues, and legal issues. Since reach operations involve DOD civilians, contractors, and joint and multinational forces, protection operations become paramount to mission accomplishment. Single individuals within an agency can carry out terrorist operations.

The benefits of reach operations must be carefully weighed against protection requirements especially when using reach assets that are not US military forces. The ability and tendency of our adversaries to use asymmetrical force against US forces increases the inherent risks of some reach operations. If local hostilities escalate, support provided by civilians or contractors may also be disrupted. Commanders must consider potential risks and develop detailed plans for compensating for sudden variances in reach support.

II. Brigade Support Battalion (BSB)/BSA

Ref: FM 4-90.7, Stryker Brigade Combat Team Logistics (Sept '07) and related sources. When published, FM 4-90.1, Brigade Support Battalion, will describe how sustainment operations take place in heavy and infantry BCT area of operations.

I. Brigade Support Area (BSA)

The BSA is the logistical, personnel, and administrative hub of the BCT. It consists of BSB, but could also include a BCT alternate CP (if formed), battalion field trains, brigade special troops battalion units, signal assets, and other sustainment units from higher HQ. The BCT operations staff officer (S-3), with the BCT S-4 and the BSB, determines the location of the BSA. The BSA should be located so that support to the BCT can be maintained, but does not interfere with the tactical movement of BCT units, or with units that must pass through the BCT area. The BSA's size varies with terrain; however, an area 4 km to 7 km in diameter is a planning guide. Usually the BSA is on a main supply route (MSR) and ideally is out of the range of the enemy's medium artillery. The BSA should be positioned away from the enemy's likely avenues of approach and entry points into the BCT's main battle area (MBA).

Brigade Support Area (BSA)

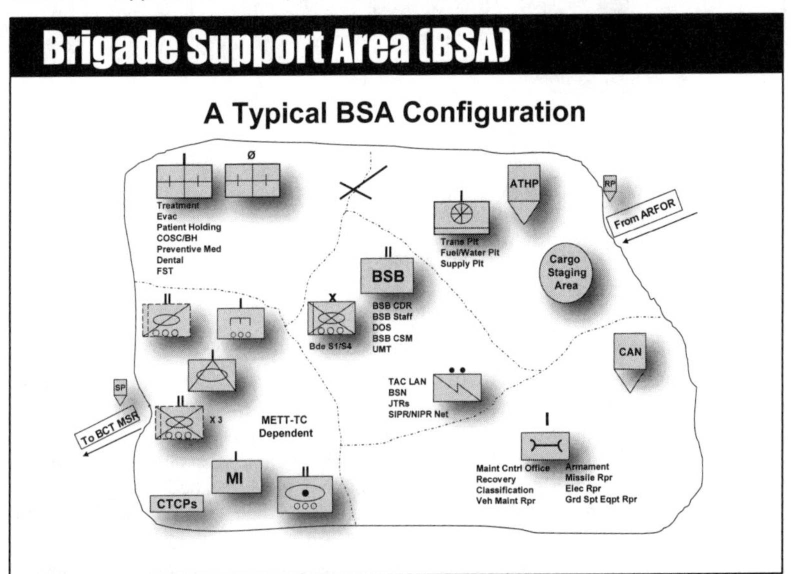

Ref: FM 4-90.7, Stryker Brigade Combat Team Logistics, fig. 3-1, p. 3-1.

The Brigade Support Battalion (BSB) is the central tenant of the Brigade Support Area (BSA) and arrays its subordinate units and other BSA tenant units to most effectively and efficiently use the terrain available for logistics operations and force protection.

The brigade support battalion (BSB) commander is responsible for the brigade support area (BSA). For security purposes, this includes the operational control (OPCON) of all elements operating within the BSA. The BSB commander might be assigned additional unassigned area functions.

See p. 3-13 for discussion of additional sustainment support areas.

II. Brigade Support Battalion (BSB) HHC

Ref: FM 4-90.7, Stryker Brigade Combat Team Logistics (Sept '07), pp. 3-1 to 3-10.

The BSB main CP centrally controls distribution-based logistics operations for the BCT. The main CP also coordinates for the protection of the BSA under direction of the BSB S2/S3.

The BSB headquarters is responsible for maintaining situational awareness using C2 information management systems. These systems provide location and configuration, total asset visibility, ITV, and overall connectivity to supported and adjacent units and higher headquarters.

The BCT administrative and logistics operations center is actually collocated within the BSB main CP and includes the BCT S1, S4, and transportation officer. The USAMC BLST provides interface with the BCT administrative and logistics operations center.

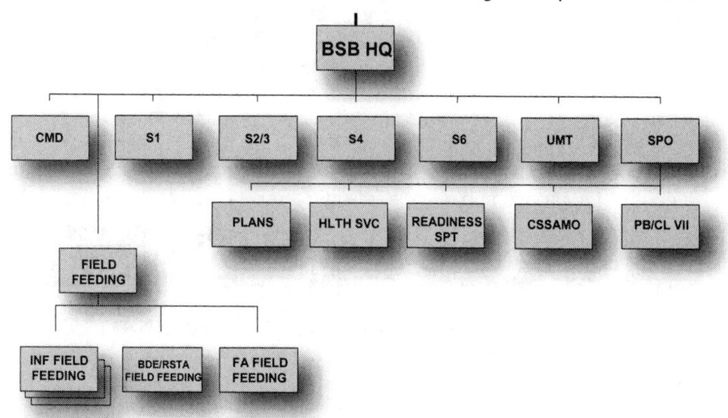

Headquarters

The HHC is one of several companies assigned to the BSB. The HHC:

- Provides the company level entity to which the BSB staff is assigned and from which they draw typical company level support
- Provides C2, administration, and logistic support
- Determines the placement of CBRNE assets in the company area and identifies elements of and plans use of base defense forces

The HHC commander provides staff supervision for field feeding and food service support for the BSB and designated personnel. Field feeding teams provide field-feeding services to assigned BCT headquarters personnel, the BSB, the infantry battalions, RSTA squadron, fires battalion, and the MI, engineer, signal, and anti-tank (AT) companies.

The headquarters contains a battalion staff organization structure with a command section, S1 section, consolidated S2/S3 section, S4 section, S6 section, UMT, and a distribution operations section. The BSB headquarters directs the C2, communications systems, and ISR functions of the BSB:

- C2 of units assigned or attached to the BSB
- C2 of all units in the BSA for security and terrain management
- Planning, direction, and supervision of sustainment administration for all units assigned or attached to the BCT
- Planning, direction, and supervision of administration and logistics for units assigned or attached to the battalion

- Limited unit-level administration and religious services for units of the battalion.
- Planning and direction of BSA security or areas as assigned by the supported brigade commander. The BSB may be designated as an alternate CP.
- Information and advice to the supported brigade commander and staff on support capabilities provided by the battalion
- Field feeding and ration storage
- Planning and execution of unit-level mortuary affairs training
- Property accountability and asset visibility for the brigade

Command Section

The command section of the BSB provides C2 for assigned and attached units and supervision for the BSB staff. It directs logistics operations for the BCT. It also provides information and advice on logistics to the BCT commander and staff.

The command section consists of the BSB battalion commander, battalion XO, command sergeant major (CSM), and coordinating, special, and personal staff officers. Staff officers supervise and coordinate the functions of subordinate sections. Command section staff officers perform duties and responsibilities common to all staff officers. FM 5-0 discusses in detail these duties and responsibilities which include:

- Preparing plans and orders
- Processing, analyzing, and disseminating information
- Preparing, updating, and maintaining estimates
- Making recommendations
- Supervising staff section and staff personnel
- Identifying and analyzing problems

Command section staff officers conduct mission analysis, develop estimates and plans, and implement policies and orders. They develop a reporting and monitoring system for operations in their area of expertise. They provide information updates to the BSB battalion commander and exchange information with other staff sections.

Battalion Commander

The brigade support battalion (BSB) commander is responsible for the brigade support area (BSA). For security purposes, this includes the operational control (OPCON) of all elements operating within the BSA. The BSB commander might be assigned additional unassigned area functions. The BSB battalion commander:

- Is the senior logistics commander and single logistics operator for the BCT
- Provides support using an array of digital information systems and a technologically competent staff that is capable of capitalizing on evolving technology
- Directs all units organic or attached to the BSB and has C2 of all elements in the BSA for security and terrain management
- Provides subordinate elements with clear missions, taskings, and statement of his/her intent

The BSB battalion commander, with the HHC staff, supervises the activities of subordinate units. They ensure that decisions, directives, and instructions are implemented and that the BCT commander's intent is being fulfilled. The BSB battalion commander and staff advise the BCT commander on logistic support as required. The BSB battalion commander's duties include:

- Providing logistics assets required to support the BCT
- Providing commander's intent and mission guidance
- Reviewing running estimates, perform course of action (COA) analysis, and recommend the COA that best supports the BCT mission
- Stating his/her estimate of the situation and announce his/her decision

Bde Cbt Tm Sustainment

III. The Field Maintenance Company (FMC)

Ref: FM 4-90.7, Stryker Brigade Combat Team Logistics (Sept '07), pp. 4-1 to 4-9.

The BSB is a centrally managed field maintenance activity. The essential maintenance tasks for this organization are to:

- Maintain BCT equipment at the Army maintenance standard before the brigade enters the operational environment
- Replace line-replaceable units (LRUs), components, and major assemblies in the operational environment

Field maintenance tasks are those that contribute to achieving and maintaining the Army maintenance standard for fully mission capable equipment. Field maintenance tasks return non-mission capable (NMC) equipment to fully mission capable status at the owning unit. The primary methods of returning systems to a mission capable status include using Class IX repair parts, battle damage assessment and repair (BDAR), controlled substitution, cannibalization, and Class VII replacement.

Organization

The FMC consists of a headquarters, a maintenance control platoon, a wheeled vehicle repair platoon, a maintenance support platoon, and five CRTs.

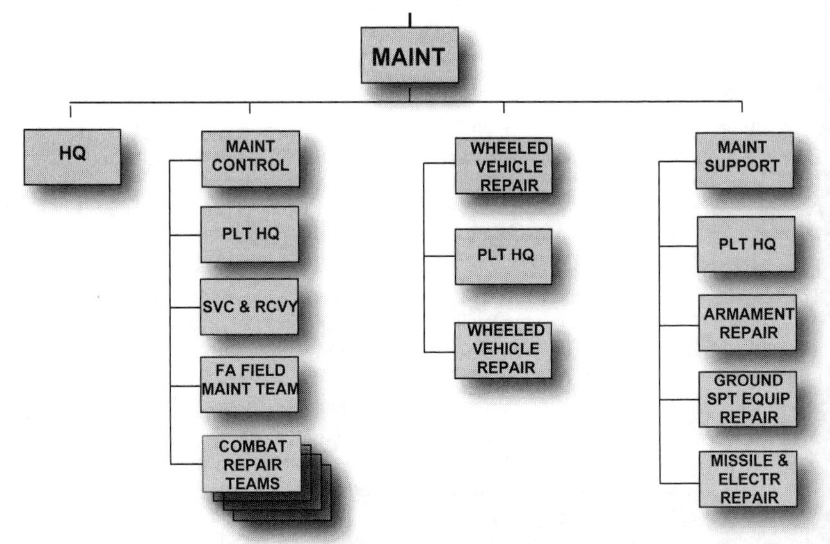

The FMC provides unscheduled field maintenance support (less medical) for the BCT. It has the maintenance capabilities to perform automotive, armament, missile, communications, special devices, and ground support equipment repair. Scheduled maintenance support for all rolling stock is provided by CLS at EAB.

Base Maintenance Operations

The FMC retains a maintenance capability in the BSA since certain pieces of test equipment are not easily transportable. Base maintenance provides dedicated field maintenance on an area basis to BSB troops as well as backup support to the CRTs and supported maneuver battalions. The MCS operates automated maintenance systems to support the BCT companies and the BSB. It also serves as the main collection point for all maintenance records prior to being sent to the BSB SPO staff. Base repair sections can perform contact maintenance missions as required.

Field Maintenance

Two-level maintenance (TLM) consists of field maintenance and sustainment maintenance. Field maintenance is focused on returning equipment to an operational status. The field maintenance level accomplishes this mission by fault isolating and replacing the failed component, assembly, or module. Field maintenance is defined by on-system repairs and the replace forward concept. The intent of this level is to replace the failed component, assembly, or module that returns the system to an operational status supporting the tactical commander. The only level of maintenance performed within the BCT is field maintenance. *See pp. 1-55 to 1-56 for further discussion.*

Replace Forward/Repair Rear

With replace forward/repair rear doctrine, the FMC uses field maintenance that quickly returns systems to a mission capable or fully mission capable status. Faults that do not render a system NMC will be deferred until augmentation arrives or the OPTEMPO permits more repair time. To be most efficient and to generate combat power, the FMC will often focus on the replacement of LRUs and major assemblies, but, when appropriate, may perform on-system repairs of components. The majority of the FMC assets are located in the BSA to reduce the burden placed on maneuver elements. The critical maintenance nodes remain in the MCPs located in the maneuver task force areas. Each of these elements will have a CRT from the FMC. Due to its limited size, the CRT will often require a daily resupply of mission critical repair parts.

Controlled Exchange

Controlled exchange is the removal of serviceable components from unserviceable but economically reparable equipment for immediate reuse in restoring another like item of equipment to combat serviceable condition. The unserviceable component must be used to replace the serviceable component or retained with the end item that provided the serviceable component. Commanders at brigade level will set guidelines for controlled exchange. Controlled exchange is managed by the BSB commander according to the set priorities and is maintained within the maintenance control section of the BSB. *Refer to AR 750-1 for more information on controlled exchange.* *See also p. 2-49.*

Cannibalization

Cannibalization is the authorized removal of components from materiel designated for disposal. It supplements supply operations by providing assets not readily available through normal supply channels. During combat, commanders may authorize the cannibalization of disabled equipment only to facilitate repair of other equipment for return to combat. Costs to cannibalize, urgency of need, and degradation to resale value of the end item should be considered in the determination to cannibalize. Cannibalization of depot maintenance candidate items, controlled exchange, or component parts by field organizations is prohibited. Exceptions will be made only in urgent cases of field operational readiness requirements and then only with the written concurrence of the AMC major subordinate command. Cannibalization is not authorized during peacetime without approval from the national inventory control point (NICP). *Refer to AR 750-1 and AR 710-2 for more information on cannibalization.* *See also p. 2-49.*

Battle Damage Assessment and Repair (BDAR)

BDAR is the procedure to rapidly return disabled equipment to the operational commander by field-expedient repair of components. BDAR restores the minimum essential combat capabilities necessary to support a specific combat mission or to enable the equipment to self-recover. BDAR is accomplished by bypassing components or safety devices, cannibalizing parts from like or lower priority equipment, fabricating repair parts, taking shortcuts to standard maintenance, and using substitute fluids, materials, or components. Depending on the repairs required and the amount of time available, repairs may or may not return the vehicle to a fully mission capable status. *See p. 2-34 and FM 4-30.31 for further discussion of BDAR. See also p. 2-34.*

IV. Distribution Company

Ref: FM 4-90.7, Stryker Brigade Combat Team Logistics (Sept '07), pp. 6-1 to 6-14 and 1-11 to 1-12.

The distribution company of the BSB provides supply and transportation support to elements of the BCT. It also provides distribution management for all classes of supply and services. The distribution company is responsible for the following:

- Planning and supervision of supply distribution points and transportation and field service support
- Daily receipt, temporarily storage, and issuance of all supplies (less Class VIII)
- Transportation of daily cargo
- Class III(B) retail fuel support to the brigade
- Water purification, storage, and distribution for the brigade

The distribution company includes a company headquarters and transportation, supply, and fuel and water platoons. The supply platoon provides support to the BCT and maintains the ASL (limited) for the BCT. The fuel and water platoon provides bulk fuel distribution and water purification and distribution. The transportation platoon provides the distribution capability to deliver supplies.

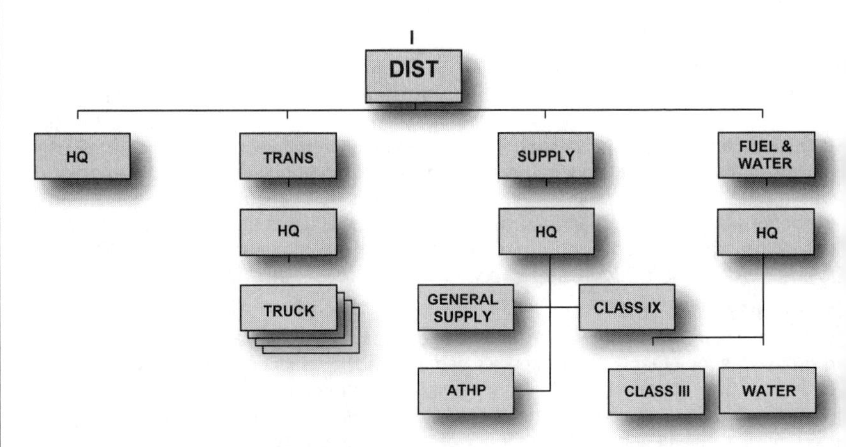

Distribution Operations

Distribution operations consist of distribution management (BSB SPO), C2 oversight (BCT S4), and physical distribution assets. Distribution operations are the processes of synchronizing all elements of the logistics system to deliver the required resources in order to generate and sustain the military capability required by joint forces. Sustainment of the BCT is planned by the BCT S4 and executed by the BSB. To ensure clear linkage between planning and execution functions, the brigade S1 and S4 may be located close to the BSB headquarters. BCT subordinate units will also locate combat trains command post (CTCP) elements in or close to the BSA.

Battalion S4s consolidate logistics status input and pass them via automated systems, such as BCS3, to the brigade S4 and BSB. Information is also provided to the BCT S1 and S3 and the BSB distribution company so they have overall knowledge of activity within their respective area. The BCT S4 and BSB SPO are then responsible for coordinating times and locations for sustainment distribution operations. Management of distribution operations may involve a combination of basic loads and regionally available sources for bulk fuel and water. Supported units are responsible for securing and conducting the distribution process as it occurs.

Support Methods

The BCT uses a number of support methods. Some of these methods are discussed in the following paragraphs.

Unit (Battalion/Company/Platoon) Distribution

In unit distribution, supplies are configured in unit sets (battalion/company/platoon, depending on the level of distribution) and delivered to one or more central locations. Heavy Expandable Mobility Tactical Truck-Load Handling System (HEMTT-LHS) fuel rack systems remain at the site to refuel unit vehicles as they cycle through the supply point. This technique makes maximum use of the capacity of BCT truck assets by minimizing delivery and turnaround time.

Supply Point Distribution

Supply point distribution requires unit representatives to move to a supply point to pick up their supplies. Supply point distribution is most commonly executed by means of a logistics release point (LRP). The LRP may be any place on the ground where unit vehicles return to pick up supplies and then take them forward to their unit. Occasionally, the LRP is the brigade support area (BSA) itself.

Refuel/Resupply on the Move (ROM)

The refuel/resupply on the move (ROM) method of replenishment is conducted by having supported unit S3 and S4 staffs coordinate with the BCT S4 and BSB support operations section to fix the time and place to conduct the ROM operations according to current unit battle rhythm. As a rule, a ROM operation is established and conducted as part of a unit movement. A ROM point is typically built to support several types of units passing through a point sequentially and provides for most classes of supply, including Class V and water.

Aerial Resupply (Deliberate, Fixed Wing, and Rotary Wing)

Aerial delivery is a viable option for distributing dry cargo to limited access or far forward areas or when delivery time is crucial. Aerial delivery is normally via containerized delivery system (CDS) or platform (463L pallet) airdrop. It may be a vital link in supporting RSTA units or other small dispersed units throughout the operation. Aerial delivery is an intensively coordinated endeavor due to requirements for drop zones and coordination lead time. The normal drop zone size for aerial delivery is 600 by 600 yards. Rigging of delivery platforms or containers usually occurs outside the brigade AO and is done by an aerial delivery unit. Flexibility is limited due to the physical nature of the drop zone and surrounding terrain and vegetation. Aerial resupply using rotary wing aviation is also an option for distributing limited quantities of supplies to remote or forward locations. Sling loading supplies is labor intensive but may be appropriate for some situations, such as Class IX delivery to combat repair teams (CRTs). This method provides ample flexibility for delivery locations as long as there is a clear area to hover the aircraft. The limiting factor is the availability of aircraft supporting the BCT. Sling load operations require aviation augmentation early in the operation. Deliberate aerial resupply may be conducted with either fixed-wing or rotary-wing aircraft.

Immediate Resupply

Immediate resupply, also referred to as emergency resupply, is the least preferred method of distributing supplies. While some emergency resupply may be required when combat losses occur, requests for immediate resupply not related to combat loss indicate a breakdown in coordination and collaboration between the logistician and customer. If immediate resupply is necessary, all possible means, including options not covered above, may be used. The battalion/squadron S4s, the BCT S4, and the BSB SPO must constantly and thoroughly collaborate to minimize the need for immediate resupply. Emergency resupply that extends beyond BSB capabilities requires immediate intervention of the next higher command capable of executing the mission. In such a case, the BCT S4 and BSB SPO immediately coordinate with the next higher echelon of support for the BCT.

V. The Brigade Support Medical Company

Ref: FM 4-90.7, Stryker Brigade Combat Team Logistics (Sept '07), pp. 5-1 to 5-11.

The mission of the BSMC is to provide AHS support to the all units subordinate to the BCT and non-brigade units operating within the BCT AO. The BSMC operates a Role 2 MTF and provides Role 1 HSS on an area basis to all BCT units that do not have organic medical assets. The company provides C2 for its organic elements and operational control of medical augmentation elements. The BSMC locates and establishes its company headquarters in the BSA and establishes a BSMC Role 2 MTF and, when required, may be augmented with a surgical capability.

For additional information on the operations and function of similar organizations, see FM 4-02.6 and FM 4-02.121.

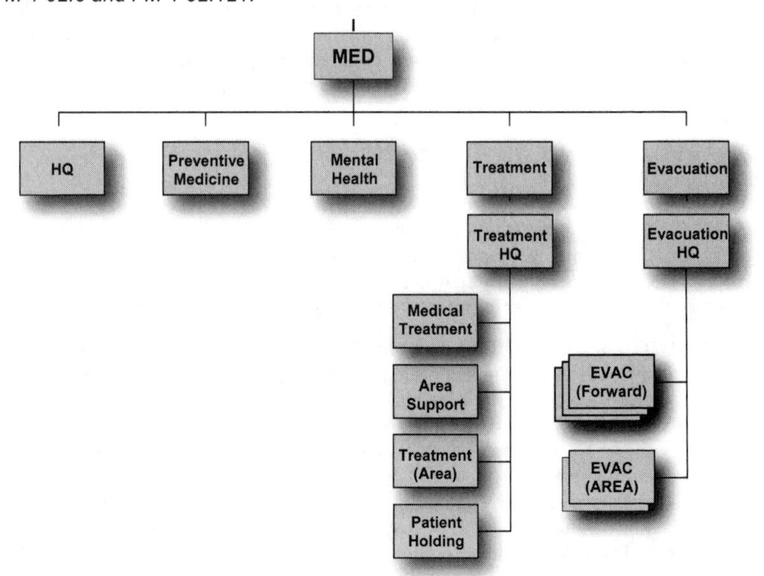

The BSMC is organized into a company headquarters, treatment and evacuation platoons, and PVNTMED and mental health (MH) sections, and performs the following :

- Emergency medical trmt and ATM for wounded and nonbattle injury patients
- Sick call services
- Ground ambulance evacuation
- Operational dental treatment that includes emergency essential dental care
- Class VIII resupply and medical equipment maintenance support
- Limited medical laboratory and radiology diagnostic services
- Outpatient consultation services for patients referred from Role 1 MTFs
- Patient holding for up to 20 patients
- Reinforcement and augmentation of maneuver battalion medical platoons
- Coordination with the UMT for required religious support
- Preventive medicine consultation and support
- Combat and operational stress control support
- Mass casualty operations (triage and management)
- Patient decontamination and treatment

III. Bde Combat Team Concept of Support

Adapted from FM 4-90.7, Stryker Brigade Combat Team Logistics (Sept '07) and related sources. When published, FM 4-90.1, Brigade Support Battalion, will describe how sustainment operations take place in heavy and infantry BCT area of operations.

BCTs are organized with the self-sustainment capability for up to 72 hours of combat. Beyond 72 hours, sustainment organizations at the division and corps levels are required to conduct replenishment of the BCT's combat loads. That replenishment is a function of the higher headquarters' sustainment brigade(s).

The Brigade Support Battalion (BSB)

The brigade support battalion (BSB) is the core of sustainment to the brigade combat team (BCT). The BSB is organic to the BCT and consists of functional and multifunctional companies assigned to provide support to the BCT.

All logistics requirements (less medical) beyond the BSB's ability are either furnished by or coordinated through the supporting sustainment brigade. The sustainment brigade supports the BCT on an area basis.

See pp. 3-1 and 3-3 to 3-10 for further discussion.

Mission Tailoring

Sustainment operations can be tailored in response to changes in tactical requirements. In most cases, the BSB will provide the supplies and services required by the supported unit at a specific point in time (scheduled delivery). For example, a typical day may include distribution to a battalion level distribution point for one customer cluster, to company/battery level for another customer cluster, and all the way to platoon/team level for a third cluster, while the fourth cluster receives no delivery (due to low/no requirements) that particular day.

Supported unit commanders coordinate through their S-3 and S-4 staffs according to current unit battle rhythm to fix the time and place for replenishment operations at a temporarily established point. Assets can be retasked if the situation demands. This approach, executed according to centralized management, optimizes the employment of personnel.

See following pages (pp. 3-12 to 3-13) for further discussion.

Support Areas

A support area is a designated area in which sustainment elements, some staff elements, and other elements locate to support a unit. The BCT S-4, BCT S-3, and BSB S-3 coordinate the location of the BCT sustainment support areas.

See pp. 3-14 to 3-17 for further discussion.

Sustainment Support

The BSB provides support in accordance with the three sustainment WFF categories: logistics, personnel services, and health service support. The BCT's organic sustainment capabilities will require augmentation to provide some of these functions and services.

See pp. 1-2 to 1-3 and 2-16 to 2-18 for further discussion.

I. Mission Tailoring of Sustainment Assets

Ref: FM 4-90.7, Stryker Brigade Combat Team Logistics (Sept '07) and related sources.

The brigade support battalion (BSB) is the core of sustainment to the brigade combat team (BCT). The BSB is organic to the BCT and consists of functional and multifunctional companies assigned to provide support to the BCT. Field Manual (FM) 4-90.1, Brigade Support Battalion, when published, will describe how sustainment operations take place in the heavy and infantry BCT area of operations.

Sustainment operations can be tailored in response to changes in tactical requirements. In most cases, the BSB will provide the supplies and services required by the supported unit at a specific point in time (scheduled delivery). For example, a typical day may include distribution to a battalion level distribution point for one customer cluster, to company/battery level for another customer cluster, and all the way to platoon/team level for a third cluster, while the fourth cluster receives no delivery (due to low/no requirements) that particular day.

Supported unit commanders coordinate through their S-3 and S-4 staffs according to current unit battle rhythm to fix the time and place for replenishment operations at a temporarily established point. Assets can be retasked if the situation demands. This approach, executed according to centralized management, optimizes the employment of personnel.

Logistical Support Teams (LSTs)

To provide necessary agility and flexibility, the BSB may use temporary task-organized logistic support teams (LSTs) allocated to maneuver units. LSTs are METT-TC dependant and used to support the commander's intent, to reduce the amount of supplies and equipment in the BSA, and to provide maneuver units with logistics assets to support battalion-level operations. As a rule, LSTs are allocated to maneuver units to perform supply distribution, transportation, and food service to support that unit. If one unit has priority for support over another based on METT-TC factors and the BCT commander's intent, it may be necessary to mix, combine, or shift assets from the BSB, LST, or CRT to another to support mission requirement. The preferred solution is to provide backup or surge maintenance capability from within the assets of the BSB, but in some cases that may not be the most efficient or effective way to support the task force.

Combat Repair Teams (CRTs)

As a rule, CRTs are allocated to maneuver units to perform field maintenance to support that unit. Sometimes one unit will have priority for maintenance over another based on METT-TC factors and the BCT commander's intent. In such cases, it may be necessary to combine or shift assets from one CRT to another to support maintenance requirements. The preferred solution is usually to provide backup or surge maintenance capability from within the assets of the FMC, but in some cases that may not be the most efficient or effective way to support the task force. Given the METT-TC situation, there may be a requirement to temporarily task organize a CRT or portions thereof.

Support to Separate Companies

The BCT has an engineer company, a military intelligence company, an antitank company, and a network support company that do not operate under a battalion. These companies, like the brigade HHC, are supported by the BSB regardless of where they are located on the battlefield. If one of these companies, or part of the company, is task organized to a maneuver battalion, it will remain under the support of the BSB. The company commander must coordinate with the maneuver battalion's S4, the BSB SPO, the distribution and maintenance company commanders, and the supporting CRT chief. Based upon the local situation and conditions, they may decide to integrate the company's logistics requirements into the battalion's logistic support structure.

Support for Attachments

Because of the BCT's austere logistics capabilities, attachments to the BCT should arrive with appropriate support augmentation. When a company, team, or detachment is attached to the BCT, the BCT S4 integrates required augmentation into the BCT support systems. The BCT S4 must clearly state who will provide medical, maintenance, and recovery services, and provide support for Classes III, V, and IX. When the BCT receives attachments, the S1 orients those units to processes that maintain personnel accountability and arranges for the necessary administrative support for those units. Logistics planners require some basic information from the sending unit's S4 to anticipate how to develop a synchronized concept of support. Some considerations are:

- Number and type of supplies, personnel, and equipment
- Current status and/or strength
- When attachment is effective and for how long
- What support assets will accompany the attached element
- When and where linkup will occur, and who is responsible

Mission Staging Operations

To maintain continuous pressure on enemy forces, the division designs operations that accommodate the cycling of BCTs to temporary bases where they rest, refit, and receive supplies. These can be sustaining operations. The BCT moves to the area established by the sustainment brigade for mission staging. While in mission staging, the BCT is not available for tactical tasks other than local security. Typically, mission staging involves the sustainment brigade, maneuver enhancement brigade, and the BCT. In offensive operations, one BCT may replace another in the attack, usually when one has a follow and assume mission. The division commander then orders a mission staging operation for the BCT that is out of the fight. After mission staging, that BCT may assume the attack while the second BCT refits continuing a tactical cycle of mission staging without relinquishing the initiative. The BCT can do this on a scheduled basis not tied to a unit's respective combat power status.

II. Sustainment Support Areas

Ref: FM 4-90.7, Stryker Brigade Combat Team Logistics (Sept '07) and related sources.

A support area is a designated area in which sustainment elements, some staff elements, and other elements locate to support a unit. The BCT S-4, BCT S-3, and BSB S-3 coordinate the location of the BCT sustainment support areas.

Sustainment Support Areas

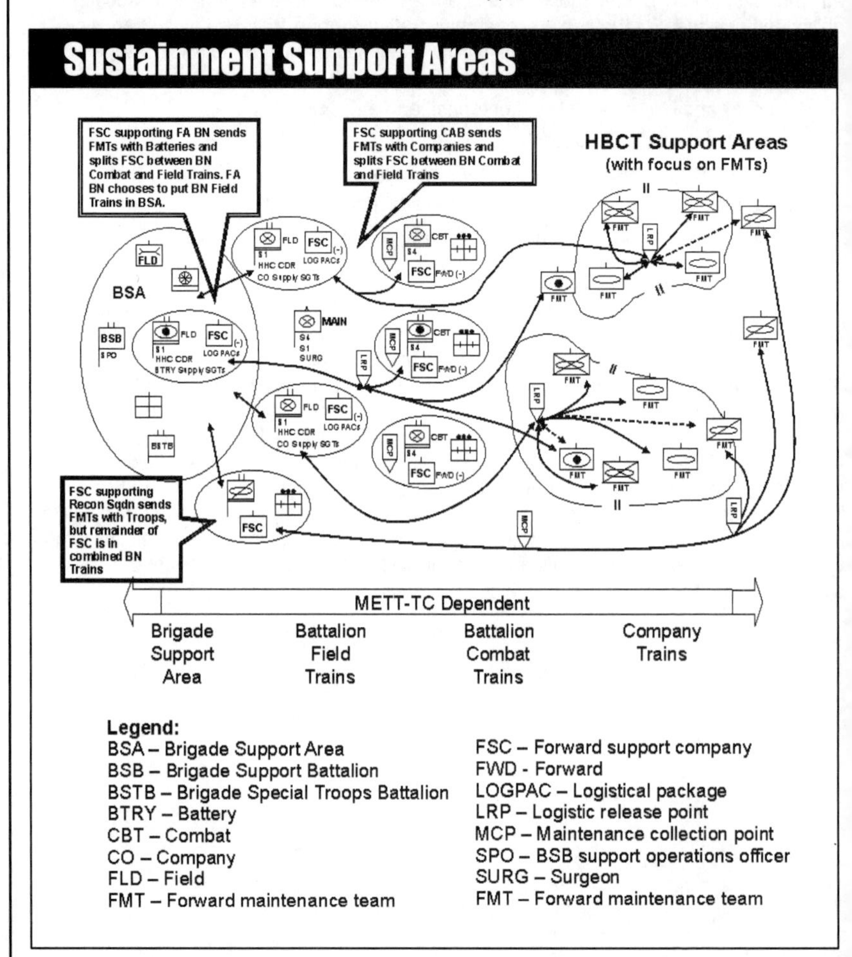

Legend:
BSA – Brigade Support Area
BSB – Brigade Support Battalion
BSTB – Brigade Special Troops Battalion
BTRY – Battery
CBT – Combat
CO – Company
FLD – Field
FMT – Forward maintenance team
FSC – Forward support company
FWD - Forward
LOGPAC – Logistical package
LRP – Logistic release point
MCP – Maintenance collection point
SPO – BSB support operations officer
SURG – Surgeon
FMT – Forward maintenance team

A. Trains

Trains are a grouping of unit personnel, vehicles, and equipment to provide sustainment. It is the basic sustainment tactical organization. Maneuver battalions use trains to array their subordinate sustainment elements including their designated forward support company. Battalion trains usually are under the control of the battalion S-4, assisted by the battalion S-1. The composition and location of battalion trains varies depending on the number of units attached to, or augmenting, the battalion.

Company Trains

Company trains provide sustainment for a company during combat operations. Company trains usually include the first sergeant, medical aid/evacuation teams, supply sergeant, and the armorer. Usually, the forward support company provides a field maintenance team with capabilities for maintenance, recovery, and limited combat spares. The supply sergeant can collocate in the combat trains if it facilitates logistics package (LOGPAC) operations. The first sergeant usually directs movement and employment of the company trains; although the company cdr may assign the responsibility to the company XO.

Battalion Trains

Battalion trains consist of two types: combat trains and field trains.

- **Combat Trains**. The combat trains usually consist of the forward support company and the battalion medical unit. The UMCP should be positioned where recovery vehicles have access, or where major or difficult maintenance is performed. The factors of mission, enemy, terrain and weather, troops and support available, time available (METT-TC) must be considered when locating combat trains in a battalion support area.
- **Field Trains**. Field trains can be located in the BSA and include those assets not located with the combat trains. The field trains can provide direct coordination between the battalion and the BSB. When organized, the field trains usually consist of the elements of the forward support company, battalion HHC, battalion S-1, and battalion S-4.

Battalion trains can be employed in two basic configurations:

- **Unit Trains**. Unit trains at the battalion level are appropriate when the battalion is consolidated, during reconstitution, and during major movements.
- **Echeloned Trains**. Echeloned trains can be organized into company trains, battalion combat trains, unit maintenance collection point (UMCP), battalion aid station (BAS), or battalion field trains.

B. Sustainment-Related Command Posts (CPs)

The battalion commander may choose to create a **combat trains command post (CTCP)** or a **field trains command post (FTCP)** to control administrative and sustainment support. Most of the time, the S-4 is the officer in charge of the CTCP. If constituted, the FTCP could be led by the HHC commander. These command posts (CPs) can be organized to accomplish specific logistical tasks.

C. Brigade Support Area (BSA)

The BSA is the logistical, personnel, and administrative hub of the BCT. It consists of BSB, but could also include a BCT alternate CP (if formed), battalion field trains, brigade special troops battalion units, signal assets, and other sustainment units from higher HQ. The BCT operations staff officer (S-3), with the BCT S-4 and the BSB, determines the location of the BSA. The BSA should be located so that support to the BCT can be maintained, but does not interfere with the tactical movement of BCT units, or with units that must pass through the BCT area. The BSA's size varies with terrain; however, an area 4 km to 7 km in diameter is a planning guide. Usually the BSA is on a main supply route (MSR) and ideally is out of the range of the enemy's medium artillery. The BSA should be positioned away from the enemy's likely avenues of approach and entry points into the BCT's main battle area (MBA).

Usually the S-4 coordinates the BCT main CP's sustainment cell which contains the BCT S-4, BCT S-1, BCT surgeon section, and the BCT UMT. The BCT commander can create an alternate CP for sustainment, should the administrative and logistics presence in the main CP become too large. The BSB or brigade special troops battalion CPs may be able to host the sustainment cell if communications links are adequate. *See also p. 3-3.*

III. Support Area Considerations

Ref: FM 4-90.7, Stryker Brigade Combat Team Logistics (Sept '07) and related sources.

The brigade support battalion (BSB) commander is responsible for the brigade support area (BSA). For security purposes, this includes the operational control (OPCON) of all elements operating within the BSA. The BSB commander might be assigned additional unassigned area functions.

Locations for Support Areas

The trains should not be considered a permanent or stationary support area. The trains must be mobile to support the battalion when it is moving, and should change locations frequently, depending on available time and terrain. The trains changes locations for the following reasons:

- Change of mission
- Change of unit AOs
- To avoid detection caused by heavy use or traffic in the area
- When area becomes worn by heavy use (e.g., wet and muddy conditions)
- Security becomes lax or complacent due to familiarity

All support areas have many similarities, including:

- Cover and concealment (natural terrain or man-made structures)
- Room for dispersion
- Level, firm ground to support vehicle traffic and sustainment operations
- Suitable helicopter landing site (remember to mark the landing site)
- Good road or trail networks
- Good routes in and out of the area (preferably separate routes going in and going out)
- Access to lateral routes
- Positioned along or good access to the MSR
- Positioned away from likely enemy avenues of approach

Security of Support Areas

Tactical logistics organizations are normally the units least capable of self-defense against a large, enemy combat force. Given the common operating environment (COE), they are also often the targets of enemy action. As the enemy threat increases, unit commanders cannot decrease logistics operations in favor of enhancing force protection. The supported commander and the logistics unit commander must have previously discussed what risks are reasonable to accept and what risk mitigation measures they should implement based on requirements and priorities including Force Health Protection. Only then can logistics commanders and staffs plan accordingly. Logisticians and unit commanders must be competent in warfighting, military decision-making, maneuver, and other tactical skills to anticipate and decide on appropriate risk mitigation measures.

Sustainment elements must organize and prepare to defend themselves against ground or air attacks. Often, they occupy areas that have been secured by maneuver elements of the BCT. The security of the trains at each echelon is the responsibility of the individual in charge of the trains. The best defense is to avoid detection. The following activities help to ensure trains security:

- Select good trains sites that use available cover, concealment, and camouflage
- Use movement and positioning discipline as well as noise and light discipline to prevent detection
- Establish a perimeter defense
- Establish observation posts and patrols

- Position weapons (small arms, machine guns, and antitank weapons) for self-defense
- Plan mutually supporting positions to dominate likely avenues of approach
- Prepare a fire plan and make sector sketches
- Identify sectors of fires
- Emplace target reference points (TRPs) to control fires and for use of indirect fires
- Integrate available combat vehicles within the trains (i.e., vehicles awaiting maintenance or personnel) into the plan, and adjust the plan when vehicles depart
- Conduct rehearsals
- Establish rest plans
- Identify an alarm or warning system that would enable rapid execution of the defense plan without further guidance; the alarm, warning system, and defense plan are usually included in the standing operating procedure (SOP)
- Designate a reaction force. Ensure the force is equipped to perform its mission. The ready reaction force must be well rehearsed and briefed unit assembly.
- Friendly and threat force recognition
- Actions on contact

Supply Routes

The BCT S-4, in coordination with the BSB support operations officer and BCT S-3, select supply routes between support areas. MSRs are routes designated within the BCT's AO upon which the bulk of sustainment traffic flows in support of operations. An MSR is selected based on the terrain, friendly disposition, enemy situation, and scheme of maneuver. Alternate supply routes are planned in the event that an MSR is interdicted by the enemy or becomes too congested. In the event of chemical, biological, radiological, and nuclear (CBRN) contamination, either the primary or alternate MSR can be designated as the —dirty MSR to handle contaminated traffic. Alternate supply routes should meet the same criteria as the MSR. Military police (MP) may assist with regulating traffic, and engineer units, if available, could maintain routes. Security of supply routes in a noncontiguous environment might require the BCT commander to commit non-sustainment resources. Route considerations include:

- Location and planned scheme of maneuver for subordinate units
- Location and planned movements of other units moving through the BCT's AO
- Route characteristics such as route classification, width, obstructions, steep slopes, sharp curves, and type of roadway surface
- Two-way, all-weather trafficability
- Classification of bridges and culverts
- Requirements for traffic control such as at choke points, congested areas, confusing intersections, or along built-up areas
- Number and locations of crossover routes from the MSR to alternate supply routes
- Requirements for repair, upgrade, or maintenance of the route, fording sites, and bridges
- Route vulnerabilities that must be protected. This can include bridges, fords, built-up areas, and choke points
- Enemy threats such as air attack, conventional and unconventional tactics, mines, ambushes, and chemical strikes
- Known or likely locations of enemy penetrations, attacks, chemical strikes, or obstacles
- Known or potential civilian/refugee movements that must be controlled or monitored

See p. 2-85 for additional discussion of main supply routes (MSRs)/alternate supply routes (ASRs). See pp. 2-79 to 2-92 for additional protection considerations.

IV. Sustainment Reporting

Ref: FM 4-90.7, Stryker Brigade Combat Team Logistics (Sept '07), pp. 1-7 to 1-9.

Support operations are fully integrated with the BCT battle rhythm through integrated planning and oversight of ongoing operations. Sustainment and operational planning occurs simultaneously rather than sequentially. Incremental adjustments to either the maneuver or sustainment plan during its execution must be visible to all BCT elements. The sustainment synchronization matrix and sustainment report are both used to initiate and maintain synchronization between operations and sustainment functions.

Accurately reporting the sustainment status is essential to keeping units combat ready. The sustainment report is the primary product used throughout the brigade and at higher levels of command to provide a logistics snapshot of current stock status, on-hand quantities, and future requirements. Battle Command Sustainment Support System (BCS3), Movement Tracking System, and Force XXI Battle Command Brigade and Below (FBCB2) are some of the fielded systems that the BSB uses to ensure effective situational awareness and sustainment support.

Company Level

At company level, the first sergeant (1SG) or designated representative is responsible for gathering the information from the platoon sergeants and submitting a consolidated report to the battalion S4. The 1SG can direct cross-leveling between platoons and forecast requirements based on current balances and upcoming mission requirements. Some possible details to include in the logistics report are systems with an operational readiness rate below 60 percent, changes to anticipated expenditure rates, Class V status, and significant incidents. The primary means of gathering this information and submitting it to the battalion S4 is through the logistics report in FBCB2.

Battalion Level

The battalion S4 is responsible for collecting reports from all companies and ensuring reports are complete, timely, and accurate. The battalion S4, with the support operations officer (SPO) and executive officer's (XO's) concurrence, determines which units receive which supplies. That decision is based on mission priority and the battalion commander's guidance. Upon receiving the logistics report, the company then validates external supplies to fulfill its requirements (where capable) and provides input to the logistics report on the adjusted balance of external supplies. The adjusted balances of external supplies are added to the logistics report and returned to the battalion S4. The company also provides a coordination copy to the BSB's SPO.

Brigade Level

The brigade S4 is responsible for collecting reports from all battalions, including the BSB logistics report on internal supplies. Prior to the brigade S4 forwarding a consolidated report to the BSB SPO, the brigade S4, with brigade executive officer's concurrence, determines which units receive which supplies. Their decision is based upon mission priority and the brigade commander's guidance. Upon receiving the logistics report, the BSB SPO conducts a brigade logistics synchronization meeting. The BSB SPO then disseminates the external supplies to fulfill battalion requirements (where capable), synchronizes distribution, and provides input to the logistics report. The updated logistics report and logistics synchronization matrix complement paragraph 4 and annex I of the operations order (OPORD) or fragmentary order (FRAGO).

Division/Corps Level

The division/corps G4 is responsible for collecting reports from all task-organized brigades and ensuring reports are complete, timely, and accurate. The division/corps may add information such as changes to theater opening and changes to anticipated expenditure rates. The division/corps G4 has a complete logistics report and forwards this report to the next higher level of command and then forwards a logistics report for coordination to the supporting TSC/Expeditionary Sustainment Command (ESC) SPO.

I. Planning Sustainment Operations

Ref: FM 4-0, Sustainment (Aug '09), chap. 4, pp. 4-2 to 4-7.

Planning begins with analysis and assessment of the conditions in the operational environment with emphasis on the enemy. It involves understanding and framing the problem and envisioning the set of conditions that represent the desired end state (FM 3-0). Sustainment planning indirectly focuses on the enemy but more specifically on sustaining friendly forces to the degree that the Army as a whole accomplishes the desired end state. There are several tools available for conducting course of action analysis. We will highlight a couple of them below.

Many of the C2 planning functions (such as battle command, the determination of end state, and sustainment staff roles) can be found on pp. 1-23 to 1-34. This chapter focuses on more specific tools and planning considerations sustainment commanders and staffs use in planning for sustainment of full spectrum operations. See chap. 5, Joint Logistics, pp. 5-7 to 5-12 for information on planning joint operations.

I. Sustainment Preparation of the Operational Environment

Sustainment preparation of the operational environment is the analysis to determine infrastructure, environmental, or resources in the operational environment that will optimize or adversely impact friendly forces means for supporting and sustaining the commander's operations plan. The sustainment preparations of the operational environment assist planning staffs to refine the sustainment estimate and concept of support. It identifies friendly resources (HNS, contractable, or accessible assets) or environmental factors (endemic diseases, climate) that impact sustainment.

Some of the factors considered (not all inclusive) are as follows:

- **Geography**. Information on climate, terrain, and endemic diseases in the AO to determine when and what types of equipment are needed. For example, water information determines the need for such things as early deployment of well-digging assets and water production and distribution units.
- **Supplies and Services**. Information on the availability of supplies and services readily available in the AO. Supplies (such as subsistence items, bulk petroleum, and barrier materials) are the most common. Common services consist of bath and laundry, sanitation services, and water purification.
- **Facilities**. Information on the availability of warehousing, cold-storage facilities, production and manufacturing plants, reservoirs, administrative facilities, hospitals, sanitation capabilities, and hotels.
- **Transportation**. Information on road and rail networks, inland waterways, airfields, truck availability, bridges, ports, cargo handlers, petroleum pipelines, materials handling equipment (MHE), traffic flow, choke points, and control problems.
- **Maintenance**. Availability of host nation maintenance capabilities.
- **General Skills**. Information on the general skills such as translators and skilled and unskilled laborers.

See also pp. 4-9 to 4-14 for discussion of logistics preparation of the battlefield.

Operations Logistics (OPLOG) Planner

The OPLOG Planner is a web-based interactive tool that assists commanders and staff from strategic through operational levels in developing a logistics estimate. It is designed to support operations typically associated with multi-phase operations plans and orders. The OPLOG Planner enables staffs to develop estimated mission requirements for supply Class I, Class II, Class III(P), Class IV, Class VI, Class VII, Class X including water, ice, and mail. The tool uses the latest Army approved planning rates and modular force structures. It is updated at least annually to stay current with force structure and rate changes.

OPLOG Planner allows planners to build multiple task organizations from a preloaded list of units and equipment or from custom built units that are generated or imported. Each task organization is assigned a consumption parameter set that establishes the rates, climate, Joint Phases (Deter, Seize Initiative, and Dominate), and Army full spectrum operations (Offense, Defense, and Stability). The planners have the option of using predefined default planning rates or customizing rates based on what a unit is experiencing.

OPLOG Planner generates the logistics supply requirements which can be viewed by the Entire Operation, each Phase of the Operation, each Task Organization, each Unit, or each Unit's Equipment. Logistics requirement reports can be printed or exported to automated spread sheets for further analysis or saved for recall to be used in course of action analysis.

See pp. 4-14 for further discussion of the Operations Logistics (OPLOG) Planner.

Planning Heath Services Support (HSS)

Medical planners determine the capabilities and assets needed to support the mission. To ensure effective and efficient support, medical plans adhere to the principles of AHS support, the commander's planning guidance, medical intelligence related to the operational area, and other planning considerations.

The theater evacuation policy, health threat, troop strength or size of the supported population, and the type, intensity and duration of the operation are some of the factors considered for determining medical requirements. The medical staff estimates and patient estimates are also developed during planning. The patient estimate is derived from the casualty estimate prepared by the G-1.

In-depth analysis is critical at every level of the operation to ensure the flexibility to quickly react to changes in the mission and continue to provide the required support. The observations of commanders, disease and nonbattle injury rates, and running estimates are the primary means of assessing an operation to ensure that the concept of operations, mission, and commander's intent are met. These factors and continuous analysis help to make certain that once developed, the plan includes the right number and combination of medical assets to support the operation.

See FM 8-55 for additional information.

Medical and nonmedical automated information systems are used to plan AHS mission. Medical commanders must know the complete COP which includes situational awareness of three areas—(1) tactical (via FBCB2), (2) medical (via MC4/DHIMS), and (3) sustainment (via GCSS-Army). The commanders and medical planners must maintain situational awareness, in-transit visibility and tracking of patients and equipment, and a common operational picture of the AO. This information is obtained through various plans, reports, and information systems available to commanders and planners to facilitate the decision making process.

See pp. 4-31 to 4-36 for discussion and listing of HSS planning considerations. Automated information systems and other medical systems are discussed on pp. 1-26 to 1-27 (FM 4-0 app A).

II. Planning Considerations for Full Spectrum Operations

Army forces plan offensive, defensive, and stability or civil support operations simultaneously as part of an interdependent joint force. The proportion and role of offensive, defensive, and stability or civil support tasks are based on the nature of the operations, tactics used, or the environment. While full spectrum operations may occur simultaneously versus sequentially, they have slightly different planning requirements.

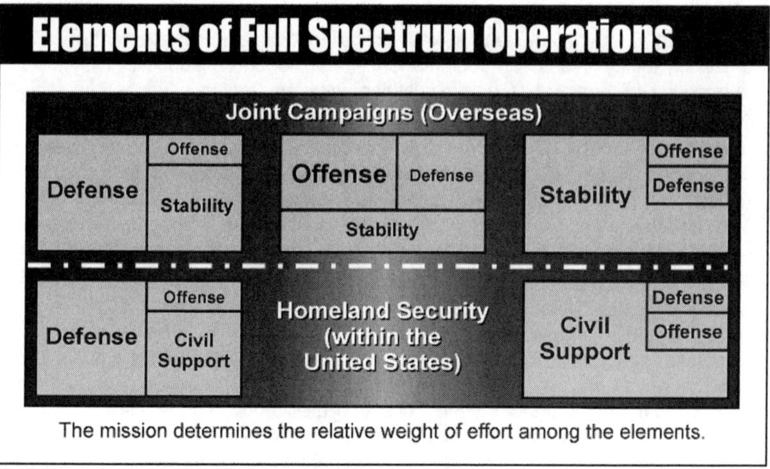

Ref: FM 4-0, Sustainment, fig. 4-2, p. 4-4.

A. Offensive Operations

Offensive operations are defined as combat operations conducted to defeat and destroy enemy forces and seize terrain, resources, and population centers. They impose the commander's will on the enemy (see FM 3-0). Sustainment of offensive operations is high in intensity. Commanders and staffs plan for increased requirements and demands on sustainment. Sustainment planners work closely with other WFF staffs to determine the scope of the operations and develop estimates for quantity and types of support required. They anticipate where the greatest need might occur and develop a priority of support. Sustainment planners may consider positioning sustainment units in close proximity to operations to reduce response times for critical support. They also consider alternative methods for delivering sustainment in emergency situations.

See following pages (4-4 to 4-5) for further discussion of supporting offensive operations.

Offensive Operations

Ref: FM 4-0, Sustainment (Aug '09), pp.4-4 to 4-5.

Offensive operations are defined as combat operations conducted to defeat and destroy enemy forces and seize terrain, resources, and population centers. They impose the commander's will on the enemy (see FM 3-0). Sustainment of offensive operations is high in intensity. Commanders and staffs plan for increased requirements and demands on sustainment. Sustainment planners work closely with other WFF staffs to determine the scope of the operations and develop estimates for quantity and types of support required. They anticipate where the greatest need might occur and develop a priority of support. Sustainment planners may consider positioning sustainment units in close proximity to operations to reduce response times for critical support. They also consider alternative methods for delivering sustainment in emergency situations.

To maintain momentum and freedom of action, coordination between staff planners must be continuous. During offensive operations, certain requirements present special challenges. The most important materiel is typically fuel (Class III Bulk) and ammunition (Class V), Class VII, movement control, and medical evacuation. Based on planning assessments, sustainment commanders direct the movement of these and other support to meet anticipated requirements.

Another challenge in planning for and sustaining an offensive operation is the lengthened lines of communication (LOCs). Widely dispersed forces, longer LOCs, and congested road networks increase stress on transportation systems. As a result, a combination of ground and aerial delivery maybe planned to accommodate the distribution. Distribution managers and movement control units synchronize movement plans and priorities according to the commander's priority of support. Distribution must be closely coordinated and tracked to ensure delivery of essential support. The routing function of movement control becomes an essential process for coordinating and directing movements on main supply routes or alternate supply routes, and regulating movement on LOCs to prevent conflict and congestion.

Higher casualty rates associated with offensive operations increase the requirement of medical resources. Plans to position medical support close to operations to facilitate treatment and evacuation are considered. If increased casualty rates overwhelm medical resources, nonmedical transportation assets may be needed for evacuation. Another planning consideration may be moving combat and operational stress control teams to support combat stress casualties following operations.

Higher casualty rates also increase the emphasis on personnel accountability, casualty reports, and replacement operations. G-1s and S-1s plan for accurate tracking of casualties and replacements through coordination with Casualty Liaison Teams (CLTs) and the HRSC.

Plans should also provide for religious support, which may become critical during offensive operations. Religious support through counseling and appropriate worship can help reduce combat stress, increase unit cohesion, and enhance performance.

Using contractors during the sustainment of offensive operations entails great risk and raises significant practical and legal considerations. However when necessary, the force commander may be willing to accept this risk and use contractors in forward areas. Commanders should seek counsel from their judge advocates when considering the use of contractors during offensive operations.

Tactical-Level Considerations

Sustainment in the offense is characterized by high-intensity operations that require anticipatory support as far forward as possible. Commanders and staffs ensure adequate support for continuing the momentum of the operation as they plan and synchronize offensive operations. Plans should include agile and flexible sustainment capabilities to follow exploiting forces and continue support (FM 4-0).

The following sustainment techniques and considerations apply to offensive planning:

- Plan for dealing with threats to sustainment units from bypassed enemy forces in a fluid, non-contiguous area of operations
- Recover damaged vehicles only to the main supply route for further recovery or evacuation
- Pre-stock essential supplies forward to minimize interruption to lines of communications
- Plan for increased consumption of petroleum, oils, and lubricants (POL)
- Anticipate increasingly long lines of communications as the offensive moves forward
- Anticipate poor trafficability for sustainment vehicles across fought-over terrain
- Consider planned/pre-configured sustainment packages of essential items
- Plan for increased vehicular maintenance, especially over rough terrain
- Maximize maintenance support teams well forward
- Request distribution at forward locations
- Increase use of meals-ready-to-eat (MREs)
- Use captured enemy supplies and equipment, and particularly support vehicles and POL. Before use, test for contamination.
- Suspend most field service functions except airdrop and mortuary affairs
- Prepare thoroughly for casualty evacuation and mortuary affairs requirements
- Select potential/projected supply routes, logistic release points, and support areas based on map reconnaissance
- Plan and coordinate enemy prisoner of war operations
- Plan replacement operations based on known/projected losses
- Consider the increasing distances and longer travel times for supply operations
- Ensure that sustainment preparations for the attack do not compromise tactical plans

B. Defensive Operations

Ref: FM 4-0, Sustainment (Aug '09), p. 4-5.

Defensive operations defeat an enemy attack, gain time, economize forces, and develop conditions favorable for offensive or stability operations (FM 3-0). The commander positions assets so they can support the forces in the defense and survive. Requirements for sustainment of forces in the defense depend on the type of defense. For example, increased quantities of ammunition and decreased quantities of fuel characterize most defensive operations. However, in a mobile defense, fuel usage may be a critical factor. Conversely, in a fixed defensive position, fuel requirements are lower.

Typically, defensive operations require more centralized control. Movements of supplies, replenishment, and troops within the AO have to be closely and continuously coordinated, controlled, and monitored. In retrograde operations (a type of defense) special care is necessary to assure that combat units receive necessary support across the depth of the AO and that the sustainment unit and stocks are not lost as the unit moves away from enemy activity.

Distribution managers direct resupply forecasted items to designated units. Increases in items such as barrier and construction materials should be pushed to designated collection points for unit retrieval whenever possible. Planners should consider the impact of increased ammunition expenditures on available transportation assets.

The task of medical units is to triage casualties, treat, and return to duty or resuscitate and stabilize for evacuation to the next higher level of medical care or out of the theater of operations. MTFs and other AHS support assets should be placed within supporting distance of maneuver forces but not close enough to impede ongoing combat operations.

Tactical-Level Considerations

The BCT commander positions sustainment assets to support the forces in the defense and survive. Sustainment requirements in the defense depend on the type of defense. For example, increased quantities of ammunition and decreased quantities of fuel characterize most area defensive operations. However, in a mobile defense, fuel usage may be a critical part of support. Barrier and fortification materiel to support the defense often has to move forward, placing increased demands on the transportation system (FM 4-0).

The following sustainment techniques and considerations apply to defensive planning:

- Preposition ammunition, POL, and barrier materiel in centrally located position well forward
- Make plans to destroy those stocks if necessary
- Resupply during limited visibility to reduce the chance of enemy interference
- Plan to reorganize to reconstitute lost CSS capability
- Use maintenance support teams in the unit maintenance collection point (UMCP) to reduce the need to recover equipment to the brigade support area (BSA)
- Consider and plan for the additional transportation requirements for movement of CL IV barrier materiel, mines, and pre-positioned ammunition, plus the CSS requirements of additional engineer units assigned for preparation of the defense
- Plan for pre-positioning and controlling ammunition on occupied and prepared defensive positions

C. Stability Operations

Stability operations encompass various military missions, tasks, and activities conducted outside the United States in coordination with other instruments of national power to maintain or reestablish a safe and secure environment, provide essential governmental services, emergency infrastructure reconstruction, and humanitarian relief (FM 3-0). Although Army forces focus on achieving the military end state, they ultimately need to create conditions where the other instruments of national power are preeminent. Sustainment of stability operations often involves supporting U.S. and MNFs in a wide range of missions and tasks. The tasks most impacted by sustainment are briefly discussed below.

1. Establish Civil Security

Civil security involves providing for the safety of the host nation and its population, including protection from internal and external threats (FM 3-0). Sustainment staffs must consider plans to support I/R operations. Sustainment may be provided to these operations until HNS, NGO, and OGOs are available.

Sustainment planners must address the sustainment of I/R operations. Sustainment to I/R involves all of the sustainment functions. Planners should consider general engineering support horizontal and vertical construction of detention centers, as well as repair and maintenance of the infrastructure (see FM 3-34.400). A more detailed discussion of sustainment support of I/R and detainee operations will be provided later in this chapter.

2. Establish Civil Control

Legal staffs should plan for supporting civil control operations. Civil control centers on rule of law, supported by efforts to rebuild the host nation judiciary and corrections systems. It encompasses the key institutions necessary for a functioning justice system, including police, investigative services, prosecutorial arm, and public defense. If transitional military authority is instituted, intervening forces will likely carry out judicial and correctional functions.

3. Restore Essential Services

Efforts to restore essential services involve developing host nation capacity to operate, maintain, and improve those services. At the tactical or local level, sustainment in support of civil authorities will restore essential civil services as defined in terms of immediate humanitarian needs (such as providing food, water, shelter, and medical and public health support) necessary to sustain the population until local civil services are restored. Other sustainment tasks associated with restoration of services include support to dislocated civilians and demining operations.

4. Support to Economic and Infrastructure Development

The role of sustainment in supporting economic stabilization and infrastructure development is significant, especially at the local level. Here the emphasis is on generating employment opportunities, infusing monetary resources into the local economy, stimulating market activity, fostering recovery through economic development, and supporting the restoration of physical infrastructure. Drawing on local goods, services, and labor through contracting, presents the force an opportunity to infuse cash into the local economy, which in turn stimulates market activity.

Restoring the transportation infrastructure in the area is central to economic recovery. General engineering is required in order to initiate immediate improvement of the transportation network. These networks enable freedom of maneuver, logistics support, and the movement of personnel and material to support ongoing operations.

Tactical-Level Considerations

Sustainment in stability operations involves supporting U.S. and multinational forces in a wide range of missions. Stability operations range from long-term sustainment-focused operations in humanitarian and civic assistance missions to major short-notice peace enforcement missions. Some stability operations may involve combat. Tailoring sustainment to the requirements of a stability operation is key to success of the overall mission (FM 4-0). The sustainment techniques and considerations that are applicable to offensive and defensive operations also apply to stability operations conducted in hostile environment. When these operations are in urban areas, the following considerations may also apply:

- Preconfigure resupply loads and push them forward at every opportunity
- Provide supplies to using units in required quantities as close as possible to the location where those supplies are needed
- Protect supplies and sustainment elements from the effects of enemy fire
- Disperse and decentralize sustainment elements with proper emphasis on communication, command and control, security, and proximity of main supply route
- Plan for extensive use of carrying parties
- Plan for and use host country support and civil resources when practical
- Position support units as far forward as the tactical situation permits
- Plan for special equipment such as rope, grappling hooks, ladders, and hand tools

D. Civil Support Operations

The Armed Forces of the United States are authorized under certain conditions to provide assistance to U.S. civil authorities. This assistance is known as civil support within the defense community. Civil support is Department of Defense support to U.S. civil authorities for domestic emergencies and for designated law enforcement and other activities. This support usually entails Class III, subsistence, medical services, and general engineering support.

Civil support will always be in support of a lead federal agency such as the Department of Homeland Security. Requests for assistance from another agency may be predicated on mutual agreements between agencies or stem from a Presidential designation of a federal disaster area or a federal state of emergency. The military typically only responds after the resources of other federal agencies, state and local governments, and NGOs have been exhausted or when specialized military assets are required.

Within the Joint Staff J–3, the Joint Director of Military Support (JDOMS) serves as the Action Agent for Assistant Secretary of Defense for Homeland Defense (ASD (HD)) who has EA responsibilities for Defense Support of Civil Authorities (DSCA). JDOMS ensures the performance of all DSCA planning and execution responsibilities. JDOMS conducts planning and prepares warning and execution orders that task DOD resources in response to specific requests from civil authorities.

The National Response Framework (NRF) is an all-discipline, all-hazards plan that establishes a single, comprehensive framework for the management of domestic incidents. It provides the structure and mechanisms for the coordination of Federal support to State and local incident managers and for exercising direct Federal authorities and responsibilities. Through the NRF, FEMA assigns emergency support functions (ESF) to the appropriate federal agencies. The Army maintains a permanently assigned Defense Coordinating Officer in each FEMA region to plan, coordinate, and integrate DSCA with local, state, and federal agencies. DOD is the Primary Coordinating Agency for ESF 3 (Public Works and Engineering) and when requested, and upon approval of the SECDEF, DOD provides DSCA during domestic incidents and is considered a support agency to all ESFs.

The Army National Guard often acts as a first responder on behalf of state authorities when functioning under Title 32 U.S. Code authority or while serving on State active duty.

II. Logistics Preparation of the Battlefield (LPB)

Ref: FMI 4-93.2, Sustainment Brigade (Feb '09), app. A. See pp. 5-7 to 5-12 for additional information on planning joint and theater Logistics. See also p. 4-1.

Logistics preparation of the battlefield (LPB) is a key conceptual tool available to personnel in building a flexible strategic/operational support plan. Logistics preparation of the theater of operations consists of the actions taken by logisticians at all echelons to optimize means (force structure, resources, and strategic lift) of supporting the joint force commander's plan. These actions include identifying and preparing ISB's and forward operating bases; selecting and improving LOC; projecting and preparing forward logistics bases; and forecasting and building operational stock assets forward and afloat. LPB focuses on identifying the resources currently available in the theater of operations for use by friendly forces and ensuring access to those resources. A detailed estimate of requirements, tempered with logistics preparation of the theater of operations, allows support personnel to advise the JTF/ASCC/ARFOR commander of the most effective method of providing responsive support.

I. Intelligence in Support of Logistics

The logistician uses intelligence to develop and execute the logistics support plan. Logistics intelligence is critical to the planning effort. Some of the areas that should be included in the intelligence analysis are listed below:

- Intent to engage in multinational operations and the extent of logistics support to be provided to non-DOD agencies and allies
- Available resources in the AO
- Conditions that alter consumption factors, such as severe climate changes or a requirement to provide support to allies
- Capabilities of local facilities to support reception and staging operations
- Foreign military logistics structure, national infrastructure capabilities, and political inclination to facilitate joint forces support
- Environmental, geographical, climatological, and topographical factors that may affect support operations
- Analysis of the capabilities of the host nation's and region's LOC's and capabilities to support the operation

Intelligence is equally critical for war and stability operations. Logisticians must have a complete logistics database or file to develop a solid plan for the LPB.

LPB is those actions (force structure, resources, and strategic lift) taken to reduce the cost of logistically supporting an OPLAN or a contingency plan. LPB minimizes or eliminates potential problems at the outbreak of hostilities, during deployment, and throughout the operations. It is a systematic tool used by logisticians and commanders to complete their mission. It becomes the basis for deciding where, when, and how to deploy limited resources (supplies, equipment, people, and money).

The ASCC of a combatant command will prepare supporting Army plans with logistics planners concentrating on the logistics plans. Once logistics planners know the contingency country or geographic region, they can begin to build a logistics information database. This applies even if the command has a small chance of being deployed to a particular area. Once completed, the information data base file can be used to develop a comprehensive plan for LPB. The relative priority given to this effort will depend on the concept of operations and other command priorities. The key point is that the logisticians cannot afford to wait until maneuver units deploy to begin the LPB. It is a complex and time-consuming function.

Any actions that can reduce the cost of moving supplies, equipment, and people into an objective or contingency area are candidates for inclusion in the LPB plan. Planning must provide for the timely arrival of sustainment assets that are balanced according to the mission. Strategic lift assets are extremely limited.

II. Relevant Logistics Information

The following paragraphs contain types of relevant logistics information. These can be added to or taken away as individual missions dictate.

1. Geography

Collect information on climate and terrain in the AO. Determine if current maps are available. Use this information to determine when various types of supplies, equipment, and field services will be needed. For example, use water information to determine the need for early deployment of well-digging assets and water production and distribution units.

2. Supply

Collect information on supply items that are readily available in the AO. Determine which of these can be used in support of US forces. Subsistence items, bulk petroleum, and barrier materials are often available in country. Collect information on the supply system of the armed forces of the supported country; determine if it is compatible with the US system. Has the host nation bought, through foreign military sales, repair parts supporting current US systems? Can contingency contracting provide resources from HNS sources or third country sources until Army capabilities arrive in the AOR? Answers to these types of questions will aid in analyzing whether HNS negotiations are possible.

3. Facilities

Collect information on warehousing and cold storage facilities, production and manufacturing plants, reservoirs, administrative facilities, sanitation capabilities, and hotels. Their availability could reduce the requirement for deployment of similar capacity.

4. Transportation

Collect information on road nets, truck availability, rail nets, bridges, ports, cargo handlers (longshoremen), petroleum pipelines, and MHE. Also collect information on traffic flow, choke points, and control problems.

5. Maintenance

Collect information on maintenance facilities that could support US or coalition equipment. Examine the supported country's armed forces. Could they supplement our capability? Is there a commonality in equipment and repair parts? Does the country have adequate machine works for possible use in the fabrication of repair parts?

6. General Skills

Collect information on the general population of the supported country. Is English commonly spoken? Are personnel available for interpreter/translator duties? Will a general labor pool be available? What skills are available that can be translated to joint forces? Will drivers, clerks, MHE operators, food service personnel, guards, mechanics, and longshoremen be available?

7. Miscellaneous

Include any other information that could prove useful. Set up other categories as needed.

III. General Information Sources

Ref: Adapted from FMI 4-93.2, Sustainment Brigade (Feb '09), app. A.

Collectors routinely provide an abundance of information on targeted theaters or likely contingency areas. Also, agencies can assist CSS personnel in building the information file. The CSS planner must not underestimate the time and resources required for these actions. The LPT is a living document that is in a continual state of review, refinement, and use. Forces should use it as the basis for negotiations, preparing the TPFDD, and the Total Army analysis process.

The following sources of information are only a few; this list is not all-inclusive.

Department of State

Department of State embassy staffs routinely do country studies. They also produce information on foreign countries, including unclassified pamphlets. These pamphlets focus on political and economic issues, not military or CSS matters.

Intelligence Preparation of the Battlefield (IPB)

The weather and terrain databases in the IPB, with its overlays, provide current information for preselecting LOC and sites for CSS facilities. The IPB event analysis matrix and template can determine the need for route improvements and bridge reinforcements. *FM 34-130 has more details. See also The Battle Staff SMARTbook.*

Special Operations Forces, to Include Civil Affairs Units

Whether in country or targeted on a specific country, SOF can provide a wealth of CSS information. They include functional specialists who focus on particular areas (such as civilian supply, public health, public safety, and transportation). Civil affairs (CA) units also can provide vital assistance when coordinating theater contract support and CUL support to NGOs.

Culturegrams

Culturegrams are a series of unclassified pamphlets published by Brigham Young University that provide general/social information on specific countries. Though not focused on governmental or military interests, they provide a variety of useful information that can be used by deploying forces.

Army Country Profiles

Army country profiles (ACPs) are produced by the Army Intelligence Threat Analysis Center. ACPs are classified country profiles providing information on logistics, military capabilities, intelligence and security, medical intelligence, and military geography. They include photos, maps, and charts.

Country Contingency Support Studies

Country contingency support studies are produced by the Defense Intelligence Agency (DIA). These classified documents contain extensive information on railways, highways, bridges, and tunnels within a given country.

Other assets or tools the CSS planner may want to consider as the LPT plan is developed include:

- Army prepositioned stocks
- Use of containerization to limit handling
- HNS agreements
- ISSAs and ACSAs
- Prearranged contracts to provide support

Logistics Prep of the Battlefield (LPB)

Ref: Adapted from FM 4-93.52 Tactics, Techniques, and Procedures for the DISCOM (Digitized), chap. 5.

Logistics preparation of the battlefield is the process of gathering data against pertinent battlefield components, analyzing their impact on sustainment, and integrating them into tactical planning so that support actions are synchronized with maneuver. It is a conscious effort to identify and assess those factors, which facilitate, inhibit, or deny support to combat forces. Just as intelligence preparation of the battlefield is important to the conduct of actual combat operations, logistics preparation of the battlefield is equally important to sustaining the combat power of the force.

The process requires tacticians to understand the data needed by logisticians to plan and provide timely, effective support. It requires TF logisticians to understand the mission, the tactical plan, and the battlefield's time and space implications for support. It is a coordinated effort to prepare the battlefield logistically. The basic steps in systematizing the process are:

- Determine battlefield data pertinent to support actions
- Determine sources from which raw data can be derived
- Gather pertinent data
- Analyze collected data elements and translate them into decision information by assessing their impact on the mission and the competing courses of action
- Integrate decision information into tactical planning by incorporating it in sustainment estimates and TF plans and orders

Sustainment Planning

When determining what battlefield data are relevant to sustainment, it's helpful to break down CSS operations into certain key elements against which data can be collected for study and analysis. These data elements are called the components of tactical sustainment. The following descriptions of the components of tactical sustainment are not intended to be all-inclusive. They are offered here, however, to stimulate thought and to facilitate an understanding of those factors which impact on tactical sustainment:

- Logistics resources are the wherewithal to effect support, including CSS organizational structures, command and control, task organizing for support, communications, information automation systems, medical facilities, and materiel such as transportation assets and supply, maintenance and field services equipment.
- Logistics capabilities include soldier and leader skills and the personnel staffing which, collectively, activate sustainment resources and bring to life the required support.
- Logistics capacities include reception and clearance capacities, carrying capacities of transportation assets, volumes of storage facilities, maintenance production output rates, and supply route characteristics such as surface composition, tunnels, overhead obstructions, bridge weight limits and traffic circulation rates.
- Materiel stocks include the quantity and status of weapon systems, ancillary equipment, ammunition, repair parts and consumable supplies required or available to sustain or reconstitute combat power of deployed units. Also included are sustainment status reports and known or projected shortfalls.
- Consumption and attrition rates include experienced or expected usages of consumable supplies and weapon systems, which must be considered to anticipate support requirements.
- Time and space factors are those requirements and restrictions of the battlefield, which influence whether logistic support is provided to deployed forces at the right place and time. Included here are plans, orders, rehearsals, priority of support,

positioning for support, tempo of support (intensity of demand), security, risk assessment, the effects of terrain, weather, contaminated areas, minefields, night time enemy threat on sustainment operations, and the battlefield signatures of logistic resources. Time and space factors, especially, impact on the synchronization and integration of sustainment on the battlefield.

Sources from which relevant battlefield data are derived include:

- Higher headquarters briefs, plans and orders
- The commander's planning guidance
- The commander's intent (or concept)
- Operations and intelligence briefings and overlays
- Modified table of equipment (MTOE) of task force units
- Sustainment status reports
- Scouts
- Engineer route reconnaissance overlays
- Traffic circulation and highway regulating plans
- Personal reconnaissance

Logisticians should treat the components of tactical sustainment as essential factors that should be assessed for each plan. By doing so, they bring a professional approach to the contributions they make in the planning process. The components are variables. Some are dynamic and change with METT-TC so they should be validated daily, even hourly, if necessary. Commanders should appreciate the unique contributions their logisticians make in the planning process and when they've done a thorough job of collecting and analyzing pertinent battlefield data. Commanders must not accept less.

The commander and staff conduct LPB. Successful LPB contributes immeasurably to the favorable outcome of battle. Logistics preparation of the battlefield is an on-going process by which logisticians analyze:

- Tactical commander's plan/concept of operation
- Tactical commander's intents
- Supported force sustainment requirements
- Available sustainment resources
- Combat service support shortfalls
- The enemy (intentions, capabilities, weaknesses, doctrine)
- Terrain and weather
- Intelligence preparation of the battlefield (IPB) products
- Transportation infrastructure
- Host nation support available
- Time/distance factors

Logistics preparation of the battlefield (LPB) products are:

- A logistics estimate
- A visualization of the pending battle and logistics activity required by phase
- Anticipated logistics challenges and shortfalls
- Solutions to logistics challenges and shortfalls
- How, when, and where to position logistics units to best support the tactical commander's plan
- A synchronized tactical and logistical effort

Operations Logistics (OPLOG) Planner

Ref. https://www.cascom.army.mil/private/CDI/FDD/Multi/PDB/OPLOGPlanner.HTM
** Requires Army Knowledge Online (AKO) log-in and password.*

The OPLOG Planner is a web-based interactive tool that assists commanders and staff from strategic through operational levels in developing a logistics estimate. It is designed to support operations typically associated with multi-phase operations plans and orders. The OPLOG Planner enables staffs to develop estimated mission requirements for supply Class I, Class II, Class III(P), Class IV, Class VI, Class VII, Class X including water, ice, and mail.

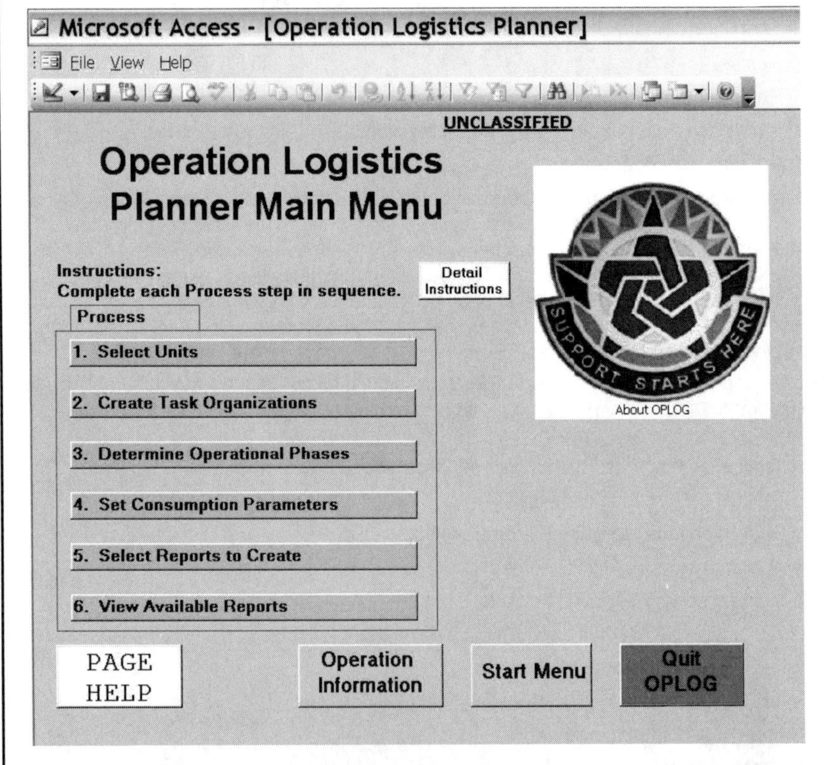

OPLOG Planner allows planners to build multiple task organizations from a preloaded list of units and equipment or from custom built units that are generated or imported. Each task organization is assigned a consumption parameter set that establishes the rates, climate, Joint Phases (Deter, Seize Initiative, and Dominate), and Army full spectrum operations (Offense, Defense, and Stability). The planners have the option of using predefined default planning rates or customizing rates based on what a unit is experiencing.

OPLOG Planner generates the logistics supply requirements which can be viewed by the Entire Operation, each Phase of the Operation, each Task Organization, each Unit, or each Unit's Equipment. Logistics requirement reports can be printed or exported to automated spread sheets for further analysis or saved for recall to be used in course of action analysis.

Chap 4

III. The Military Decision Making Process (MDMP)

Ref: Adapted from FM 5-0, Army Planning and Orders Production (Jan '05) and FM 4-93.52 Tactics, Techniques, and Procedures for the DISCOM (Digitized), chap. 4.

The military decision making process is a planning tool that establishes procedures for analyzing a mission, developing, analyzing, and comparing courses of action against criteria of success and each other, selecting the optimum course of action, and producing a plan or order. The MDMP applies across the spectrum of conflict and range of military operations. Commanders with an assigned staff use the MDMP to organize their planning activities, share a common understanding of the mission and commander's intent, and develop effective plans and orders.

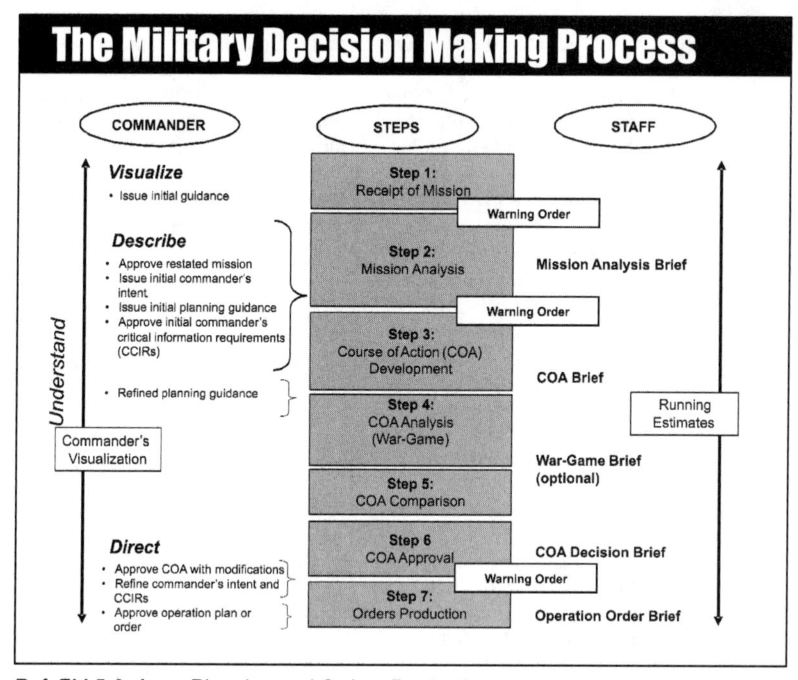

Ref: FM 5-0, Army Planning and Orders Production.

The MDMP helps organize the thought process of commanders and staffs. It helps them apply thoroughness, clarity, sound judgment, logic, and professional knowledge to reach decisions.

Sustainment Planning

MDMP - Sustainment Considerations

Ref: Adapted from FM 4-93.52 TTP for the DISCOM (Digitized), chap. 4.

Refer to The Battle Staff SMARTbook and FM 5-0, Planning, Chapter 5, for more detailed information reference MDMP.

I. Mission Analysis

See pp. 4-19 to 4-26 for discussion of personnel/logistics estimates.

The MDMP begins when a mission is received from higher headquarters. Very rarely will this be in the form of a complete operations order. More likely it will begin after a verbal or written warning order is received. The commander, upon receiving a mission, should provide his staff with guidance as to how they should proceed with their analysis, and a warning order, in five-paragraph field order format, should be issued to subordinate units to allow them to begin to prepare for a new mission.

The staff begins mission analysis by developing their initial staff estimates based upon the higher headquarters order and their commander's guidance. Mission analysis also determines what the mission of higher headquarters is, what this equates to as a mission for their unit, and the situation/circumstances that may impact upon their unit's ability to execute a particular course of action that will be proposed to accomplish the mission. Each staff officer produces an estimate in his or her area. The results of mission analysis should include completed staff estimates, including an initial intelligence preparation of the battlefield (IPB) by the S-2. The staff will also produce a proposed restated mission for their unit. These products will be presented to the commander and he will provide additional planning guidance to include: number of courses of action he wants the staff to develop, initial commander's critical information requirements (CCIR), timeline, risk guidance, and rehearsals to be conducted. Additionally, the commander will provide his initial intent that will include the method and end state for the operation. A second warning order should be issued to subordinate units no later than the end of mission analysis.

The Logistics Estimate

The logistician's input during mission analysis primarily comes from the logistics estimate. The logistics estimate is a continuous process that begins during mission analysis and is continually refined and updated through mission completion.

II. COA Development

During COA development, the logistics planners must refine the logistics estimate they developed during mission analysis. Facts and assumptions developed during mission analysis must be verified and updated. Sustainment planners must identify any significant Sustainment considerations and requirements that have a major impact on mission accomplishment. Additionally, the sustainment planner must develop a draft concept of support during this phase of the MDMP.

COA Sustainment Feasibility

For each course of action, the logistics planner must access its sustainment feasibility. The sustainment feasibility is determined by whether or not the unit possesses the required resources to sustain the unit throughout the tactical operation. Tailoring your logistics estimate for each course of action can help make this determination. If requirements do not exceed capabilities, the sustainment of the course of action will generally be feasible. If any requirements do exceed capabilities you must again determine its significance and potential impact upon the mission. If the shortfall is a "WAR-STOPPER", and there are no workable solutions to the problem, then sustainment of the COA is not feasible. Ensure you have exhausted all possible means to solve the problem, to include support from higher headquarters, before you deem the COA not feasible.

Synchronization Requirements

The synchronization of sustainment during COA analysis is critical to ensure continuous support during the operation. During the war game, the logistical planner will determine, based on the scheme of maneuver, what supplies and services must be where at a given time. This will generate critical Sustainment actions that must be accomplished to sustain the mission. He must consider time-distance factors and determine which support activity will be available to provide the required support. This is where the logistical planner begins to directly link the actions of task force logistics assets with the support battalion sustainment activities and division/corps resupply activities. He must ensure that all critical Sustainment activities are included in the synchronization matrix to successfully synchronize all levels of support.

Critical Sustainment Requirements

During this phase of MDMP, the courses of action are compared using the synchronization matrices and notes taken for each evaluation criteria used. A decision matrix with the evaluation criteria and some type of weighting factor (e.g., numbers, +/-, etc.) should be used to record the results of the course of action comparison. A decision matrix can be used as an aid to obtain a decision from the commander as to what course of action will be selected.

III. COA Comparison

In order to compare COAs and determine which is more supportable, logistical planners must calculate estimated attrition rates, project battle losses for critical weapons systems, and project personnel battle losses. The RSR for each COA must be refined and compared to any CSR that may be in effect. Quantities of supplies required, demands on transportation assets, and reconstitution requirements must be compared to determine which COA stresses the units' logistical system the most. An analysis of the risks to Sustainment assets and operations must be compared and considered.

IV. Orders Production

See pp.4-27 to 4-30 for further discussion of the concept of support (para. 4a).

The brigade logistic planners are responsible for paragraph 4 of the OPORD as well as the logistics annex (annex I). These products must be complete, concise and synchronize all levels of logistics support from top to bottom with the tactical plan.

Final Concept Of Support

The paragraph four consists of the final approved concept of support. Remember that this paragraph is written primarily to inform the commanders how they will be supported throughout the tactical operation, so do not include details on how the support elements are to execute the plan. Save all those details for the logistics annex. The concept of support should include a brief synopsis of the support command mission, support area locations to include the locations of the next higher logistics bases, the next higher levels support priorities, the commander's priority of support, significant or critical Sustainment activities, any significant risks, and the major support requirements in each tactical logistics function. If the tactical concept of operation is phased, the concept of support should also be phased to facilitate changes of priorities and logistics focus during each phase.

Complete/Concise

It is essential that the OPORD be complete, concise, and include all critical tasks that must be accomplished to support the tactical mission. Ensure you consider the command and support relationships of all units within your area of operation and ensure all elements receive support. Ensure you address all of the tactical logistics functions and properly phase the support concept if the tactical concept of operation is phased.

Synchronized Top-to-Bottom

To ensure proper synchronization, include all critical tasks and coordination requirements that were developed during the war gaming phase. Consider developing a logistics synchronization matrix, if possible. Ensure that all priorities of support are in agreement with the scheme of maneuver and weight the main effort. Coordinate with the other battlefield operating system elements to ensure that there are no inconsistencies in logistics information within the maneuver, engineer, artillery, and Sustainment annexes. As subordinate OPORDs are developed, you must ensure that their support plans are consistent and executable within your support framework. Synchronization of resupply operations from the corps/division, to the support battalion, to the task force level is critical.

The Military Decision-Making Process: A Sustainment Planner's Perspective

Ref: Adapted from FM 4-93.52, TTP for the DISCOM (Digitized), chap. 4.

Appendix A, FM 4-93.50 provided an overview of the military decision making process (MDMP) used by the Sustainment battle staff providing support to brigade and below echelons and is not intended as an all-inclusive description of the process. Refer to The Battle Staff SMARTbook and FM 5-0, Army Planning & Orders Production, Chapter 5, for more detailed information on the MDMP.

The MDMP must be integrated from top to bottom and from bottom to top in order to produce a synchronized concept of support that effectively supports the brigade tactical operation. The support battalion should have a Sustainment planner (liaison officer-LNO) who actively participates with the brigade S-1 and S-4 in the MDMP. Information must flow continuously between the brigade S-1/S-4, the support battalion, and the battalion task force S-1/S-4. At each level, the logistics estimate process should assess Sustainment capabilities, develop detailed requirements, and identify shortfalls as well as possible solutions. The logistics estimate process must be continuous and communication between the many Sustainment planners is essential. An integrated Sustainment concept of support must provide, at each level, the details of how a unit will both receive and provide support throughout an operation. It must provide enough detail so commanders know how they will be supported as well as how they and their subordinate units will execute the Sustainment

Sustainment planners at all echelons must actively participate during each stage of the MDMP, and these planners must not only participate, but they must communicate with each other throughout the process.

The courses of action are compared, and one is recommended to the commander as the best option for providing support to the task force, support battalion, and the brigade. The commander selects the course of action which he feels best supports his concept of the operation. Throughout this stage of the MDMP, information must flow between the brigade, support battalion, and task force Sustainment planners. After the commander has made his decision, warning orders to subordinate units must be issued. Staffs at each echelon now produce a complete operations order. For brigade level Sustainment planners, this includes paragraph four (concept of support), a Sustainment annex/overlay and possibly a Sustainment matrix. For task force level Sustainment planners, this includes paragraph 4 (concept of support) and possibly a Sustainment annex/matrix with additional information on support arrangements. The support battalion should produce a full five-paragraph field order. Paragraph four for the support battalion should discuss the concept of internal Sustainment support. Additionally, this paragraph should be expanded upon in a Sustainment annex and possibly a Sustainment matrix. External Sustainment support to the brigade should be discussed in paragraph three of the support battalion base order, in an external Sustainment support annex, and possibly a Sustainment matrix. The support battalion order must also contain in the base order appropriate annexes on how the BSA and Sustainment assets will be supported by the battlefield operating systems (BOS) of fire support, air defense, intelligence (to include a reconnaissance and surveillance plan) and mobility/survivability (to include NBC).

IV. Running Estimates & Mission Analysis

Ref: Adapted from FM 5-0, Operations Process (2009) and FM 4-0, Sustainment (Apr '09). See FM 5-0 or The Battle Staff SMARTbook for detailed discussion of the operations process and the military decision making process (MDMP)

Running Estimates

A running estimate is a staff section's continuous assessment of current and future operations to determine if the current operation is proceeding according to the commander's intent and if future operations are supportable (FM 3-0). Building and maintaining running estimates is a primary task of each staff section. The running estimate helps the staff provide recommendations to commanders on the best course of action to accomplish their mission. Running estimates represent the analysis and expert opinion of each staff section by functional area.

Mission Analysis

Mission analysis helps commanders to understand the situation to include their mission. This enables commanders to issue the appropriate guidance that drives the rest of the planning process. Commanders—supported by their staffs—gather, analyze, and synthesize information to orient themselves on current conditions in the AO. Such orientation helps commanders to better understand the relationships among the operational and mission variables. Mission analysis helps commanders understand the problem they have been called upon to resolve and how their units fit into the higher headquarters' plan.

During mission analysis, the staff conducts intelligence preparation of the battlefield and updates running estimates in relation to the new mission. The commander and staff analyze the higher headquarters' order to completely understand the higher headquarters commander's intent, mission, and concept of operations. They develop facts and assumptions about the upcoming operations and determine specified, implied, and essential tasks. They identify forces available for the mission, resource shortfalls, and any constraints placed on them from the higher command.

The logistician's input during mission analysis primarily comes from the logistics estimate. The logistics estimate is a continuous process that begins during mission analysis and is continually refined and updated through mission completion. The logistics estimate does not have a doctrinal format at the brigade level.

Mission analysis considerations feed information into the estimate process. The estimates are as thorough as time permits. Personnel/logistics estimates are kept current. As factors that influence operations change, new facts are developed and assumptions become facts or become invalid.

Ideally, the commander holds several informal meetings with key staff members before the mission analysis briefing. These meetings increase a common understanding, pass information to the staff, and issue guidance.

The duration of the mission analysis briefing may vary. It may be with only a few staff briefing the commander, or it may take be several days in the form of a conference that includes commanders, subordinate commanders, staff, and other partners.

See following pages (pp. 4-20 to 4-21) for a listing of suggested sustainment mission analysis considerations by sub-function.

Mission Analysis Considerations

Ref: Adapted from FM 5-0 and FM 4-0.

A comprehensive mission analysis briefing helps the commander, staff, subordinates, and other partners develop a shared understanding of the requirements of the upcoming operation. Time permitting, the staff briefs the commander on its mission analysis using the following outline:

- Mission and commander's intent of the headquarters two levels up
- Mission, commander's intent, and concept of operations of the HQs one level up
- Review of the commander's initial guidance
- Initial IPB products, including modified combined obstacle overlays and situation templates
- Civil considerations that impact the conduct of operations
- Specified, implied, and essential tasks (to include minimum essential stability tasks)
- Pertinent facts and assumptions
- Constraints
- Forces available and resource shortfalls
- Initial risk assessment
- Recommended initial information themes and messages
- Recommended initial critical commander's critical information requirements (CCIRs) and essential elements of friendly information (EEFIs)
- Initial intelligence, surveillance, and reconnaissance (ISR) plan
- Recommended timelines
- Recommended collaborative planning sessions
- Recommended restated mission

Sustainment Planning

Sustainment Sub-Functions

I. Logistics

A. Supply

1. General Supply

a. Facts
- Classes I, II, III(p), IV, VI, VII, X, and water status
- Distribution system
- Critical shortages

b. Assumptions
- Resupply rates
- Host nation/multinational support
- Other

c. Conclusions
- Projected supply levels and status
- Shortfalls and critical sustainment risks/events

2. Class III (B)

a. Facts
- Class III(b) status
- Restrictions
- Distribution Systems
- Critical shortages

b. Assumptions
- Resupply rates
- Host nation/multinational support
- Other

c. Conclusions
- Projected supply status
- Projected distribution system
- Shortfalls and critical risks/events

3. Class V

a. Facts
- Class V status
- Distribution system
- Restrictions
- Critical shortages

b. Assumptions
- Resupply rates
- Host nation/multinational support
- Other

c. Conclusions
- Projected supply status
- Projected distribution status
- Shortfalls and critical risks/events

B. Field Services

Life support functions, including field laundry, showers, light textile repair, food preparation, mortuary affairs, aerial delivery support, food services, billeting, and sanitation.

a. Facts

b. Assumptions

c. Conclusions

C. Maintenance

Field and sustainment maintenance.

a. Facts

- Maintenance status (equip readiness)
- Class IX status
- Repair times, evacuation policy, assets
- Critical shortages

b. Assumptions

- Host nation/multinational support
- Other

c. Conclusions

- Projected maintenance status
- Shortfalls and critical CSS risks/events
- Recommendations

D. Transportation

Movement control, mode operations, terminal operations, container management.

a. Facts

- Status of transportation assets
- Critical LOC and MSR status
- Critical shortages

b. Assumptions

- Host nation/multinational support
- Other

c. Conclusions

- Projected status of transportation assets
- Projected status of LOCs and MSRs
- Shortfalls and critical risks/events
- Recommendations

E. Operational Contract Support

Operational contract support is the process of planning for and obtaining supplies, services, and construction from commercial sources along with associated contract management functions.

A. Facts

B. Assumptions

C. Conclusions

II. Personnel Services

A. Human Resource (HR) Support

a. Facts

- Personnel strengths and morale
- Replacements and medical RTD
- Critical shortages

b. Assumptions

- Replacements
- Host nation/multinational support
- Other

c. Conclusions

- Projected personnel strength
- Projected critical MOS status
- Shortfalls and critical risks/events

B/C/D/E. Finance/Legal/Religious/Band

a. Facts

b. Assumptions

c. Conclusions

III. Health Services Support (HSS)

See pp. 4-31 to 4-36 for discussion of considerations for HSS in support of joint and multinational operations. Include medical logistics considerations (class VIII supply).

a. Facts

b. Assumptions

c. Conclusions

IV. Other Sustainment-Related Functions

As a result of the movement from battlefield operating systems to the Warfighting Functions (WFF) construct, some tasks are realigned. Two of those tasks include:

A. Explosive Ordnance Disposal (EOD)

From a WFF perspective, EOD falls under the Protection WFF

a. Facts

b. Assumptions

c. Conclusions

B. Internment/Resettlement (I/R) Opns

Internment/Resettlement (I/R) operations are included under the Sustainment WFF (FM 3.0). While not a major sub-function of the sustainment WFF; I/R are supported by logistics, personnel services, and HSS.

Recommendations

- Sustainment unit and system capabilities, limitations, and employment
- Risk identification and mitigation
- Organization for operations, allocations to subordinate units, command and support relationships among subordinate units
- Resource allocation and employment synchronization of organic and supporting units (including other joint assets)
- General locations and movements of units

I. The Logistics Estimate

Ref: Adapted from FM 5-0, FM 4-0 and FM 4-93.50, app. A.

A running estimate is a staff section's continuous assessment of current and future operations to determine if the current operation is proceeding according to the commander's intent and if future operations are supportable (FM 3-0). Building and maintaining running estimates is a primary task of each staff section. The running estimate helps the staff provide recommendations to commanders on the best course of action to accomplish their mission. Running estimates represent the analysis and expert opinion of each staff section by functional area.

Staff representatives do more than collect and store information; they process it into knowledge and apply judgment to get that knowledge to those requiring it. The staff section's running estimate is one product of this effort. By synthesizing relevant information from a variety of sources into a prescribed format, the staff and commander can quickly identify and summarize how a given functional area impacts other functional areas and the operation as a whole.

1. Requirements

The first step in the log estimate process is to determine the logistical requirements for the mission. To determine the requirements, you may use a number or combination of methods. Automated systems such as OPLOG PLANNER are good tools to use to estimate requirements. If you prefer to do the number crunching yourself, planning factors from the CGSC ST 101-6 (G1/G4 BATTLEBOOK) or the FM 101-10-1/2 may be used to determine estimates; historical data from previous missions should also be used.

2. Capabilities

To correctly determine the logistics capability of your unit, you must consider the capabilities of all the available Sustainment assets at your disposal. This includes all available Sustainment units assigned, attached, or OPCON, and the sustainment capability organic to the maneuver units themselves. Be sure to consider the unit's current status in terms of personnel and equipment, as well as the projected status at mission execution.

3. Comparison/Shortfall

Once you have determined your estimated requirements and the unit's sustainment capability, compare them to determine any logistical shortfalls. If there are no shortfalls, go to the analysis step of this methodology. Shortfalls may occur in terms of storage, distribution, and transportation capability or may be caused by personnel or equipment shortfalls based on current on hand shortages or maintenance status. A shortfall may also occur if required facilities or terrain are not available or the plan does not provide enough time to prepare. If there is a shortfall, determine what the shortfall is in terms of short tons, gallons, or other units of measurement and when or where during the operation the shortfall occurs.

4. Analysis

Whether or not there is a shortfall, the analysis process must occur for all support operations. The sustainment planner must determine: when the support operation must begin, how much time there is to prepare for the mission, the purpose of the support mission, the duration of the mission, and whether the mission can be supported from a fixed location or whether to echelon support forward in some way. If there is a shortfall identified in the comparison of requirements and capabilities, you must also determine its cause, its significance and its potential impact on the tactical operation.

5. Solutions

Determine the most workable solutions based on your analysis. Ensure that all solutions are integrated into the MDMP to enhance continuity between the tactical decision making and logistical planning.

A comprehensive running estimate will address all aspects of operations and contain both facts and assumptions based on the staff's experience within a specific area of expertise. A base document for a running estimate is provided below. Each staff section modifies it to account for their specific functional area. All running estimates should cover essential facts and assumptions, a summary of the current situation by the mission variables, conclusions, and recommendations. The mission variables of METT-TC are mission, enemy, terrain and weather, troops and support available, time available, civil considerations.

1. Mission.
Show the restated mission resulting from mission analysis

2. Situation and Considerations

a. Characteristics of the Operational Environment. Identify area structures, capabilities, organizations, people, and events that impact or affect functional area considerations.

b. Characteristics of the Area of Operations.

(1) Weather. State how weather affects staff functional area's capabilities

(2) Terrain. State how terrain affects staff functional area's capabilities

(3) Enemy Forces. Describe enemy disposition, composition, strength, capabilities, systems, and possible courses of action (COAs) with respect to their effect on functional area.

(4) Friendly Forces. List current functional area resources in terms of equipment, personnel, and systems. Identify additional resources available for functional area located at higher, adjacent, or other units. Compare requirements to current capabilities and suggest solutions for satisfying discrepancies.

(5) Civilian Considerations. Describe additional personnel, groups, or associations that cannot be categorized as friendly or enemy. Discuss possible impact these entities may have on functional area.

(6) Assumptions. List all assumptions that affect the functional area. Best practice is to prioritize assumptions based on operational requirement or to divide assumptions (if many) into categories such as —friendly assumptions and —enemy assumptions.

3. Courses of Action (COAs)

a. List friendly COAs that were war-gamed.

b. List enemy actions or COAs that were templated that impact functional area.

c. List the evaluation criteria identified during COA analysis. All staff use the same criteria.

4. Analysis.
Analyze each COA using that evaluation criteria from COA analysis. Review enemy actions that impact functional area as they relate to COAs. Identify issues, risks, and deficiencies these enemy actions may create with respect to functional area.

5. Comparison.
Compare COAs. Rank order COAs for each key consideration. Use a decision matrix to aid the comparison process.

6. Recommendations

a. Recommend the most supportable COAs from the perspective of the functional area.

b. Prioritize and list issues, deficiencies, and risks and make recommendations on how to mitigate them

See pp. 4-24 to 4-25 for a listing of suggested mission analysis considerations.

Sustainment Planning

II. The Personnel Estimate

Ref: FM 1-0, Human Resources Support (Feb '07), pp. I-3 to I-5. An alternate (macro) format for the Personnel Estimate can also be found in App. A, Joint Pub 1-0.

G-1s and S-1s at all levels, HRSC divisions, Expeditionary Sustainment Command (ESC) and Sustainment Brigade SPO HR Operations Cells, HR Company headquarters, TG R5 Teams, MMT Teams, R5 Plans and Operations Teams, and Postal Plans and Operations Teams are responsible for executing the core competency of HR Planning and Operations to ensure that effective and efficient HR support is provided to both units and individuals.

Mission analysis is a critical early step in the MDMP. The HR planner/G-1/S-1 has specific responsibilities that establish favorable conditions for the remainder of the MDMP from the HR perspective. The following are G-1/S-1 mission analysis responsibilities:

- Analyze personnel strength data to determine current capabilities and project future requirements
- Analyze unit strength maintenance, including monitoring, collecting, and analyzing data affecting Soldier readiness
- Prepare estimates for personnel replacement requirements, based on estimated casualties, non-battle losses, and foreseeable administrative losses to include critical MOS requirements
- Determine personnel services available to the force (current and projected)
- Determine personnel support available to the force (current and projected)
- Determine HR specified, implied, and essential tasks (if any)
- Determine any HR constraints which may impact successful completion of the mission
- Identify critical HR facts and assumptions which will impact the mission
- Determine any HR Commander's Critical Information Requirements (CCIR) and Essential Elements of Friendly Information (EEFI) which may impact the mission

One of the primary tools used by HR planners is the Personnel Estimate, and like any staff estimate it: "Is an assessment of the situation and an analysis of those COAs a Commander is considering that best accomplishes the mission. It includes an evaluation of how factors in the HR staff section's functional area influence each course of action and includes conclusions and a recommended COA to the Commander" from the perspective of the HR supporter. The personnel estimate is a continuous process that evaluates current and future operations from the functional perspective of the HR provider. Once the commander has selected a COA, the Personnel Estimate becomes an assessment tool to evaluate the relative success of the COA and a means for HR operations cells to track the progress of the operation from an HR support perspective. Format is not the critical element of the personnel estimate. It is a tool that facilitates the collection and processing of key HR information, and may be adapted by HR planners for their own organization or purposes. The following are general characteristics of the personnel estimate:

- Prepared as thoroughly as time allows in either a simple form or a preformatted digital product
- Revised continuously as planning factors, manning levels, facts and assumptions change
- Prepared at all levels of command from Battalion to Army/ASCC by S-1s/G-1s and HR Operations Cell planners
- Not necessarily prepared in a fixed sequence
- Provides a thorough, clear, unemotional analysis of all pertinent data

1. MISSION: Commander's restated mission resulting from mission analysis

2. SITUATION AND CHARACTERISTICS:

a. Characteristics of the AO: Discuss how the weather, terrain, civil considerations, and other AO-specific conditions affect the HR support to the planned operation

b. Enemy Forces: Discuss any affect enemy dispositions, composition, strength, capabilities and COAs may have on the HR support to the planned operation

c. Friendly Forces: Discuss current status of resources available from the HR perspective and other resources which impact HR support and compare requirements and capabilities to develop recommendations to solve discrepancies.

1) Current subordinate unit manning levels/critical MOS shortages (consider all manning influences – Task Organization, R&R flow, Boots On Ground data, etc.)
2) New personnel requirements resulting from the operation (language skills, ASI, etc.)
3) Casualty estimates (as developed)
4) Forecasted replacement availability
5) Evacuation policy for the operation
6) Supporting HR organizations (location, command/support relationship, controlling element)
7) Supporting medical elements
8) Replacement fill priority (Coordinate with S-3/G-3)
9) Crew/key leader replacement
10) Projected postal flow/limitations
11) Change to established PASR reporting flow/times
12) Change to casualty reporting flow (CLT change, reporting changes)
13) RSOI Reception HR impacts
14) R&R schedule/flow operational impact (during operations conducted during sustained operations)
15) Civilian and JIM manning impacts on the operation focusing on strengths and personnel service requirements
16) Specific Army G-1 Personnel Planning Guidance (PPG) impacts on the operation
17) Status of other personnel service support factors (postal, MWR, legal, MILPAY support)
18) Connectivity and NIPR/SIPR and HR system access requirements and availability

d. Sustainment Situation: Discuss the sustainment supporting organizations (TSC, ESC, Sustainment Brigades) and possible impacts to include: supported/supporting relationship, reporting requirements, transportation planned, MSR status, etc.

e. Assumptions: List any assumptions that affect the delivery of HR support.

f. Other Considerations: List everything not covered elsewhere in the estimate.

3. COURSES OF ACTION:

a. COAs: List friendly COAs to be wargamed

b. Evaluation Criteria: List HR evaluation criteria determined during COA analysis

4. ANALYSIS: Analyze each COA using the HR evaluation criteria established during the COA analysis determining advantages and disadvantages of each COA

5. COMPARISON: Compare COAs and rank order from the perspective of HR supportability based on Step 4 (Analysis).

6. RECOMMENDATION and CONCLUSIONS:

a. Recommendation: Recommend the most supportable COA from HR support perspective.

b. Issues/Risks/Mitigations: List the issues, risks, or deficiencies associated with the recommended COA and proposed mitigation to reduce the impact.

APPENDIXES (Use appendixes when information/graphs is of such detail and volume that inclusion in the body makes the estimate too cumbersome. Appendixes should be lettered sequentially as they occur through the estimate.

III. Casualty Estimation

FM 1-0, Human Resources Support (Feb '07), pp. I-6 to I-7.

The Army's functional proponent for casualty estimation is the DA G-1. DA G-1 oversees policy on Army casualty estimation (to include methods and procedures) to ensure forces-wide planning coherence. DA G-1 is proponent for battle casualties (killed, captured, missing, wounded—total battle casualties, TBC). The Surgeon General is proponent for disease and nonbattle injury (DNBI) casualties.

Casualty estimates support operations planning, future force planning, and training at all levels. Operations' planning includes deliberate and contingency plans. Force planning describes options for out-year force structure and force capability designs. Training includes the Battle Command Training Program and other unit exercises. Supported functions include:

- Commander's evaluation of Courses of Action, by assessment of force strength for missions within the concept of operations and scheme of maneuver
- Personnel replacements and flow planning, and allocation among forces
- Medical support planning, for both force structure and logistics support
- Transportation planning, including both inter- and intra-theater requirements, to deliver medical force structure and to evacuate and replace personnel
- Evacuation policy options to sustain the force by balancing minimal support force footprint, maximum in-theater returns-to-duty, and stable personnel rotation.

At all echelons, the G-1/S-1, as the principal staff officer for manning and personnel readiness requirements projections and recommendations, prepares the casualty estimate as part of the Personnel Estimate. The G-1 (S-1) estimates battle casualties and administrative losses, and combines this with the medical staff's DNBI estimate.

The G-1/S-1 and Surgeon (medical staff) coordinate with the battle staff so estimated casualties reasonably reflect projected force activity in the planning scenario. The G-1/S-1:

- Consolidates the overall casualty estimate, stratifies the projected losses by skill and grade, projects personnel readiness requirements, and recommends and plans the support for appropriate replacements and evacuation flows.
- Coordinates with other staff elements that use casualty projections to guide their planning process. Examples include the C1/J1 at both component and COCOM (and higher) levels for a range of force planning needs, the Surgeon (who uses the wounded-in-action estimate, and possibly an EPW or other estimate, to help size medical support) and Army and/or Joint lift planners regarding transportation of the replacements and evacuation flows.

Achieving Reasonable Casualty Estimates

Ensuring reasonable casualty estimates requires more than a numeric estimating procedure or set of rates; a rates frame of reference is critical to show which rates relate to which variables, and how. The Benchmark Rate Structure (BRS) orients the planner by showing how rates vary as forces (size, type), time (duration of rate application), and operational settings vary. The BRS describes rate ranges and patterns seen in actual operations for both maneuver forces (Battalions, brigades, divisions) and support forces across the spectrum of conventional operations, including major combat and stability operations. The operating environments described range from peer or near-peer confrontations, to overwhelming dominance by one side, to isolated asymmetric events. It is critical to note that numerous non-standard and non-authorized casualty estimation tools are available to various planners, and that these various other estimators do not provide accurate casualty data.

V. The Concept of Support (para. 4a)

Ref: Adapted from FM 4-90.7, Stryker Brigade Combat Team Logistics (Sept '07) and related sources (ST 101-6, chap. 3 and app. C-G). See also p. 4-17.

After the commander selects a specific COA, the staff communicates this decision by publishing the operation plan/operation order (OPLAN/OPORD). The G4, with input from the other logistic staff elements (G1, G5, surgeon, finance and personnel officers, and the support command), will prepare paragraph 4 of the plan.

Paragraph 4a is the support concept. This concise, but comprehensive, paragraph tells the maneuver commander and his primary staff those critical or unusual logistic actions that will occur by phase or before, during, and after the battle to support the concept of the operation.

Additional subparagraphs can be used to provide more detailed sustainment information by functional area. Usually, these subparagraphs are omitted, and this detailed information is published as part of the service support annex to the plan. The G4 prepares this order with input from the other logistic staff elements. The G4 can also prepare a Sustainment overlay to show supported units' supply route locations and supporting logistic organizations. Finally, routine, doctrinal, or constant information is incorporated into the unit tactical standing operating procedures (TSOP) to avoid repetition.

I. Developing the Sustainment Concept

The logistician actively participating in the decisionmaking process facilitates the support concept's development. Specifically, during mission analysis, the Sustainment planner determines the units' current materiel and personnel posture before the operation begins. This, with the commander's priorities, determines which units and items of equipment should receive priority before the operation.

The wargaming and quantitative analysis portions of COA analysis highlight critical and/or unusual logistic requirements and determine support priorities during each phase of the operation. By its very nature, wargaming facilitates logistic synchronization with the concept of the operation.

There are numerous other information sources for the support concept:

- Commander's guidance and intent
- Concept of the operation
- Higher HQ support concept, service support order or plan (if applicable), and Sustainment overlay
- Maneuver control system screens and/or other locally generated status charts
- Lessons learned data and historical perspectives to view how others successfully, or unsuccessfully, supported other similar operations
- The unit's battle book

II. The Sustainment Overlay

The sustainment overlay is a graphic representation of the tactical array of support areas and units. Ideally, it accompanies copies of the OPLAN and/or OPORD distributed to subordinate HQ and is used as a graphic backdrop to the support concept briefing.

The sustainment overlay should include (as a minimum):

- Locations of current and proposed support areas
- Boundaries for sustainment responsibilities
- MSRs
- Locations of major HQ
- Locations of sustainment installations and units
- Locations of critical resources [potable water, maintenance collection points, ATPs, mortuary affairs (MA) collection points, ambulance exchange points (AXPs), etc.]

The sustainment overlay will not only depict the tactical array of sustainment units/nodes, but it is also an integral part of the overall OPLAN/OPORD graphics and must be synchronized with the operations overlays.

The **BRIGADE sustainment overlay** would include (as a minimum)—

- The brigade support area (BSA) location and, using type unit symbols, the sustainment units and HQ located therein
- Locations of alternate/proposed BSAs
- Locations of forward logistics elements (FLEs)
- The supply routes from the BSA to the logistic release points and/or maintenance collection points
- The MSR from the division support area (DSA) to the BSA

The **DIVISION sustainment overlay** would include (as a minimum):

- The division support area (DSA) location and using type unit symbols, the sustainment units and HQ contained there-in, whether they are divisional or nondivisional
- Locations of alternate and/or proposed DSAs
- The MSRs from the corps rear area to the DSA and from the DSA to each BSA

The **CORPS sustainment overlay** may have to encompass the entire corps area of operation (AO) as well as a part of the communication zone (COMMZ) and, as a minimum, would depict:

- The logistic support areas (LSAs) and, using type unit symbols, the Sustainment units and HQ located therein, and the locations of any other critical sustainment nodes not located in an LSA
- The MSRs leading into the corps rear area from the COMMZ and the MSRs leading from the corps rear area to each DSA (or, as a minimum, to the division rear boundary) and to other critical logistic nodes
- Locations of alternate and/or proposed LSAs
- Locations of corps sustainment units operating forward of the divisional rear boundaries

III. The Sustainment Matrix

The oral support concept briefing will allow the commander and his subordinates to visualize how the operation will be logistically sustained. The sustainment planners' oral briefing, using the sustainment overlay, is useful in communicating the support concept to the commander. In addition, a support concept matrix can be used to make complex logistic concepts more easily understood.

The sustainment matrix's design is aligned with the support concept format. The logistic functions are in the "by phase" context. The matrix can also be modified to reflect before, during, and after phases. The matrix will highlight those critical aspects of each sustainment function. It can also depict other critical information such as priorities, shifts in priorities, problem areas, critical events, and other critical action. The matrix is not intended to stand alone or to replace the support concept briefing. It should complement and supplement the support concept briefing.

Concept of Support (Format & Briefing)

Ref: Adapted from ST 101-6, chap. 3, pp. 3-5 to 3-7. See also p. 4-17.

The logistician's role in the overall OPLAN/OPORD briefing is to brief the support concept, but he must first understand the general concept of the operation and the commander's intent. The support concept briefing should address the critical, non-SOP, or unusual aspects of logistic support by phase of an operation by critical Sustainment functions. Doctrinal, usual, or SOP matters should not be addressed unless there is a deviation in support relationships or normal methods.

4. Concept of Support (Paragraph 4a)

a. Support Concept. Paragraph 4a will provide an overall view of the support concept. Its intent is to provide the non-Sustainment commanders and their primary staffs an image of how the operation will be logistically supported. If the information pertains to the entire operation, or if it pertains to more than one unit, include it in the introductory portion of paragraph 4a. Change it in the ensuing subparagraphs when needed. This could include: a brief synopsis of the support command mission; support command headquarters and/or support area locations, including locations of next higher logistics bases if not clearly conveyed in the Sustainment overlay; the next higher level's support priorities and where the unit fits into those priorities; priorities that remain unchanged throughout the operation; units in the next higher Sustainment organization supporting the unit; significant and/or unusual Sustainment issues that might impact the overall operation; the use of host nation support; and any significant sustainment risks

(1) PHASE I (repeat for each phase)

- Logistics focus.
- Priorities: By unit; for personnel replacements; maintenance and/or recovery and evacuation priorities (by unit and equipment type); mvmt; class of supply.
- Critical events. Use the Sustainment functions for information.
- Sustainment risks.

(4) Paragraphs 4b through 4e are normally more detailed and are included in the service support annex. They are not part of the support concept.

(5) Concept of support written before, during, and after format. Follow the same guidance as by phase.

Concept of Support Briefing Format

1. Introduction (overview of the support concept and orientation to the map, if required). Orientation to the map is not required if another briefer has done so previously. Do not assume the commander totally knows the terrain. Focus on locating critical Sustainment nodes, MSRs, etc.

2. Brief the support concept starting with critical actions that must be accomplished in the first phase of the operation and concluding with critical actions to be accomplished in the last phase. This will prepare for future operations using the Sustainment functions as a guide.

3. Identify which units have priorities for each critical sustainment function (this should correlate with the commander's priorities; e.g., main effort).

4. Identify the next higher echelon unit providing support and/or backup support.

5. Identify any critical shortages/problem areas for each sustainment function and solution. For example, this can be supported, but _____ , or it can be done, but not without risk in_____ .

6. Identify any other sustainment problem areas, arrangements, special requirements, or any other critical aspects addressed elsewhere in the briefing.

BCT Sustainment Planning (The BCT S-4)

Ref: Adapted from FM 4-90.7, Stryker Brigade Combat Team Logistics (Sept '07) and related sources.

The lead planner for sustainment in the BCT is the BCT S-4, assisted by the BCT S-1, the BCT surgeon, and the BSB support operations officer. Representatives from these and other sections form a sustainment planning cell at the BCT main CP to ensure sustainment plans are fully integrated into all operations planning. The unit SOP should be the basis for sustainment operations, with planning conducted to determine specific requirements and to prepare for contingencies. BCT and subordinate unit orders should address only specific support requirements for the operation and any deviations from SOP.

Although the sustainment planners at the BCT main CP control and coordinate sustainment for specific BCT operations, routine operations usually are planned in the BSA. The BCT S-1 may have representatives at or near the BSB to handle various personnel functions (e.g., replacement or monitor casualty operations). The BCT S-4 might choose to locate the property book or movement sections with the BSB support operations section.

To provide effective support, sustainment planners and operators must understand the mission statement, intent, and concept of the operation which in turn will lead to developing a concept of support that will be provided in the BCT operation order (OPORD). The BCT S-4 is responsible for producing the sustainment paragraph and annexes of the OPORD, which should include the following:

- Commander's priorities
- Class (CL) III/ V resupply during the mission, if necessary
- Movement criteria
- Type and quantities of support required
- Priority of support, by type and unit
- Sustainment overlay
- Supply routes
- Logistic release points (LRPs)
- Casualty evacuation (CASEVAC) points
- Maintenance collection points

The remainder of this section provides a description of the concepts that apply to planning sustainment support within the BCT. It also describes how the BCT organizes the sustainment staff and organizations during full spectrum operations.

BCT Concept of Support

The concept of support (Paragraph 4 of the Operations Order) establishes priorities of support (by phase or before, during, and after) for the operation. These priorities are set by the BCT commander at each level in his intent statement and the concept of the operation. Priorities include such things as personnel replacements; maintenance and evacuation, by unit and by system (aviation and surface systems would be given separate priorities); fuel and/or ammunition; road network use by unit and/or commodity; and any resource subject to competing demands or constraints. To establish the concept of support, the BCT sustainment planners must know:

- Subordinate units' missions
- Times missions are to occur
- End states
- BCT scheme of movement and maneuver
- Timing of critical events

VI. (HSS) Health Service Support Considerations

Ref: FM 4-02, Force Health Protection (Feb '03), app. F and app. I.

I. HSS Planning Considerations for Joint Operations

Future operations will be joint in nature. HSS planners must anticipate this fact and plan accordingly. *See following pages (4-31 to 4-32) for further discussion and specific considerations.*

- What C2 infrastructure will be established for the operation?
- What is the nature of the operation and its anticipated duration?
- What is the anticipated level of violence to be encountered?
- What are the capabilities of all Service component HSS assets in theater?
- Are communications systems and automation equipment interoperable?
- What are the rules of engagement (ROE)?
- What are the security requirements/force protection measures for HSS?
- Is the contracting for HNS feasible for medical activities?

II. HSS Planning Considerations for Multinational Operations

Multinational operations present new challenges to the HSS planner. In addition to ensuring the rapid, effective, and efficient delivery of health care on the battlefield for US forces, the planner must coordinate support with the health authorities of all participating nations. Thorough coordination is required to ensure that a duplication of services does not occur and that maximum use and benefit is achieved from scarce medical resources. Health service support in multinational operations is a national responsibility (Joint Pub 4-02). *See following pages (4-31 to 4-32) for further discussion.*

- What is the mission of the force and how does it effects HSS operations?
- What is the composition of the force?
- What are the HSS capabilities of the force?
- Who has been designated to provide HSS to the multinational force?
- Has a command surgeon been identified within the multinational force?
- Are there any ISAs among the participating nations?
- What is the anticipated level of compliance with the provisions of the Geneva Conventions (friendly and enemy)?
- Will all nations have interoperable communications and automation systems?
- Has a determination of eligible beneficiaries (in conjunction with the SJA) been made for care in US facilities?
- If (when) members of the participating nations are treated in US facilities, what is the mechanism for returning them to their parent nation for continuing medical care?
- What is the anticipated level of violence to be encountered?
- What are the ROE?
- What are the mechanisms for reimbursement of services?

Joint Operations (HSS Considerations)

Ref: FM 4-02, Force Health Protection (Feb '03), pp. F-1 to F-7. See also p. 4-31.

Future operations will be joint in nature. HSS planners must anticipate this fact and plan accordingly. This checklist addresses some interagency and/or HN considerations. Section II provides a checklist for multinational operations. Section III provides an eligibility for medical care matrix.

Specific doctrine for HSS in joint operations is contained in Joint Pubs 4-02, 4-02.1, and 4-02.2.

Preventive Medicine and the Medical Threat

- Do all Services have PVNTMED assets deployed in the theater?
- What is the medical threat in the AO?
- Have site surveys been conducted for areas to be inhabited by US forces?
- Is it anticipated that refugee, internally displaced persons (IDP), retained/detained persons, and/or EPW operations will be required?
- Do units have field hygiene and sanitation supplies and equipment on hand?
- How will medical waste be collected and disposed of?
- Do service members have personal protective supplies and equipment available and/or issued?
- If continuous operations are anticipated, have work/rest schedules (sleep plans) been developed and implemented when appropriate?
- Is a command policy established and disseminated on water discipline?

Medical Treatment (Area Support)

- What units will provide Level I and Level II medical care?
- Will troop clinics/dispensaries be established in areas of troop concentrations?
- Do any operations security (OPSEC) requirements exist which must be accommodated?

Hospitalization

- What hospital resources will be in the theater?
- What hospitals will be designated for the care of retained/detained persons and EPW?
- Has an eligibility determination been made for care in US facilities?
- Are there any hospital resources within the theater that can operate as shared resources with hospitals from the other Services?

Medical Evacuation and Medical Regulating

- What is the theater evacuation policy?
- What are the specific responsibilities for each Service component?
- Will a TPMRC or a GPMRC be activated for the operation?
- Will a MASF or aeromedical staging facility (ASF)/ASTS be established for staging patients awaiting medical evacuation aircraft?
- What other USAF aeromedical evacuation resources will be available in theater?
- How will patient movement items be handled?
- Are US Army aeromedical evacuation unit personnel deck-landing qualified for USN ships?

Health Service Logistics (to Include Blood Management)

- Has the combatant commander designated a SIMLM for the operation?
- How will medical equipment maintenance and repair be accomplished?
- What units/organizations will provide optical fabrication support?
- Are there donated medical supplies and equipment for use?
- Are there any Service specific HSL requirements?
- How are blood management functions/activities conducted?

Dental Service

- What dental resources are deployed in theater?
- Is it anticipated that dental personnel will be required to perform their alternate wartime role during the operation?
- Where will dental resources be located?

Veterinary Service

- Although the US Army is the Executive Agent for Veterinary Support for all Services, will the USAF conduct its own subsistence inspection on USAF installations?
- What types of rations are to be used by the forces in theater?
- Will military working dogs (MWD) and/or other government-owned animals be used in the operation?
- Does a command policy exist on unit mascots or pets?
- How will animals requiring evacuation be managed?

Combat Operational Stress Control/Mental Health Activities

- Do all the Services have MH personnel deployed to the theater?
- During the operation is it likely that a mass casualty situation will develop?
- What is the likelihood of a terrorist attack?

Medical Laboratory Support

- What medical laboratory assets will be deployed to the theater?
- What medical laboratory will provide the identification of suspect BW and CW agents?
- Will a near-patient testing capability be present in any of the in-theater medical units?
- Will any intheater medical laboratory assets have a split-base operating capability?

Operations in a Nuclear, Biological, and Chemical Environment

- Is the use of NBC weaponry anticipated?
- What is the potential for accidental contamination?
- What medical units have the capability to perform patient decontamination operations?
- What are the reporting and notification requirements in the event of a suspect NBC incident?
- Is collective protection available for MTFs?
- Are veterinary personnel available to inspect NBC contaminated subsistence?
- Are PVNTMED personnel available to inspect NBC contaminated water supplies?
- Are immunizations, chemoprophylaxis, antidotes, pretreatments, and barrier creams available?

Multinational Opns (HSS Considerations)

Ref: FM 4-02, Force Health Protection (Feb '03), pp. F-8 to F-16. See also p. 4-31.

Multinational operations present new challenges to the HSS planner. In addition to ensuring the rapid, effective, and efficient delivery of health care on the battlefield for US forces, the planner must coordinate support with the health authorities of all participating nations. Thorough coordination is required to ensure that a duplication of services does not occur and that maximum use and benefit is achieved from scarce medical resources. HSS in multinational operations is a national responsibility (Joint Pub 4-02).

Preventive Medicine and the Medical Threat

- What are the diseases (endemic and epidemic) in the AO and/or in the separate national contingents?
- Are immunizations or chemoprophylaxis available to counter the disease threat?
- Have site surveys been conducted in areas US forces will inhabit?
- What PVNTMED support will US forces provide other national contingents?
- What is the level of training in field hygiene and sanitation for US forces and other national contingents?
- Are refugee, IDP, retained/detained persons, and/or EPW operations required?
- How will medical waste be collected and disposed of?
- Do units have required field hygiene and sanitation supplies and equipment on hand?
- Are personal protective supplies and equipment available and/or issued?

Medical Treatment (Area Support)

- Are interpreters available to translate patient complaints to the attending medical personnel?
- What units are providing Levels I and II medical care?

Hospitalization

- What hospitals are established in the AO?
- What ancillary services are offered by the hospitals?
- What is the surgical capability of in-theater hospitals?
- What procedures/notifications are required when a non-US soldier is admitted to a US facility?
- Will non-US physicians be permitted to treat patients in a US facility?
- Has a formulary been established for prescription drugs?
- What outpatient services will be provided?
- How will patients be transferred from one hospital to another within the theater?

Medical Evacuation and Medical Regulating

- What is the theater evacuation policy?
- What units are conducting medical evacuation operations?
- What types of evacuation assets are available?
- How are requests for evacuation transmitted?
- How will units requesting medical evacuation be located and identified?
- Do medical evacuation vehicles/aircraft require armed escort while performing their mission?
- How will patient movement items (PMI) be managed?
- Will MASFs/ASFs/ASTSs [or similar organizations] be established at airheads to sustain patients awaiting evacuation from the theater?
- What nation will provide the medical regulating function?

Health Service Logistics (to Include Blood Management)

- What is the Class VIII stockage level?
- What is the impact of multinational operations on blood management?
- Is the US tasked to provide HSL support to the multinational force?
- Are there donated medical supplies and equipment?
- How will resupply be affected?
- What reports are required to be submitted to the supporting HSL facility?
- Can medical supplies and equipment from non-US sources be used for US forces?
- If operations are conducted under the auspices of an international organization (such as the UN) how do their supply/resupply procedures and requirements impact on US Class VIII operations?

Dental Service

- What units will provide dental services for the multinational force?
- What is the scope of dental services to be provided within theater?
- Do all members of the multinational force have panographs on file?
- Will a preventive dentistry program be implemented for US forces and/or multinational forces in theater?
- What dental conditions will necessitate the evacuation of patients from the theater?

Veterinary Service

- What type of rations are to be used in theater?
- Will Class I operations be consolidated for the multinational force?
- Will US forces provide veterinary inspection of subsistence for food safety and quality assurance for multinational forces?
- Will government-owned animals be used in the operation?
- Has a command policy been disseminated on unit mascots/pets?
- How will animals be evacuated?
- Will the operation involve nation assistance activities?
- What veterinary PVNTMED activities will be implemented in-theater?

Combat Operational Stress Control/Mental Health Activities

- Is each national contingent responsible for its own MH programs and treatment?
- How will NP and/or COSC patients be evacuated?
- What preventive programs will be implemented in theater?
- Who will conduct critical event debriefings?

Medical Laboratory Services

- What laboratory capability exists within the national contingents?
- How are specimens and samples collected, handled, stored, and transferred?

Nuclear, Biological, and Chemical Environments

- What is the potential threat for use of NBC weaponry during the operation?
- What is the level of protection for each national contingent?
- Is collective protection available to the MTFs?
- Have patient decontamination teams been identified from supported units?
- What are the reporting and notification requirements?
- Are veterinary personnel available to inspect NBC contaminated subsistence?
- Are PVNTMED personnel available to inspect NBC contaminated water supplies?
- Are treatment protocols established for the treatment of NBC casualties?
- Immunizations, chemoprophylaxis, antidotes, pretreatments, barrier creams?

III. Special Medical Augmentation Response Teams (SMARTs)

Ref: FM 4-02, Force Health Protection (Feb '03), pp. I-1 to I-8.

SMART teams provide a rapidly available asset to compliment the need to cover the full spectrum of military medical response locally, nationally, and internationally. These teams are organized by the USAMEDCOM and its subordinate commands; they are not intended to supplant TOE units assigned to US Army Forces Command (USAFORSCOM) or other major commands. The USAMEDCOM, RMCs, USACHPPM, USAMRMC, and US Army Veterinary Command (USAVETCOM) commanders organize SMARTs using their TDA assets. These teams enable the commander to field standardized modules in each of the SMART functional areas to meet the requirements of the mission.

The types of SMARTs include:

- Trauma/Critical Care (SMART-TCC)
- Nuclear/Biological/Chemical (SMART-NBC)
- Stress Management (SMART-SM)
- Medical Command, Control, Communications, and Telemedicine (SMART-MC3T)
- Pastoral Care (SMART-PC)
- Preventive Medicine/Disease Surveillance (SMART-PM)
- Burn (SMART-B)
- Veterinary (SMART-V)
- Health Systems Assessment and Assistance (SMART-HS)
- Aeromedical Isolation (SMART-AIT)

These teams provide military support to civil authorities during disasters, CMO, and humanitarian and emergency services incidents

Requests for Assistance

Requests for assistance may be generated from numerous sources to MEDCOM. These requests are received using appropriate, recognized, and approved channels. These sources may include:

- Director of Military Support (DOMS)
- United States Joint Forces Command

Team Composition and Specialty-Specific Equipment

The USAMEDCOM determines the composition of each team and identifies the specialty-specific equipment required to accomplish the mission. The composition of the team is task-organized based on the METT-TC and medical risk analysis in order to provide the appropriate level of response and technical augmentation to civil and military authorities. This information is provided to its subordinate commands through appropriate command policy statements, directives, or SOPs. These teams may be comprised of active duty military, DOD civilians, or contractors as determined by the commander.

Deployability and Continuous Operations

Within 12 hours of notification, the SMARTs will be alerted, issued a warning order (WARNO), and assembled; within 12 hours of the WARNO the SMARTs will be capable of deploying.

The SMARTs are not capable of 24-hour continuous operations. To conduct continuous operations the deployed SMARTs require augmentation/reinforcement of both personnel and materiel or support from follow-on medical specialty personnel.

Chap 5

I. Joint Logistics Overview

Ref: JP 4-0, Joint Logistics (Jul '08), chap. I. For more information on joint operations, see The Joint Forces Operations & Doctrine SMARTbook.

The Nation's ability to project and sustain military power depends on effective joint logistics. Joint logistics delivers sustained logistic readiness for the combatant commander (CCDR) and subordinate joint force commanders (JFCs) through the integration of national, multinational, Service, and combat support agency (CSA) capabilities. The integration of these capabilities ensures forces are physically available and properly equipped, at the right place and time, to support the joint force. Joint logisticians coordinate sustained logistic readiness through the integrating functions of planning, executing and controlling joint logistic operations.

I. Sustainment as a Joint Function

The six joint functions described in Joint Publication (JP) 3-0, Joint Operations, include Command and Control (C2), Intelligence, Fires, Movement and Maneuver, Protection, and Sustainment. Sustainment is the provision of logistics and personnel services necessary to maintain and prolong operations until successful mission completion. Sustainment in joint operations provides the JFC flexibility, endurance and the ability to extend operational reach. Effective sustainment determines the depth to which the joint force can conduct decisive operations, allowing the JFC to seize, retain, and exploit the initiative. Sustainment is primarily the responsibility of the supported CCDR and subordinate Service component commanders in close cooperation with the Services, CSAs, and supporting commands. Key considerations include employment of logistic forces, facilities, environmental considerations, health service support (HSS), host-nation support (HNS), contracting, disposal operations, legal support, religious support, and financial management. JP 4-0 concentrates on the logistic function of sustainment; the personnel services function can be found in the JP 1-0 series.

II. Joint Logistics

Joint logistics is the coordinated use, synchronization, and sharing of two or more Military Departments' logistic resources to support the joint force. From a national perspective, it can be thought of as the ability to project and sustain a logistically ready joint force through the sharing of Department of Defense (DOD), interagency, and industrial resources. In today's operating environment this will include coordination and sharing of resources from multinational partners, intergovernmental organizations (IGOs) and nongovernmental organizations (NGOs). This provides the JFC the freedom of action necessary to meet mission objectives. It is an essential component of joint operations because the Services, by themselves, seldom have sufficient capability to independently support a joint force. By purposefully combining capabilities, the commander can optimize the allocation of limited resources to provide maximum flexibility to the joint force. It is this kind of interdependence, focused on common outcomes, that delivers sustained logistic readiness.

III. The Joint Logistics Environment (JLE)

Ref: JP 4-0, Joint Logistics (Jul '08), pp. I-4 to I-5.

Political and military leaders conduct operations in a complex, interconnected, and increasingly global operational environment. This environment is characterized by uncertainty and surprise. Operations are also distributed and conducted rapidly and simultaneously across multiple joint operations areas (JOAs) within a single theater or across boundaries of more than one geographic combatant commander (GCC) and can involve a large variety of military forces and multinational and other government organizations. The joint logistics environment (JLE) exists within this operational environment and consists of the conditions, circumstances and influences that affect the employment of logistic capabilities. It exists at the strategic, operational, and tactical levels of war; and includes the full range of logistic capabilities, stakeholders, and end-to-end processes.

Physical Domains

Joint logistics takes place within the physical domains of air, land, maritime, and space. Service components provide the expertise within these domains and the JFC and staff focus on leveraging and integrating those capabilities.

Information Environment

The global dispersion of the joint force and the rapidity with which threats arise have made real-time or near real-time information critical to support military operations. Joint logistic planning, execution, and control depend on continuous access to make effective decisions. Protected access to networks is imperative to sustain joint force readiness and allow rapid and precise response to meet JFC requirements.

Levels of War

Joint logistics spans all levels of war. It is, however, at the tactical level where the principal outcome - sustained logistic readiness - of joint logistics must be measured.

1. Strategic

At the strategic level, joint logistics is characterized by the vast capacity of the Nation's industrial base, both government and commercial. The Nation's ability to project and sustain military power comes from the strategic level; it enables sustained military operations over time and represents one of our Nation's greatest strengths. At this level, modern, clearly defined, well-understood and outcome-focused processes should drive effectiveness across joint, Service, agency, and commercial organizations. These global processes combined with agile force positioning are fundamental to optimizing joint logistics and critical to the Nation's ability to maintain flexibility in the face of constantly changing threats.

2. Operational

At the operational level, joint logistics has its most significant impact. It is at the operational level that strategic and tactical capabilities, processes, and requirements intersect, and it is here where the essence of joint logistics resides. Joint logisticians at this level must integrate or coordinate national, DOD, combatant command, Service and functional components, multinational, interagency and other partner capabilities, and HNS, with the JFC's tactical requirements. Joint and Service logistics fuse at the operational level. Logisticians face their greatest challenge at the operational level because of the difficulty of coordinating and integrating capabilities from many providers to sustain logistically ready forces for the JFC.

3. Tactical

The tactical level represents that part of the operational environment where outcomes are realized. At the tactical level, logistic support is Service-oriented and executed. Organizations operating at the tactical level are focused on planning and executing those operations, engagements, and activities to achieve assigned military objectives. Tactical units require sustained logistic readiness to meet assigned objectives. Sustained logistic readiness results from the cumulative efforts of Service, agency, and other providers across the entire JLE.

Global Relationships

The JLE is bound together by a web of relationships among global logistic providers, supporting and supported organizations and units, and other entities. The key global providers in the JLE are the Services, the Defense Logistics Agency (DLA), United States Joint Forces Command (USJFCOM) and United States Transportation Command (USTRANSCOM).

See JP 4-0, Appendix A, Joint Logistic Roles and Responsibilities, for a listing of logistic-related roles and responsibilities of the Services, CSAs, and commands.

Integrating Functions

Sustained logistic readiness is driven by the effective and efficient delivery of joint logistics through coordinating and integrating Service, agency, and other capabilities to meet the supported commander's requirements. To achieve this level of integration, commanders and their staffs, especially logisticians, must be able to: effectively and efficiently plan, execute, control, and assess joint logistic operations.

1. Planning

Planning joint logistic support links the mission, commander's intent, and operational objectives to core logistic capabilities, procedures and organizations. Joint logistic planning defines joint processes to establish an effective concept for logistic support. Effective planning among the combatant commands, Services, CSAs, and other government and nongovernment agencies is essential to enable integration and visibility across the operational environment. Obtaining and understanding joint requirements for supplies and services is vital to supporting the deployment, employment, and redeployment of forces and equipment.

See pp. 5-7 to 5-12 for further discussion of planning joint logistic support.

2. Executing

Executing joint logistics involves the employment of capabilities and resources to support joint and multinational operations. The joint logistician must be able to assess and respond to requirements by monitoring dynamic situations and providing accurate feedback to subordinates and decision makers. The joint logistician must determine the proper balance of efficiency and effectiveness in processes being executed, and remain flexible to employ new methods as the environment changes.

See pp. 5-13 to 5-16 for further discussion of executing joint logistic operations.

3 Controlling

Effective control of joint logistic operations results from the exercise of authority and direction for the sustained logistic readiness of the joint force. This integrating function includes choosing organizational options to best execute effective joint logistic operations.

See pp. 5-17 to 5-24 for further discussion of controlling joint logistic operations.

IV. Joint Logistics Imperatives

Ref: JP 4-0, Joint Logistics (Jul '08), pp. I-7 to I-8.

The value of joint logistics can be determined by how well three imperatives are attained: unity of effort, JLE-wide visibility, and rapid and precise response. These imperatives define the desired attributes of a federation of systems, processes, and organizations that effectively adapt within a constantly changing environment to meet the emerging needs of the supported CCDR.

1. Unity of Effort

In accordance with JP 1, Doctrine for the Armed Forces of the United States, unity of effort is the coordination and cooperation toward common objectives, even if the participants are not necessarily part of the same command or organization - the product of successful unified action. For joint logisticians this is the synchronization and integration of logistic capabilities focused on the commander's intent and is the most critical of all joint logistic outcomes. To achieve unity of effort, joint logisticians must develop a clear understanding of how joint and multinational logistic processes work; know the roles and responsibilities of the providers executing tasks in those processes; build agreement around common measures of performance (MOPs) (process outcomes); and ensure appropriate members of the JLE have visibility into the processes.

2. JLE-Wide Visibility

JLE-wide visibility is having assured access to logistic processes, resources, and requirements to gain the knowledge necessary to make effective decisions. JLE-wide visibility provides the means to optimize logistic capabilities to maximize outcomes, increase readiness, and build confidence in joint logistics. It provides access to authoritative information and enables the user to respond quickly to the CCDR's changing needs. Visibility fundamentally answers the CCDR's questions, "Where is it?" "How will it get there?" and "When will it get there?"

3. Rapid and Precise Response

Rapid and precise response is the ability of the core logistic capabilities, military and commercial, to meet the constantly changing needs of the joint force. The effectiveness of joint logistics can be measured by assessing the following attributes, or key performance indicators:

- **Speed** is at the core of responsiveness. Speed does not mean everything moves at the same rate or fastest rate, but everything moves according to priority at the rate that produces the most effective support to the joint force.
- **Reliability** is reflected in the dependability of the global providers to deliver required support when promised. Reliability is characterized by a high degree of predictability, or time-definite delivery of support. Time-definite delivery is the consistent delivery of requested logistic support at a time and destination specified by the requiring activity.
- **Efficiency** is directly related to the amount of resources required to deliver a specific outcome. In the tactical and operational environments, inefficiency increases the logistic footprint and increases force protection requirements and risk. At the strategic level, inefficiency increases the cost for a unit of process outcome.

The joint logistics imperatives enable the measurement of our ability to provide sustained logistic readiness. The essence of these imperatives guide joint logisticians in the performance of the three integrating functions needed for successful joint logistic operations.

Chap 5

II. Core Logistics Capabilities

Ref: JP 4-0, Joint Logistics (Jul '08), chap. II and pp. I-10 to I-12.

Joint logistics can be described in terms of the capabilities it delivers. These capabilities enable achievement of objectives (ends) through combinations of functions (ways) executed by the people and processes (means) within a broad range of conditions and to a specified set of standards. Joint logistics, in the larger sense, can best be understood as a joint capability area. The ways of joint logistics are its core logistic capabilities: supply, maintenance operations, deployment and distribution, HSS, engineering, logistic services, and operational contract support. Each of these capabilities includes the people, processes, and resources required to conduct joint logistics.

Core Logistic Capabilities

Core Capabilities	Functional Capabilities
Supply	▪ Manage Supplies and Equipment ▪ Inventory Management ▪ Manage Supplier Networks
Maintenance Operations	▪ Depot Maintenance Operations ▪ Field Maintenance Operations ▪ Manage Life Cycle Systems Readiness
Deployment and Distribution	▪ Move the Force ▪ Sustain the Force ▪ Operate the Joint Deployment and ▪ Distribution Enterprise
Health Service Support	▪ Casualty Management ▪ Patient Movement ▪ Medical Logistics ▪ Preventive Medicine and Health Surveillance ▪ Theater Medical Information
Engineering	▪ Combat Engineering ▪ General Engineering ▪ Geospatial Engineering
Logistic Services	▪ Food Service ▪ Water and Ice Service ▪ Base Camp Services ▪ Hygiene Services
Operational Contract Support	▪ Contract Support Integration ▪ Contract Management

Ref: JP 4-0, Joint Logistics, fig. I-3, p. I-10.

Core logistic capabilities provide a framework to facilitate integrated decisionmaking, enable effective synchronization and allocation of resources, and optimize joint logistic processes. The challenges associated with support cut across all core logistic capabilities – especially when multiple joint task forces (JTFs) or multinational partners are involved. The core logistic capabilities are supply, maintenance operations, deployment and distribution, HSS, engineering, logistic services, and operational contract support. The core logistic capabilities must be integrated within a complex operational environment, bridging the strategic sustaining base of the Nation to the tactical environment where outcomes are measured. An important objective for logisticians at the operational level is to set the conditions for tactical level logisticians to achieve success.

Core Logistics Capabilities

Ref: JP 4-0, Joint Logistics (Jul '08), pp. I-10 to I-12.

1. Supply

Operations that include identifying requirements, selecting supply sources, scheduling deliveries, receiving, verifying and transferring product, inspection and acceptance, and authorizing supplier payments. It includes the following functions: management of supply operations, inventory management and the mgmt of DOD's supplier networks.

2. Maintenance Operations

Operations that encompass key functions executed by the Services to deliver systems readiness and enable the JFC's freedom of action. Field maintenance operations are focused on rapidly returning systems to the user. Depot maintenance operations are focused on rebuilding/repairing systems and components to sustain long-term life cycle readiness. Total life cycle systems management is focused on the readiness and the integrated control of systems' long-term health by maximizing availability and reliability of systems at best value to the Military Departments.

3. Deployment and Distribution

Deployment and Distributions operations include planning, coordinating, synchronizing, moving forces, and sustainment, and operating the Joint Deployment and Distribution Enterprise (JDDE) in support of military operations. Distribution capabilities are a part of joint logistics, while the full range of deployment activities are a series of operational events enabled by logistics. The portion of deployment that falls within the logistics capabilities is the movement of forces and materiel.

4. Health Service Support

Services that promote, improve, conserve, or restore the mental or physical well-being of personnel. These services include, but are not limited to, the management of health services resources, such as manpower, monies, and facilities; preventive and curative health measures; evacuation of the wounded, injured, or sick; selection of the medically fit and disposition of the medically unfit; blood management; medical supply, equipment, and maintenance thereof; combat stress control; and medical, dental, veterinary, laboratory, optometry, nutrition therapy, and medical intelligence services.

5. Engineering

Operations that assure mobility, provide the infrastructure necessary to position, project, protect and sustain the joint force, and enhance visualization of the operational area across the full range of military operations. Operational engineering is the integration of combat, general, and geospatial engineering to meet requirements.

6. Logistic Services

Operations that are essential to the technical management and support of the joint force. Logistics Services includes food, water and ice, base camp, and hygiene services in an expeditionary environment.

7. Operational Contract Support

Operations that provide the ability to orchestrate and synchronize the provision of integrated contract support and management of contractor personnel providing that support to the joint force in a designated operational area. As indicated, the major functional capabilities are support integration and contractor (personnel) management. Contract support integration gives the CCDR the ability to synchronize and integrate contract support in support of mission requirements. Contractor management provides the CCDR with the ability to manage and maintain visibility of the associated contractor personnel in the designated operational area.

Chap 5

III. Planning Joint Logistics

Ref: JP 4-0, Joint Logistics (Jul '08), chap. III. See also chap. 4, Sustainment Planning.

The demands and complexities of global operations require that joint logistic planning be an integral part of all planning activities to deliver adaptive, integrated, and synchronized joint logistic support. Effective planning enables logisticians to anticipate requirements, and validate, synchronize and integrate them with available resources to minimize duplication of effort, resolve shortfalls, mitigate risk and ensure effective support of CCDR requirements. Joint logistic planning includes the identification of roles, responsibilities, key tasks and resources, along with the sequencing of logistical capabilities to meet the commander's intent. Effective joint logistics planning identifies future requirements and proposes solutions; it requires joint logisticians to clearly understand the commander's intent and concept of operation. The objective of joint logistic planning is to fully integrate support planning and operations. The more integrated the logistics plan is with the operational concept, the more effective the overall operation will be.

I. Planning Functions

Planning translates strategic guidance and direction into executable OPLANs and operation orders (OPORDs) for contingency or crisis action response. Planning is initiated from a continuous awareness of global events, recognition of the need for a prepared military response to support the National Security Strategy (NSS), and follows a collaborative, iterative planning process. From a logistician's perspective it is important for the operations planners to understand the capabilities and limitations of their apportioned core logistic capabilities. The joint logistician is deeply involved in each of the planning functions and can use the principles of logistics to assist in preparing the logistic plan to support the CCDR's mission.

A. Strategic Guidance

At the CCDR level, planning begins with the receipt of strategic guidance or a planning directive and continues as the CCDR develops a mission statement. This JOPES planning function relates to the first two JOPP steps (Initiation and Mission Analysis). The staff's planning activities initially focus on mission analysis, which develops information to help the commander, staff, and subordinate commanders understand the situation and mission. Planning activities include identifying assumptions, planning forces, mission and desired end state. Logisticians identify critical logistical assumptions. During mission analysis joint logisticians must provide critical information to operations planners on the guidance contained in strategic logistical documents such as the Joint Strategic Capabilities Plan (JSCP) (Mobility and Logistic supplements) and related or supplemental publications. Additionally, detailed information on airfields, seaports, road, rail, and bridging capabilities and other critical infrastructure.

B. Concept Development

This JOPES planning function relates to the following JOPP steps: COA Development, COA Analysis and Wargaming, COA Comparison, and COA Approval. The staff develops, analyzes, and compares valid COAs and develops staff estimates that are coordinated with the CSAs, and JPEC when applicable. The output is an approved COA along with common understanding of the enemy situation, interagency coordination requirements, multinational involvement (if applicable), and capability requirements. Logistic planners coordinate and integrate planning efforts with operational planners so that sustainment requirements are an integral part of COA development. The logistician identifies requirements and critical items and services needed and must be fully aware of force structure planning, time-phased force and deployment data (TPFDD) development and joint reception, staging, onward movement, and integration (JRSOI) requirements as they commence development of concepts of support (addressing supply, maintenance operations, deployment and distribution, HSS, engineering, logistic services, and operational contract support) to meet sustainment requirements from theater entry and operations, to redeployment and reset.

C. Plan Development

During the plan development function, the CCDR's staff will create a detailed OPLAN, OPORD, or operation plan in concept format (CONPLAN), with required annexes. The supported CCDR, staff and subordinate commanders, and supporting commanders conduct a number of different planning activities to include: force planning, support planning, nuclear strike, deployment planning, shortfall identification, feasibility analysis, refinement, documentation, plan review and approval, and supporting plan development. A clear understanding of the concept of operations (CONOPS) is essential to the joint logistician's ability to meet joint force requirements. Because logistic support is provided through a variety of different organizations, joint logistic planning must provide the integration mechanism to unify all sources of support. The joint logistic concept of support specifies how capabilities will be delivered over time, it identifies who is responsible for delivering a capability, and it defines the critical logistical tasks necessary to achieve objectives during the phases of the operation. The logistic concept of support coordinates the capabilities of joint, multinational, host nation, interagency, IGO, NGO, plus Active Component and Reserve Component forces.

Two planning activities logisticians are most involved in are:

1. Support Plan Development

The purpose of support planning is to determine the sequence of the personnel, logistics, and other support required to provide supply, maintenance operations, deployment and distribution, HSS, engineering, logistic services and operational contract support in accordance with the CONOPS.

See following pages (pp. 5-10 to 5-11) for discussion of support plan development.

2. Joint Logistic Concept of Support

The deliverable product at the conclusion of plan development is a completed joint logistic concept of support which resides in the base plan and logistic annex to an OPLAN/OPORD.

CCDRs follow the guidelines for contingency planning in Enclosure CJCSM 3122.03C, JOPES, Volume II, Planning Formats.

D. Assessment (Plan Refinement, Adaptation, Termination, or Execution)

The supported commander extends and refines planning while supporting and subordinate commanders complete their plans. Branch plans and other options continue to be developed. The CCDR and staff continue to evaluate the situation for any changes that would trigger plan refinement, adaptation, termination, or execution.

II. Principles of Logistics

Ref: JP 4-0, Joint Logistics (Jul '08), pp. III-3 to III-4. See also p. 1-4.

Logisticians use the principles of logistics as a guide for analytical thinking when assessing COAs or plans/orders. These principles are not a set of rigid rules, nor do they apply in every situation.

1. Responsiveness

Responsiveness is providing the right support when it is needed and where it is needed. Responsiveness is characterized by the reliability of support and the speed of response to needs of the joint force. Responsiveness is enhanced by visibility commanders need to see where their support is and when it will arrive.

2. Simplicity

Simplicity is defined as a minimum of complexity in logistic operations. Complexity introduces confusion into an already chaotic environment. Simplicity fosters efficiency in planning and execution, and allows for more effective control over logistic operations. Clarity of tasks, standardized and interoperable procedures, and clearly defined command relationships contribute to simplicity. Simplicity is a way to reduce the "fog of war" or the friction caused by combat.

3. Flexibility

Flexibility is the ability to improvise and adapt logistic structures and procedures to changing situations, missions, and operational requirements. Flexibility is reflected in how well logistics responds in an environment of unpredictability. Where responsiveness is a commander's view of logistic support, flexibility is a logistician's view of being responsive.

4. Economy

Economy is achieved when support is provided using the fewest resources within acceptable levels of risk. At the tactical and operational levels, economy is reflected in the number of personnel, units and equipment required to deliver support. Among the key elements of the logistic principle of economy is the identification and elimination of unnecessary duplication and redundancy.

5. Attainability

Attainability is the assurance that the minimum essential supplies and services required to execute operations will be available. Attainability is the point at which the CCDR or subordinate JFC judges that sufficient supplies, support, distribution capabilities, and LOC capacity exist to initiate operations at an acceptable level of risk. It is also that point at which logistic capabilities exist at a level that will allow the transition of operations between phases.

6. Sustainability

Sustainability is the ability to maintain the necessary level and duration of operational activity to achieve military objectives. Sustainability is a function of providing for and maintaining those levels of ready forces, materiel, and consumables necessary to support military effort. Sustainability is focused on the long-term objectives and requirements of the supported forces.

7. Survivability

Survivability is the capacity of an organization to prevail in the face of potential threats. To ensure continuity of support critical logistic infrastructure must be identified and plans developed for its protection. Survivability is directly affected by dispersion, design of operational logistic processes and the allocation of forces to protect critical logistic infrastructure.

III. Support Plan Development

Ref: JP 4-0, Joint Logistics (Jul '08), pp. III-7 to III-10.

The purpose of support planning is to determine the sequence of the personnel, logistics, and other support required to provide supply, maintenance operations, deployment and distribution, HSS, engineering, logistic services and operational contract support in accordance with the CONOPS. Support planning is conducted in parallel with other planning, and encompasses such essential factors as EA identification; assignment of responsibility for base operating support; airfield operations; HSS; aeromedical evacuation; personnel services; handling of prisoners of war and detainees; theater general engineering policy; logistic-related environmental considerations; support of noncombatant evacuation operations and other retrograde operations; disposal; and nation assistance.

Support planning is primarily the responsibility of the Service component commanders and begins during CONOPS development. Service component commanders identify and update support requirements in coordination with the Services, USJFCOM, DLA, and USTRANSCOM. They initiate the procurement of critical and low-density inventory items; determine HNS availability; develop plans for asset visibility; and establish phased delivery plans for sustainment in line with the phases and priorities of the CONOPS. They develop plans for battle damage repair; retrograde of repairables; container management; force and LOC protection; and transportation and support that are aligned with the CONOPS. Service component commanders continue to refine their sustainment and transportation requirements as the force providers identify and source force requirements.

During distribution planning, the supported CCDR and USTRANSCOM resolve gross distribution feasibility questions impacting intertheater and intratheater movement and sustainment delivery. USTRANSCOM and other transportation providers identify air, land, and sea transportation resources to support the approved CONOPS. These resources may include apportioned intertheater transportation, GCC-controlled theater transportation, and transportation organic to the subordinate commands. USTRANSCOM and other transportation providers develop transportation schedules for movement requirements identified by the supported commander. A transportation schedule does not necessarily mean that the supported commander's CONOPS is transportation feasible; rather, the schedules provide the most effective and realistic use of available transportation resources in relation to the phased CONOPS. Mobilization planning includes two processes: the military mobilization process by which the Nation's Armed Forces are brought to an increased state of readiness, and the national mobilization process of mobilizing the national economy to meet non-defense needs as well as sustaining the Armed Forces across the range of military operations.

A. Transportation Feasibility

Transportation refinement simulates the planned movement of resources that require lift support to ensure that the plan is transportation feasible. The supported commander evaluates and adjusts the CONOPS to achieve end-to-end transportation feasibility if possible, or requests additional resources if the level of risk is unacceptable. Transportation feasibility determination will require concurrent analysis and assessment of available strategic and theater lift assets, transportation infrastructure, and competing demands and restrictions.

B. Logistics Supportability Analysis (LSA)

The LSA, as outlined in CJCS Instruction (CJCSI) 3110.03C, Logistics Supplement to the Joint Strategic Capabilities Plan (JSCP), provides a broad assessment of core logistic capabilities required to execute the CCDR plans. The LSA is a critical plan assessment tool that seeks to define the total unconstrained logistical requirement for execution of a CONOPS. The LSA findings should highlight deficiencies and their associated risk to supporting theater operations.

The LSA assesses each core logistic capability, and is usually accomplished as part of plan development and updated during plan assessment:

- **Critical Items.** Critical supplies and materiel must be identified early in the planning process. Critical items are supplies vital to the support of operations that are in short supply or are expected to be in short supply. Critical items may also be selected mission-essential items that are available but require intense management to ensure rapid resupply for mission success.
- **Limitations.** Logistic planners must understand the limiting factors affecting deployment, sustainment, and redeployment degrading the ability to support a campaign or OPLAN. Identifying limitations en route to or within the theater is the first step in coordinating activities to avoid overloading LOCs. Traditionally, limited unloading capacities at ports and airfields, lack of asset visibility, and limited inland transportation have constrained the operational reach of combat forces.
- **Logistic Outsourcing**. Planning for the use of contracted capabilities is a complex undertaking. It must address both contracting capability and the management of contractor personnel. Planning for contract support is complicated by the fact that support flows from inside and outside the theater. Detailed planning should be done for both contracting support (contracting support plan) and contractor (personnel) integration (either integrated into appropriate functional areas of the plan or in a separate contractor integration plan annex). Such plans need to be at a level of detail appropriate to ensure contract support is fully integrated and on par with forces planning (e.g., in TPFDD). Planning should identify sources of supplies and services from civilian sources and integrate them with operational requirements. Contract support in a JOA is provided by theater support, external support (e.g., Navy fleet husbanding support and Defense Energy Support Center [DESC] fuels contracts), and systems support contracts. Note: Refer to JP 4-10, Operational Contract Support, for additional information.
- **Threat.** Logistic units and installations are also high-value assets that must be safeguarded by both active and passive measures. Active measures must include a defense plan for logistics with provisions for reinforcement and fire support. Passive measures include dispersion, physical protection of personnel and equipment, deception, and limiting the size of an installation to what is essential for the mission. Although the physical environment will most often only degrade logistic capabilities rather than destroy them, it must be considered when planning. Logistic operations are particularly vulnerable to weapons of mass destruction (WMD) that deny, temporarily hamper, or restrict the use of critical infrastructure (e.g., aerial ports and seaports of embarkation, APODs/SPODs) and prepositioned assets. Survivability in a CBRN environment presents additional challenges and will dictate planning for dispersion and the allocation of protective forces at critical nodes of the logistic infrastructure – particularly within the theater. Decentralization and redundancy are critical to the safety of the logistic system.

C. Logistics Synchronization Matrix

One product of detailed planning is a synchronization matrix, which allows the CCDR and his staff to display many of the known activities of the operation by phases, functional areas and operating systems. It also allows the CCDR to assign responsibility for task accomplishment and identify metrics for future execution monitoring. The joint logistic concept of support is synchronized with the OPLAN. Particular attention is given to linking critical joint logistical tasks and responsibilities to key operational objectives and vice versa. The joint logistician develops his own logistic synchronization matrix (or decision support tool) as part of joint logistic detailed planning, which can assist in identifying logistical requirements matched to force deployment and sustainment actions, operational phasing, scheme of maneuver, and the generation of logistic theater capabilities.

IV. Joint Logistic Planning Considerations

Ref: JP 4-0, Joint Logistics (Jul '08), pp. III-11 to III-14.

Military operations require specific logistical support starting at the strategic level in the national industrial base and ending at the tactical level where required sustainment is delivered on time, at the right place, in the right quantity. The principal focus of joint logistic planning is at the operational level. The challenge for logisticians is to link strategic resources to tactical unit requirements. Joint logistic planning is the accurate identification of future requirements and the development of a scheme or method of meeting those requirements through the synchronization of logistic capabilities and resources in time and space. The objective of joint logistic planning is to fully integrate and coordinate support and operational execution to ensure sustained operational readiness of the joint force.

Organizing for Joint Logistic Planning

Operations and logistics are inseparable. After the execution of a joint operation, the CCDR's planning generally occurs in three distinct but overlapping timeframes and organizational elements: future plans, future operations, and current operations. Logisticians may not be on any planning cell on a full time basis, however, a coordinated staff battle rhythm, information technology, and staff management may facilitate support to the numerous planning or coordination cells. Operations and logistics are most effectively integrated as part of a collaborative planning process that includes subordinate component commands, supporting commands and global providers. Collaborative, inclusive planning helps prevent unnecessary duplication or overlap of logistical functions among the Service component commands, and ensures early identification of risks associated with shortfalls in support capabilities.

Demands of an Expanding Force

Execution of an OPORD or campaign plan or response to a crisis may be accompanied by general expansion of the Armed Forces of factors based on the mission and environment and maintain the flexibility to adjust planning factors and resupply methods as circumstances dictate. Historically, demand for items increases faster than the supply system can provide, and special management actions might become necessary. To anticipate campaign priorities, planners must: provide instructions or guidance for redistributing common-use assets from low to high-priority organizations within the command; obtain assets from external sources with lower priority needs; control the allocation of new assets in short supply; and provide efficient means to retrograde, repair, and then reissue critical items.

Balancing Push and Pull Resupply

Automatic (push) resupply works best for commodities and classes of materiel with constant usage rates (e.g., rations). It is particularly useful for establishing and maintaining the stocks of common-user items, which may then be distributed within the theater. Requisitioning (pull) is preferable for variable usage rate requirements (e.g., repair parts). Properly used and regulated, a combination of push and pull resupply will reduce unused or wasted space by adding predictability as well as combining compatible loads, thus resulting in a more effective as well as more efficient use of transportation assets and the logistic footprint in-theater. Current logistic initiatives are designed to further reduce the logistic footprint, increase the velocity and visibility of resupply, and emphasize pull resupply for maximum efficiency. In this regard, planners must realize that for certain commodities such as repair parts and major end-items, the Services have oriented their logistic system to a pull system, heavily reliant on information systems and a rapid, time-definite distribution system.

IV. Executing Joint Logistics

Ref: JP 4-0, Joint Logistics (Jul '08), chap. IV.

The term "executing joint logistics" is used to describe actions and operations conducted by joint logistic forces in support of the JFC mission. Force reception, theater distribution, and MA are examples of joint logistic operations. Since joint logistic operations span the strategic, operational, and tactical levels, the transition from planning to execution is critical. In today's ever-changing operating environment, planning and executing operations often occur simultaneously. Even though these two functions may be running concurrently, it is critical that planning outputs serve as inputs to the execution function.

I. Joint Logistic Execution

JFCs must be able to adapt to evolving mission requirements and operate effectively across a range of military operations. These operations differ in complexity and duration. The joint logistician must be aware of the characteristics and focus of these operations and tailor logistical support appropriately. This range of military operations extends from shaping activities to major operations and campaigns. Joint logisticians must have a general understanding of the diversity, range, and scope of military operations and understand their role in each type of operation.

A. Military Engagement, Security Cooperation, and Deterrence

Shaping activities include military engagement, security cooperation, and deterrence. Developing mutually supportive relationships to enhance coordination between regional partners and CCDRs is an important enabler for joint logistic operations. The US and multinational partners collaborate in order to expand mutual support and leverage each others' capabilities to quickly respond to future contingencies. Effective joint logistic operations in peacetime provide the foundation for an expanded role in later crisis and provide additional warfighting flexibility. Specific issues that can be addressed in peacetime include:

- Securing interagency approvals and permissions, normally through the country team
- Address partner nation (PN) and regional sensitivities, changing politics, and overall stability
- Determining optimal presence and posture: Persistent DOD presence in other nations is generally less supported by both country teams and partner nation governments. Maintaining a low visibility signature to US DOD presence and activities is often the only way we can secure requisite interagency and PN permissions. In some instances, interagency and/or PN desires/mandates not only limit/restrict US military presence, but also apply to US civilian contractors.
- Developing formal agreements/permissions between the US and many developing nations (e.g., status-of-forces agreements, ACSAs, etc.): US law and military regulations often involve long approval processes and restrictions on the types of funding authorized.

B. Crisis Response and Limited Contingency Operations

Crisis response and limited contingency operations are usually single, small-scale, limited-duration operations. Many of these operations involve a combination of military forces and capabilities in close cooperation with other government agencies, IGO and NGO elements. Logisticians must understand multinational and interagency logistical capabilities and coordinate mutual support, integrating them into the joint operation when appropriate. Efforts during peacetime shaping operations to develop partner capacities can pay dividends in these types of operations. Many crisis response missions, such as foreign humanitarian assistance and disaster relief operations, require time-sensitive sourcing of critical commodities and capabilities, and rapid delivery to the point of need. In these operations, joint logistics is most often the main effort. Civil support refers to the unique DOD ability to provide support to civil authorities. DOD responds to requests for support under the National Response Framework to civilian authorities. Upon approval by the SecDef, or at the direction of the President, DOD resources may be used to support federal, state, local, and tribal authorities. These operations frequently involve supplying food and water, providing medical support, aeromedical evacuation, creating temporary shelter, providing contracting support, conducting distribution operations and assisting in the evacuation of the populace. In the event of an incident involving CBRN or high-yield explosives, joint logistics operations may also involve providing specific consequence management support, such as emergency clearance of debris and restoration of essential public services. For other capabilities, such as MA, state and local medical examiners or coroners normally maintain jurisdiction. When this is the case, DOD should be prepared to support as needed.

C. Major Operations or Campaigns

Major operations or campaigns typically involve the deployment, sustainment, and retrograde of large combat forces. Joint logisticians develop support plans for the duration of the operation, as well as the return of equipment to CONUS or other locations. These plans often leverage contractor support to alleviate logistical capability shortfalls. The primary challenges for logisticians during these types of operations are gaining visibility of the requirements, sensing competing priorities and adjusting continuously as the situation unfolds to ensure sustained readiness over time. A critical planning requirement during major operations is to plan for the transition to phase IV (Stability), and phase V (Enable Civil Authority), where logisticians will have competing requirements to include supporting stability operations, providing basic services and humanitarian relief, and assisting reconstruction efforts, while redeploying a large number of forces and equipment. The retrograde of contaminated materiel will require special handling to control contamination and protect the force and mission resources from CBRN hazards.

D. Concluding Joint Logistic Operations

Joint logistic operations are always ongoing, but it is possible that some logistic operations could be complete before the operation has been completed. For example, force reception operations could be complete when forces have moved to the tactical assembly area, have been placed under the control of the commander for integration and employment, and no other forces are flowing into the JOA. It is important for joint logisticians to monitor these transitional activities and ensure logistical resources used for the completed actions are either given new tasks, or the resources are redeployed back to home station. When operations are complete, joint logisticians should participate in the lessons learned process to review processes, roles, authorities, and the execution of the operation.

II. Framework for Joint Logistics

Ref: JP 4-0, Joint Logistics (Jul '08), pp. IV-3 to IV-5.

The CCDR's logistic staff must be able to rapidly and effectively transition from peacetime/planning activities to monitoring, assessing, planning, and directing logistic operations throughout the theater. This transition may occur through the directed expansion of the joint logistics operations center (JLOC) and/or the CCDR's joint deployment and distribution operations center (JDDOC). The CCDR's or JFC's staff is augmented (either physically or virtually) with representatives from Service components, USTRANSCOM and other supporting CCDRs, CSAs, and other national partners or agencies outside the command's staff. For example, each GCC has established a JDDOC to synchronize and optimize the flow of arriving forces and materiel between the inter-theater and intratheater transportation. As the operational tempo increases during a contingency or crisis, additional joint logisticians and selected subject matter experts (maintenance, ordnance, supply, etc.) can augment JDDOCs and use established networks and command relationships instead of creating new staffs with inherent startup delays and inefficiencies. This expanded organization must be organized and situated to ensure increased coordination and synchronization of requirements in the deployment and distribution process. This organization must have clear roles and responsibilities between the various elements and clearly understood relationships between the logistical elements and the combatant command staff.

- **Technology.** Logisticians use a variety of automated tools to assist in planning and execution.
- **Achieving Situational Awareness**. A role of the joint logistician is to support the CCDR in achieving situational awareness in order to make decisions, and disseminate and execute directives. Maintaining situational awareness requires maintaining visibility over the status and location of resources, over the current and future requirements of the force, and over the joint and component processes that deliver support to the joint force.
- **Battle Rhythm.** The combatant commander will establish a battle rhythm for the operation along with mechanisms for establishing and maintaining visibility for all functional areas, to include logistics. The joint logistician must develop a supporting battle rhythm for the sustainment staff that builds off the JFC battle rhythm.
- **Joint Logistic Boards, Offices, Centers, Cells, and Groups**. *See following page (p. 5-16) for further discussion.*
- **Synchronization Matrix.** A synchronization matrix or decision support tool/template serves to establish common reference points to help assess the progress of an operation. Joint logisticians may use a matrix to assess expected progress against actual execution and recommend adjustments as needed. A logistic synchronization matrix is built around the concept of the operation, and normally contains the phasing of the operation over time along the horizontal axis. The vertical axis normally contains the functions that the joint logistician is responsible to integrate as part of a concept of support.
- **Commander's Critical Information Requirements (CCIR).** Joint logisticians must ensure that CCIRs are a part of every operational update, and must ensure that functions, resources or processes directly linked to CCIRs are given highest priority. Operational adjustments or branch plans may be necessary if CCIRs cannot be collected to ensure mission success. Joint logisticians will most often use friendly forces information requirements to guide decision making, those requirements are often a direct reflection of resources (force availability, unit readiness, or materiel availability). *See p. 1-29 for further discussion.*

III. Joint Logistic Boards, Offices, Centers, Cells, and Groups

Ref: JP 4-0, Joint Logistics (Jul '08), app. C.

There are a number of logistic boards, offices, centers, cells, and groups that reside at the strategic and operational levels that can be used to resolve joint logistic issues during operations. These enduring or temporary organizations may be staffed on a permanent, full time basis, such as the JLOC at the Joint Staff J-4, or on a temporary basis, to resolve specific strategic and operational gaps, shortfalls, or the impact of competition with another supported commander's concurrent operations.

1. Strategic-level Boards, Offices, and Centers

Strategic-level joint logistic boards, offices, and centers provide advice or allocation recommendations to the CJCS concerning prioritizations, allocations, policy modifications or procedural changes.

- Joint Materiel Priorities and Allocation Board (JMPAB)
- Joint Transportation Board (JTB)
- Joint Logistics Operations Center
- Deployment and Distribution Operations Center (DDOC)
- Defense Health Board (DHB)
- Defense Medical Standardization Board (DMSB)
- Global Patient Movement Requirements Center (GPMRC)
- Armed Services Blood Program (ASBP)

2. Operational Joint Logistic Boards, Centers, and Cells

Operational level joint logisticians must provide advice and recommendations to the supported CCDR concerning prioritizations, allocations, or procedural changes based upon the constantly changing operational environment.

- Joint Logistics Operations Center
- Joint Deployment and Distribution Operations Center
- Combatant Commander Logistic Procurement Support Board (CLPSB)
- Joint Acquisition Requirements Board (JARB)
- Joint Civil-Military Engineering Board (JCMEB)
- Joint Environmental Management Board (JEMB)
- Joint Facilities Utilization Board (JFUB)
- Logistics Coordination Board
- Theater - Joint Transportation Board (T-JTB)
- Joint Movement Center (JMC)
- Theater Patient Movement Requirements Center (TPMRC)
- Joint Patient Movement Requirements Center (JPMRC)
- Joint Blood Program Office (JBPO)
- Joint Petroleum Office (JPO)
- Sub-area Petroleum Office
- Joint Mortuary Affairs Office (JMAO)
- Explosive Hazards Coordination Cell (EHCC)

Chap 5

V. Controlling Joint Logistics

Ref: JP 4-0, Joint Logistics (Jul '08), chap. V.

Control of joint logistics involves organizing the joint staff and operational level logistic elements and their capabilities to assist in planning and executing joint logistic support operations, integrating and synchronizing responsibilities, designating lead Service responsibilities and developing procedures to execute the CCDR's directive authority for logistics (DAFL) when required. While logistics remains a Service responsibility, there are processes and tasks that must be considered when developing a concept of support in order to optimize joint logistic outcomes.

Logistic Control Options

The need for rapid and precise response under crisis action, wartime conditions, or where critical situations make diversion of the normal logistic process necessary in the conduct of joint operations, the CCDR's logistic authority enables him to use all logistic capabilities of all forces assigned as necessary for the accomplishment of the mission. The President or SecDef may extend this authority to attached forces when transferring those forces for a specific mission and should specify this authority in the establishing directive or order. The CCDR may elect to control logistics through a tailored and augmented J-4 staff or through a subordinate logistics organization.

I. Authorities and Responsibilities

Title 10, USC, and DODD 5100.1, Functions of the Department of Defense and Its Major Components, describe the statutory requirements for each Military Department to provide logistical support to assigned forces. Title 10, USC, also describes the basic authority to perform the functions of command that include organizing and employing commands and forces, assigning tasks, designating objectives, and "giving authoritative direction to subordinate commands and forces necessary to carry out missions assigned to the command." This authority includes all aspects of military operations, joint training, and logistics.

Combatant Command Authority (COCOM)

COCOM over assigned forces is vested only in the commanders of combatant commands by Title 10, USC, and cannot be delegated or transferred. This authority over assigned forces includes DAFL, which gives the CCDR the authority to organize logistic resources within theater according to the operational needs.

Directive Authority for Logistics (DAFL)

Commanders of combatant commands exercise authoritative direction over logistics, commonly referred to as DAFL, in accordance with Title 10, USC, Section 164. The CCDR may delegate directive authority for as many common support capabilities to a subordinate JFC as required to accomplish the assigned mission. For some commodities or support services common to two or more Services, one provider may be given DOD EA responsibility by the SecDef or the Deputy SecDef. However, the CCDR must formally delineate this delegated authority by function and scope to the subordinate JFC or Service component commander. The exercise of DAFL by a CCDR includes the authority to issue directives to subordinate commanders, including peacetime measures necessary to ensure the following: effective execution of approved OPLANs; effectiveness and economy of operation; and prevention or elimination of unnecessary duplication of facilities and overlapping of functions among the Service component commands.

See p. 1-11 for further discussion.

Joint Logistics

The President or SecDef may extend this authority to attached forces when transferring forces for a specific mission, and should specify this authority in the establishing directive or order.

A CCDR's directive authority does not: discontinue service responsibility for logistic support; discourage coordination by consultation and agreement; or disrupt effective procedures or efficient use of facilities or organizations. Unless otherwise directed by the SecDef, the Military Departments and Services continue to have responsibility for the logistic support of their forces assigned or attached to joint commands, subject to the following guidance.

- Under peacetime conditions, the scope of the logistic authority exercised by the CCDR will be consistent with the peacetime limitations imposed by legislation, DOD policy or regulations, budgetary considerations, local conditions, and other specific conditions prescribed by SecDef or CJCS.
- Under crisis action, wartime conditions, or where critical situations make diversion of the normal logistic process necessary, the logistic authority of CCDRs enables them to use all facilities and supplies of all forces assigned to their commands as necessary for the accomplishment of their missions. The President or SecDef may extend this authority to attached forces when transferring those forces for a specific mission and should specify this authority in the establishing directive or order.

Administrative Control (ADCON)

Administrative control (ADCON) is the direction or exercise of authority over subordinate or other organizations with respect to administration and support, to include the organization of Service forces and control of resources and equipment. ADCON is synonymous with the administration and support responsibilities identified in Title 10, USC, as previously mentioned.

Executive Agent (EA)

The SecDef or Deputy Secretary of Defense may designate a DOD EA and assign associated responsibilities, functions, and authorities within DOD. The head of a DOD component may be designated as a DOD EA. The DOD EA may delegate to a subordinate designee within that official's component the authority to act on that official's behalf for any or all of those DOD EA responsibilities, functions, and authorities assigned by the SecDef or Deputy SecDef. EA designations are related to, but not the same as, CCDR lead Service designations discussed below.

For additional information on EA, refer to JP1, Doctrine for the Armed Forces of the United States, for supply commodity related EAs, refer to JP 4-0 Appendix B, Supply Commodity Executive Agents, and for logistic-related EAs, refer to JP 4-0 Appendix D, Department of Defense Logistics-Related Executive Agents.

Lead Service

The CCDR may choose to assign specific common user logistic functions, to include both planning and execution to a lead Service. These assignments can be for single or multiple common logistical functions, and may also be based on phases and/or locations within the AOR. GCC lead Service assignments are normally aligned to Office of the Secretary of Defense-level EA designations, but this may not always be the case. For example, in circumstances where one Service is the predominant provider of forces and/or the owner of the preponderance of logistic capability, it may be prudent to designate that Service as the joint logistic lead. It would be rare for one Service logistic organization to have all the capabilities required to support an operation, so the CCDR may augment the lead Service logistic organization with capabilities from another component's logistic organizations as appropriate.

II. Logistics Directorate, J-4

The CCDR must have the right resources to plan, execute, and control logistics within the joint operational area. The J-4 is the CCDR's principal staff organization responsible for integrating logistics planning and execution in support of joint operations. The J-4 staff executes its' responsibilities by integrating, coordinating, and synchronizing Service Component logistic capabilities in support of joint force requirements. The J-4 is also responsible for advising the JFC of the logistic support that can be provided, and for optimizing available resources to provide the most effective joint outcomes by fusing information to facilitate integrated, quality decision-making. Although the organizational considerations outlined below could apply to a CCDR's J-4 staff, they will most frequently be applied to subordinate joint force J-4 organizations. In addition, the J-4 and other logisticians support the J-3 lead in the planning and executing of requirements for the JRSOI process and base support installations planning and sustainment.

The J-4 is responsible for executing and controlling joint logistics, and should organize to respond to anticipated or on-going operations:

Joint Logistics Operations Center (JLOC)

The J-4 should establish a JLOC to monitor and control the execution of logistics in support of on-going operations. The JLOC is an integral part of the CCDR's operations element and provides joint logistics expertise to the J-3 operations cell. The JLOC is tailored to the operation and staffed primarily by the J-4 staff.

Joint Deployment Distribution Operations Center (JDDOC)

USTRANSCOM, as the DPO, through its Deployment and Distribution Operations Center (DDOC), collaborates with JDDOCs to link strategic deployment and distribution processes to operational and tactical functions in support of the warfighter. The geographic CCDRs are responsible for implementing their JDDOC core structure. The JDDOC is an integral organization of the GCC's staff, normally under the direction of the J-4 and collocated with the JLOC during operations. However, the GCC can place the JDDOC at any location required or under the operational control of other command or staff organizations. The JDDOC can reach back to the national partners to address and solve deployment and distribution issues for the CCDR and can have the capability to develop deployment and distribution plans, integrate multinational and/or interagency deployment and distribution, and coordinate and synchronize the movement of sustainment in support of the CCDR's priorities.

Fusion Cell

Synchronizing and integrating the many joint logistics functional capabilities, multinational and interagency capabilities and operational contract support may require the J-4 to establish a location or center where the requirements, resources, and processes can come together in a way that provides knowledge to effect quality decision-making. This fusion of information is essential to effective logistics support and critical to enabling the J-4 to "see the logistics battlefield" with clarity.

Joint Logistics Boards, Centers and Offices

The CCDR may also establish boards, centers and offices to meet increased requirements and to coordinate logistic efforts (e.g., subarea petroleum office, joint facilities utilization board, joint mortuary affairs office).

See p. 5-16 for further discussion.

III. Logistics Execution Organizations

The fundamental role of joint logistics is to integrate and coordinate logistic capabilities from Service, agency and other providers of logistic support, and to facilitate execution of the Services' Title 10, USC responsibilities while supporting the ever-changing needs of the JFC.

A. Service Logistic Control Structures

The Services' operational-level logistic control structures form the basis for joint operations, thus it is important understand how each Service and US Special Operations Command (USSOCOM) conducts logistics at the operational level. The following paragraphs briefly describe those logistic C2 capabilities and are fundamental to understanding Service and special operations forces (SOF) logistics capabilities.

See following pages (pp. 5-22 to 5-23) for further discussion.

B. US Transportation Command (USTRANSCOM)

USTRANSCOM serves as the single coordination and synchronization element on behalf of and in coordination with the JDDE community of interest to establish processes to plan, apportion, allocate, route, schedule, validate priorities, track movements, and redirect forces and supplies per the supported commander's intent. This coordination and synchronization does not usurp the supported CCDR's Title 10, USC, responsibilities but drives unity of effort throughout the JDDE to support CCDRs. The supported CCDR is responsible to plan, identify requirements, set priorities, and redirect forces and sustainment as needed to support operations within the respective AOR. USTRANSCOM exercises responsibility for planning, resourcing, and operating a worldwide defense transportation system in support of distribution operations, to include reviewing taskings and analyzing supported CCDR's requirements for transportation feasibility, and advising on changes required to produce a sustainable force deployment. During the deployment, sustainment, and redeployment phases of a joint operation, CCDRs coordinate their movement requirements and required delivery dates with USTRANSCOM, and supported CCDRs are responsible for deployment and distribution operations executed with assigned/attached force in their respective AORs.

C. US Joint Forces Command (USJFCOM)

USJFCOM is DOD's primary joint force provider and is involved in deployment operations under its charter to provide trained and ready joint force assets to geographic CCDRs as they require them. USJFCOM fulfills this responsibility by performing four distinct functions: joint force trainer, integrator, primary joint force provider, and joint deployment process owner.

D. Defense Logistics Agency (DLA)

DLA manages, integrates and synchronizes suppliers and supply chains to meet the requirements of the Armed Forces of the US, military allies and coalition partners. When directed, DLA also supports interagency and non-DOD organizations by providing humanitarian and natural disaster relief. DLA provides the equipment, supplies and services needed for sustained logistic readiness, and supports major aviation, ground and maritime systems by providing tailored logistic support, by optimizing investment strategies and by capitalizing on commercial business practices. In addition, DLA has EA responsibilities for subsistence, bulk fuel, construction and barrier materiel, and medical materiel. DLA provides a continuous forward presence through its regional commands in the Pacific, Europe, and Southeast Asia, and has liaison officers attached to every combatant command staff to assist with planning, exercises and current operations. DLA's contingency support teams are also deployed to enhance theater distribution to meet the warfighter's needs.

E. Defense Contract Management Agency (DCMA)

DCMA is the combat support agency responsible for ensuring major DOD acquisition programs (systems, supplies, and services) are delivered on time, within projected cost or price, and meet performance requirements. DCMA is a combat support agency whose major role and responsibility in contingency operations is to provide contingency contract administration services for external and theater support contracts and for selected weapons system support contracts with place of performance in the operational area and theater support contracts when contract administration services is delegated by the procuring contracting officers.

F. Defense Security Cooperation Agency (DSCA)

DSCA serves as the DOD focal point and clearinghouse for the development and implementation of security assistance plans and programs. DSCA manages major weapon sales and technology transfer issues, budgetary and financial arrangements, legislative initiatives and activities, and policy and other security assistance matters. DSCA has oversight responsibilities for DOD elements in foreign countries responsible for managing security assistance programs, and oversees the DOD Humanitarian Assistance Program that provides nonlethal property to authorized recipients. DSCA arranges DOD funded and space available transportation for nongovernmental organizations for delivery of humanitarian goods to countries in need; coordinates foreign disaster relief missions; and, in concert with DLA, procures, manages, and arranges for delivery of humanitarian daily rations and other humanitarian materiel in support of US policy objectives.

IV. Technology

The rapid advance of technology, if leveraged effectively, can enable the CCDR to effectively control logistics within the operational area. Technology, in the form of information systems, decision support tools and evolving communications capabilities can improve visibility of logistic processes, resources and requirements and provide the information necessary to make effective decisions. These technologies can also contribute to a shared awareness that enables the CCDR to focus capabilities against the joint warfighter's most important requirements, and can be used to more effectively capture source data, make data more accessible within the public domain, and integrate data into tools or applications that enable effective decision-making. Logistics operations rely on a variety of Service and agency information systems to gather the data necessary for planning, decision-making, and assessment.

The Global Combat Support System – Joint (GCSS-J)

GCSS-J is the primary information technology application used to provide automation support to the joint logistician. In order to deliver visibility over resources, requirements and capabilities, GCSS-J uses a services-oriented architecture to link the joint logistician to component, Service, multinational, and other agencies allowing all concerned to use shared data to plan for, execute and control joint logistic operations.

See JP 3-35, Deployment and Redeployment Operations, Appendix A, Enabler Tools, for more information on deployment and redeployment enabler technologies.

A. Service Logistic Control Structures

Ref: JP 4-0, Joint Logistics (Jul '08), pp. V-7 to V-9.

The Services' operational-level logistic control structures form the basis for joint operations, thus it is important understand how each Service and US Special Operations Command (USSOCOM) conducts logistics at the operational level. The following paragraphs briefly describe those logistic C2 capabilities and are fundamental to understanding Service and special operations forces (SOF) logistics capabilities.

1. Army

The overarching theater-level headquarters is the Army Service component command (ASCC). The ASCC is responsible for providing support to Army forces and common-user logistics to other Services as directed by the CCDR and other authoritative instructions. The theater sustainment command (TSC) is the logistic C2 element assigned to the ASCC and is the single Army logistic headquarters within a theater of operations. The TSC is responsible for executing port opening, theater opening, theater surface distribution and sustainment functions in support of Army forces, and provides lead Service and EA support for designated common user logistics to other government agencies, multinational forces, and NGOs as directed. Additionally the TSC, working with the JFC's J-4, or as designated or directed, is responsible for establishing and synchronizing the intratheater segment of the surface distribution system in coordination with the JDDOC with the strategic-to-theater segment of the global distribution network. The TSC rapidly establishes C2 of operational level logistics in a specified area of operations/JOA by employing one or more expeditionary sustainment commands (ESCs). The ESC provides a rapidly deployable, regionally focused, forward-based C2 capability that mirrors the organizational structure and functionality of the TSC.

2. Marine Corps

The Marine expeditionary force (MEF) is the largest force the Marine Corps employs at the operational level. The logistics combat element (LCE) that supports the MEF is the Marine logistics group (MLG). The MLG is the principal and largest Marine logistics element. The MLG is organized to provide multifunctional direct support and functional general support to logistic units and combat arms and tactical units. A standing and experienced command and control capability, as well as an operations and planning capability, are organic to the MLG. It can rapidly and seamlessly task organize, and deploy to meet MEF mission requirements. While the Marine Corps does not possess an organic capability to execute operational-level logistics, the Marine component commander may be augmented and/or may task elements of an LCE to perform limited operational-level functions. Integration with strategic level logistical support is coordinated through the operational-level Marine component commander.

3. Navy

Numbered Fleet Commanders (e.g., FIFTH FLEET, SIXTH FLEET, and SEVENTH FLEET) have operational logistics responsibilities within a CCDR's geographic boundaries. Fleet operational forces are normally organized into task forces under the command of a task force commander. The commander, logistics forces, executes tactical logistics based on numbered fleet policy, guidance and direction. The logistics task force commander normally exercises operational control (OPCON) of assigned combat logistics forces and is responsible for coordinating the replenishment of forces at sea.

4. Air Force

The air and space expeditionary task force (AETF) is the organizational structure for deployed US Air Force forces. The AETF presents a scalable, tailorable organization with three elements: a single commander, embodied in the Commander, Air Force Forces

Joint Logistics

(COMAFFOR); appropriate C2 mechanisms; and tailored and fully supported forces. The Air Force air and space operations center provides operational-level C2 of Air Force forces and is the focal point for planning, executing, and assessing air and space operations. Although the Air Force provides the core manpower capability for the Air Force air and space operations center, other Service component commands contributing air and space forces, as well as any multinational partners, may provide personnel in accordance with the magnitude of their force contribution. The Air Force air and space operations center can perform a wide range of functions that can be tailored and scaled to a specific or changing mission and to the associated task force the COMAFFOR presents to the JFC. The Air Force forces staff is the vehicle through which the COMAFFOR fulfills operational and administrative responsibilities for assigned and attached forces, and is responsible for long-range planning that occurs outside the air tasking cycle. The director of logistics (A-4) is the principal staff assistant to the COMAFFOR for JOA-wide implementation of combat support capabilities and processes, to include the coordination and supervision of force bed-down, transportation, supply, maintenance, logistic plans and programs, and related combat support activities. In general, the A-4 formulates and implements policies and guidance to ensure effective support to all Air Force forces.

5. Coast Guard (USCG)

USCG deployable units are capable of providing combat and combat support forces and as such, must be able to react rapidly to worldwide contingencies. In order to accomplish the many missions, deployable units and assets consist of high endurance cutters, patrol boats, buoy tenders, aircraft, port security units, maritime safety and security teams, maritime security response teams, tactical law enforcement teams, and the National Strike Force. Logistical support is provided through the USCG maintenance and logistic commands and their subordinate elements. When USCG units operate as part of a JTF, Coast Guard units may draw upon the logistic support infrastructure established by/for the JTF. These general support functions normally include but are not limited to the following: berthing, subsistence, ammunition, fuel, and accessibility to the naval supply systems. The Navy logistic task force commander is responsible for coordinating the replenishment, intratheater organic airlift, towing, and salvage, ship maintenance, and material control, as well as commodity management for the task force group.

6. Special Operations Forces

SOF in the US are normally under the COCOM of Commander, US Special Operations Command (CDRUSSOCOM). When directed, CDRUSSOCOM provides US based SOF to a GCC. The GCC normally exercises COCOM of assigned and OPCON of attached SOF through the commander of a theater special operations command (TSOC), a sub-unified command. When a GCC establishes and employs multiple JTFs and independent task forces, the TSOC commander may establish and employ multiple joint special operations task forces (JSOTFs) to manage SOF assets and accommodate JTF/task force special operations (SO) requirements. Accordingly, the GCC, as the common superior, normally will establish supporting or tactical control command relationships between JSOTF commanders and JTF/task force commanders. When directed, CDRUSSOCOM can establish and employ a JSOTF as a supported commander. The CDRTSOC and JSOTF J-4s are the primary logistic control authorities for SOF. Responsibilities include oversight of the core logistic capabilities. The JSOTF J-4 must ensure that JSOTF forces are supported by the Services, which is required by Title 10, USC. The JSOTF J-4 is dependent on Service and joint logistic support as the primary means of support. For rapid response operations, USSOCOM component commands will maintain the capability to support SOF elements for an initial period of 15 days. Services and/or executive agents should be prepared to support special operations as soon as possible but not later than 15 days after SOF are employed. *For additional guidance on SOF logistic operations, refer to JP 3-05, Doctrine for Joint Special Operations, and JP 3-05.1, Joint Special Operations Task Force Operations.*

V. Multinational & Interagency Arrangements

Multinational and interagency operational arrangements regarding joint logistics are bound together by a web of relationships among global providers. These relationships are critical to joint logistics success because logistical capabilities, resources, and processes are vested in a myriad of organizations which interact across multiple physical domains and the information environment, and span the range of military operations.

Multinational Operations

In today's operational environment, logisticians will likely be working with multinational partners. While the United States maintains the capability to act unilaterally, it is likely that the requirement, and the desire, to operate with multinational partners will continue to increase. Multinational logistics is a challenge; however, leveraging multinational logistical capabilities increases the CCDR's freedom of action. Additionally, many multinational challenges can be resolved or mitigated by having a thorough understanding of the capabilities and procedures of our multinational partners before operations begin. Integrating and synchronizing logistics in a multinational environment requires developing interoperable logistic concepts and doctrine, as well as clearly identifying and integrating the appropriate logistical processes, organizations, and command and control options. Careful consideration should be given to the broad range of multinational logistic support structures.

For further guidance on multinational logistics, refer to JP 4-08, Joint Doctrine for Logistic Support of Multinational Operations.

Other Government Agencies (OGAs), IGO, and NGO Coordination

Integration and coordination among military forces, OGAs, and NGOs and IGOs is different from the coordination requirements of a purely military operation. These differences present significant challenges to coordination. First, the OGA/NGO/IGO culture is different from that of the military. Their operating procedures will undoubtedly differ from one organization to another and with the Department of Defense. Ultimately, some OGAs, NGOs and IGOs may even have policies not in consonance with those of DOD. In the absence of a formal command structure, the joint logistician will need to collaborate and elicit cooperation to accomplish the mission. As in multinational operations, the benefit of leveraging the unique skills and capabilities that NGOs and IGOs possess can serve as a force multiplier in providing the joint warfighter more robust logistics.

For additional guidance on interagency, intergovernmental organization, and nongovernmental organization coordination, refer to JP 3-08, Interagency, Intergovernmental Organization, and Nongovernmental Organization Coordination During Joint Operations, Volumes I and II. For additional guidance on civil-military operations, refer to JP 3-57, Civil-Military Operations.

I. Deployment Operations Overview

Ref: FMI 3-35, Army Deployment and Redeployment (June '07), chap. I and FM 4-01.011, Unit Movement Operations (Oct '02).

The Army is transforming in an environment characterized by a wider spectrum of potential contingencies, increased uncertainty, and a more complex range of operational conditions. The situation demands swift action by the United States; through rapid strategic response, the geographic combatant commander immediately begins to neutralize the early advantages of the adversary. This race against time to establish a dominant, employable, complete spectrum military capability in the theater of operations is aimed at tipping the balance from defense to offense.

Deployment encompasses all activities from origin or home station through destination, including predeployment events, as well as intra-continental United States, inter-theater, and intra-theater movement legs. This combination of dynamic actions supports the combatant commander's concept of operations for employment of the force.

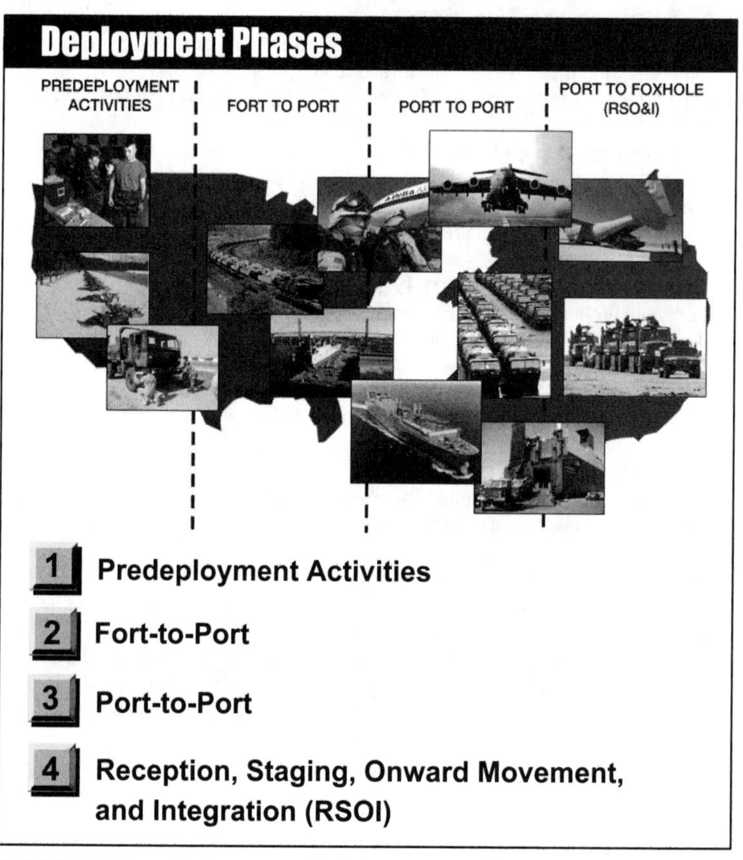

Ref: FMI 3-35, Army Deployment and Redeployment, fig. 1-2, p. 1-3.

A deploying unit undergoes a series of transformations during their movement to an area of operations (AO). Initially at its home station, personnel and equipment are separated in preparation for movement by different strategic lift modes—typically personnel via airlift and equipment by surface. Both personnel and equipment may arrive at different ports of debarkation and come together again as a combat-ready unit following the RSOI process. Experience has shown RSOI is difficult, and unprepared/untrained units find this to be the Achilles heel of their movement to the new area of operations. The manner in which the unit conducts predeployment and prepares for fort-to-port movement will have a direct impact on RSOI. Bringing the right equipment, marking and tagging equipment, proper sequencing of personnel, having and following a deployment plan, setting and meeting established timelines, and marking and packaging of hazardous materials are all elements that set the stage for a seamless transit into the new theater of operations. A detailed and integrated plan, along with a well-organized and trained team is fundamental to the success of RSOI.

I. Deployment Planning

Successful deployment planning requires knowledge of the unit's deployment responsibilities, an understanding of the total deployment process, and an intellectual appreciation of the link between deployment and employment. Steps used in planning and preparation during predeployment activities include:

1. Analyze the Mission

The mission is examined and courses of action (COA) are developed bearing in mind that the employment considerations are paramount. The primary purpose of a deployment is to provide the right force at the right place and at the right time.

2. Structure Forces

The COAs outline the ways (employment) and the means (forces) to accomplish the mission. Initially, capabilities are identified; however, as the COAs are further defined, the requirements are being translated into type units.

3. Refine Deployment Data

As forces are identified, the development of the time-phased force and deployment data (TPFDD) is begun. The supported combatant commander defines the intent for deployment which may be very specific and direct the sequence of units or just identify a general deployment timeline. In any case, the intent should clearly express how the deployment postures the force for employment.

4. Prepare the Force

Force packages are developed, ensuring the right capabilities are in the proper combinations to meet the intentions of the supported combatant commander.

5. Schedule the Movement

The supporting combatant commands must clearly and completely define their mobility requirements and milestones based on the concept of operations. The right sequencing of forces will provide the commander with the capabilities required to achieve the desired objectives. Once the strategic lift schedule is put in motion, it is difficult to change without losing the identified transportation asset and its position in the lift schedule.

Force Projection

Ref: FMI 3-35, Army Deployment and Redeployment (June '07), pp. 1-1 to 1-2.

Force projection is the military element of national power that systemically and rapidly moves military forces in response to requirements across the spectrum of conflict. It is a demonstrated ability to alert, mobilize, rapidly deploy, and operate effectively anywhere in the world. The Army, as a key member of the joint team, must be ready for global force projection with an appropriate mix of combat forces, support units, and sustainment units. Moreover, the world situation demands that the Army project its power at an unprecedented pace. Therefore, the Army must be able to defuse crises early to prevent escalation through the employment of flexible, rapidly deployable forces with sufficient depth and strength to sustain multiple, simultaneous operations.

Force Projection Processes

Force projection encompasses a range of processes including mobilization, deployment, employment, sustainment, and redeployment. These processes have overlapping timelines, are continuous, and can repeat throughout an operation. Force projection operations are inherently joint and require detailed planning and synchronization. Decisions made early in the process directly impact the success of the campaign.

1. Mobilization

Mobilization is the process of assembling and organizing resources to support national objectives in time of war and other emergencies. Mobilization includes bringing all or part of the industrial base and the Armed Forces of the United States to the necessary state of readiness to meet the requirements of the contingency.

2. Deployment

Deployment is the movement of forces to an operational area in response to an order.

3. Employment

Employment prescribes how to apply force and/or forces to attain specified national strategic objectives. Employment concepts are developed by the combatant commands (COCOM) and their component commands during the planning process. Employment encompasses a wide array of operations—including but not limited to—entry operations, decisive operations, and post-conflict operations.

4. Sustainment

Sustainment is the provision of personnel, logistics, and other support necessary to maintain and prolong operations or combat until successful accomplishment or revision of the mission or national objective.

5. Redeployment

Redeployment involves the return of forces to home station or demobilization station.

Each force projection activity influences the other. Deployment and employment cannot be planned successfully without the others. The operational speed and tempo reflect the ability of the deployment pipeline to deliver combat power where and when the joint force commander requires it. A disruption in the deployment will inevitably affect employment.

II. Deployment Phases

Ref: FMI 3-35, Army Deployment and Redeployment (June '07), pp. 1-2 to 1-5.

Deployments consist of four distinct but interrelated deployment phases. A successful deployment requires smooth and rapid implementation of each phase with seamless transitions and interactions among all of them. The four phases are not always sequential and could overlap or occur simultaneously.

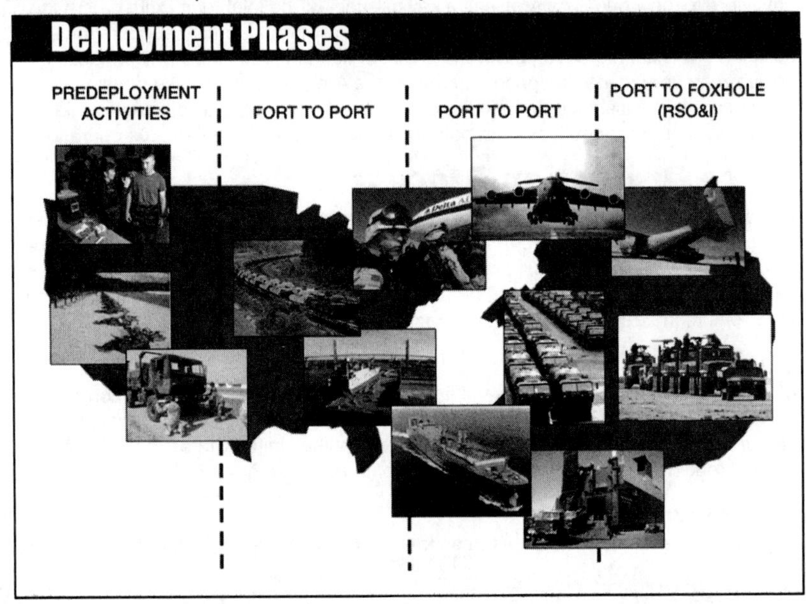

Ref: FMI 3-35, Army Deployment and Redeployment, fig. 1-2, p. 1-3.

1. Predeployment Activities

Predeployment activities are actions taken to prepare forces for deployment. An expeditionary Army demands a vigilant watch over unit preparedness to deploy beginning with predeployment activities. When units train and exercise their predeployment activities, those activities become second-nature tasks and are accomplished efficiently. Not only should units be trained, personnel must be nearly 100 percent SRP compliant. This level of readiness and training requires school-trained, long-term unit movement officers (UMO), hazardous cargo certifiers, and air load planners as directed by Army Commands and Army Service Component Commands (ASCC). In addition, units must acquire movement expertise, knowledgeable deployment support teams, joint deployment process improvement tools, and an understanding of the Joint Operation Planning and Execution System (JOPES) to enable seamless deployment operations. The Mobility Officer (specialty 882A) Program was established to embed deployment expertise in the brigade combat teams (BCT) and these officers have demonstrated the value added from the outset.

See pp. 6-9 to 6-20 for further discussion.

2. Fort-to-Port

USTRANSCOM supports the COCOM for deployment. USTRANSCOM operates the DTS and provides common user strategic transportation. The receipt of the Air Mobility Command (AMC) air tasking order and SDDC call forward order initiates port of embarkation (POE) operations and specifies the dates on which units must arrive at the POE.

Data is verified and equipment is inspected and configured for movement at the installation staging areas. Equipment is typically moved to the POEs by commercial surface transport. The installation provides supporting units to assist the deploying force, and the installation is often supplemented with the resources of a garrison support unit (GSU). A GSU is a U.S. Army Reserve (USAR) organization whose primary mission is to support mobilizing forces on an installation. Support may include load teams, pusher vehicles, maintenance teams, A/DACGs, for embark port operations, and deployment support teams. Support requirements should be identified during deployment exercises and then written into installation deployment plans.

Deploying units immediately configure for deployment with required documentation (for example, properly prepared passenger manifest, air load plans, military shipping labels, shippers' declarations of hazardous cargo, custody documentation for sensitive cargo, copy of unit equipment list, and so forth), reduce/prepare vehicles and rotary aircraft for movement, properly stow and tie down secondary loads, and construct 463L pallets and stuff containers.

The POE initiates operations to receive, stage, process, and load personnel, equipment, and vehicles onto the strategic lift platforms. SDDC verifies the resources required to accomplish the port support activity (PSA) mission and provides them from their own resources or contracts for them. Unit support requests beyond standard port operations (aircraft assembly/disassembly, convoy reception) are the responsibility of the requestor.

See pp. 6-21 to 6-28 for further discussion.

3. Port-to-Port

The port-to-port deployment phase begins with the departure from the POE and ends with arrival at the POD. During a typical deployment, personnel move by airlift, and unit equipment, vehicles, and sustainment move by sealift. Fundamental to the success of the port-to-port movement is the ability of the theater to receive and clear transiting materiel and personnel at the required rate; that is, the flow should not exceed the throughput capacity. Sufficient in-flight refueling must be available to support the flights. Also, there must be fuel at the APOE and aerial port of debarkation (APOD). Procedures for country clearance and over-flight rights, if not the agreements themselves, must be in place ready for submission at the POE.

4. Reception, Staging, Onward Movement, and Integration (RSOI)

RSOI is the process in the operational theater that supports generating combat power and delivering it to the joint force commander. The very nature of seizing the initiative demands expeditious processing of personnel and equipment throughout the deployment pipeline. Consequently, facilities for personnel and equipment reception and areas for unit staging and preparation (to include fueling) must be available on or near APOD and seaport of debarkation (SPOD) capable of 24 hour operations. Two of the essential requirements for an APOD are adequate aircraft parking and working maximum (aircraft) on ground (MOG) to meet the throughput requirements. Whether provided by theater support contracts, external support contracts (primarily the Army Logistic Civil Augmentation Program [LOGCAP]), or regionally available commercial host nation support, and/or military assets, reception support must be sufficient to immediately support the arrival of deploying units. Effective reception operations will link up personnel with their equipment while minimizing sustainment requirements for units transiting the POE. In a perfect scenario, personnel arrive and remain at the POE no longer than required, fall in on their equipment, stage, and begin onward movement. A plan to accomplish integration and maintain combat readiness must be understood, trained, and ready to implement upon arrival.

See pp. 6-29 to 6-40 for further discussion.

III. Deployment Principles

Ref: FMI 3-35, Army Deployment and Redeployment (June '07), pp. 1-1 to 1-2.

Four principles apply to the broad range of activities encompassing deployment:

1. Precision

Precision applies to every activity and piece of data. Its effect is far-reaching, and the payoff is speed. For example, precise UDLs ensure that correct lift assets are assigned against the requirement. Precision includes accurate weights, dimensions, and quantities. This degree of precision eases loading requirements and improves departure speed and safety. Precision allows units to meet the COCOM's timeline and support the COCOM's concept of employment. Bringing the application of current doctrine, realistic training, adequate support structures, and enablers together provide the framework for precision.

2. Synchronization

Just as a commander arranges activities in time and space to gain the desired effect during employment, deployment activities must be synchronized to successfully close the force. Effective synchronization of scarce lift assets and other resources maximizes their use. Synchronization normally requires explicit coordination among the deploying units and staffs, supporting units and staffs, a variety of civilian agencies, and other services. Realistic exercises and demanding training are paramount to successful synchronization.

3. Knowledge

There is a short period of time during which the deploying commander must make crucial decisions on employment. These decisions set the tone for the remainder of the deployment. Many decisions are very hard to change and have significant adverse impacts if changed; others are irrevocable. For example, knowledge and understanding of the TPFDD is imperative when making decisions on high-priority items, their sequencing, use of available time, and prioritization. Moreover, commanders must understand the entire deployment process, as well as allocating training time and placing emphasis on predeployment activities to achieve the knowledge level needed to be effective upon arrival in a new theater of operations.

4. Speed

Speed is more than a miles-per-hour metric. The proper focus is on the velocity of the entire force projection process, from planning to force closure. Critical elements of speed associated with force projection include:

- Agile (state-of-the-art) ports
- Submission of accurate information
- Safe and efficient loading
- Embedded competency
- Trained UMOs
- Timely arrival of throughput enablers
- Maintaining unit integrity
- Delivering capability rather than entire units
- Force tracking information

IV. Army Force Generation (ARFORGEN)

Army Force Generation (ARFORGEN) is the structured progression of increased unit readiness over time resulting in recurring periods of availability of trained, ready, and cohesive units. ARFORGEN uses personnel, equipment, and training to generate forces to meet current and future requirements of combatant commanders. This cyclical readiness process allows commanders to recognize that not all units have to be ready for war all the time, and units must build their Units progress through the three operational readiness cycles.

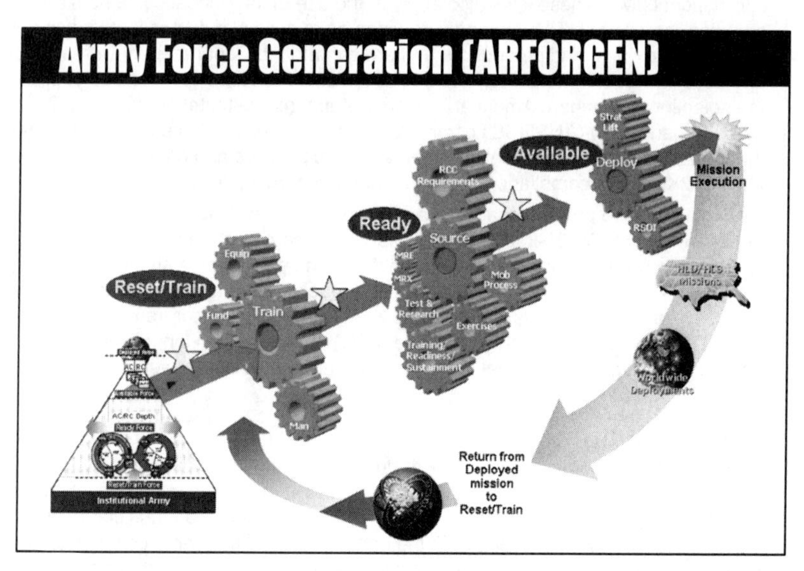

Army units will proceed through the Reset and Train, Ready and Available force pools to meet operational requirements with increased predictability. Units in the Reset and Train force pool redeploy from operations, receive and stabilize personnel, reset equipment, and conduct individual and collective training. Unit collective training is focused on core Mission Essential Task List (METL) tasks, such as offensive and defensive operations.

1. Reset and Train Pool

The Reset and Train phase culminates in a brigade-level collective training event. Units in the Reset and Train force pool are not ready or available for major combat operations. However, they should be ready to respond to homeland defense requirements and provide defense support to civil authorities at all times.

2. Ready Force Pool

Units in the Ready force pool continue mission-specific collective training and are eligible for sourcing if necessary to meet joint requirements. Their collective training is designed to focus on its directed METL, such as stability operations.

3. Available Force Pool

Units in the Available force pool are in their planned deployment windows and are fully trained, equipped, and resourced to meet operational requirements. In this way, ARFORGEN enables units to be fully-trained to conduct full-spectrum operations.

Hazardous, Classified, and Protected Sensitive Cargo

Ref: FM 4-01.011, Unit Movement Operations (Oct '02), app. D.

1. Hazardous Material (HAZMAT)

Packaging, shipping, handling, and inspecting of HAZMAT is mandated by US and international laws. These laws also apply to the use of intermodal containers and container equipment. This appendix provides an overview of doctrinal guidance and tactics, techniques, and procedures that are common to Department of Defense (DOD) and other US government agencies and organizations. This appendix also applies to the selection of standard American National Standards Institute/International Standards Organization (ANSI/ISO) commercial- or military-owned intermodal containers that meet the standards for shipment of Class I explosives and other HAZMAT. (See MIL-HDBK 138 for compliance with container standards criteria.)

HAZMAT must be properly prepared and documented IAW DOD Regulation 4500.9-R, Volume II and III; TM 38-250; and other service or command regulations. Documentation must include the total HAZMAT quantity and a certification statement stating that the HAZMAT is properly classified, described, packaged, marked, and labeled. Only specially trained individuals have authority to certify HAZMAT for transportation. Contact the Installation Transportation Officer (ITO) or Movement Control Team (MCT) for assistance in determining what certification requirements apply to each HAZMAT item being prepared for shipment.

2. Ammunition

Ammunition shipments are usually scheduled through military ammunition ports. Designated military ammunition ports serve the strategic purpose of routinely handling shipments of ammunition. To meet deployment requirements, ammunition may be moved through a commercial port. If the unit is deployed through a commercial seaport and must carry basic load ammunition with them, the MTMC manager for the port must first be notified of the intent to ship ammunition. The Joint Munitions Transportation Coordinating Activity (JMTCA) consolidates all containerized munitions movement requests for OCONUS shipment aboard common-use sealift.

3. Classified Cargo

Classified cargo is cargo that requires protection in the interest of national security. The nature of classified cargo requires that shippers and transporters handle it in a way that it be identified, accounted for, secured, segregated, or handled in a special way to safeguard it. Detailed instructions are included in DTR 4500.9R. Do not identify classified cargo on the outside of the shipping containers.

When transporting classified material, enclose it in two sealed containers, such as boxes or heavy wrappings. *Detailed instructions for packing classified material are contained in AR 380-5.*

When traveling by motor convoy, escorts must ensure constant surveillance of classified material. Classified material must stay within the escort's personal possession and observation at all times. Larger pieces of secret shipments, such as missiles, may require outside storage. If so, take special protective measures to include constant and continuous surveillance by at least one or more escorts in the area.

4. Sensitive Cargo

Sensitive cargo is cargo that could threaten public safety if compromised. Sensitive cargo must be properly secured and identified to port personnel so sufficient security can be provided. Do not identify security cargo on the outside of the shipping containers. *Detailed instructions are included in DTR 4500.9R.*

II. Predeployment Operations

Ref: FMI 3-35, Army Deployment and Redeployment, chap. 2 and FM 4-01.011, Unit Movement Operations (Oct '02).

The Army's effort to become more strategically responsive begins at home station. Predeployment activities are actions taken to prepare forces for deployment and are not limited to the deploying unit but include supporting units and the installation staff. The units' primary predeployment activities include planning, document preparation, and training. It has been noted during Deployment Excellence Award team visits that the units achieving the most success in deploying are those which the commanders are directly involved in the preparations; transportation documentation is maintained, and training is ongoing. Deployments are a function of dedicated personnel attention to detail and following the guidelines in regulations, doctrine, and other related source material.

I. Planning

The deliberate deployment planning process begins with the COCOM validating the capabilities necessary to support the contingency plan. The Secretary of Defense and the joint staff review the requirements to determine if they are valid. Joint Forces Command passes the requirements to the appropriate Service provider to task units to furnish the requested capabilities. Unit commanders and staffs analyze the requirement and determine the personnel and equipment resources necessary to accomplish the stated mission. The process may transpire over the period of several months if time allows for deliberate planning to a matter of days or even hours for crisis action planning.

A. Deployment Planning

Deliberate planning is typically used where the deployment and employment of forces is in response to anticipated contingencies. It is designed to produce a detailed operational plan for a hypothetical situation and relies heavily on a number of assumptions ranging from the threat to anticipated host nation support. Conversely, crisis action planning is accomplished in response to time-sensitive, imminent threats that may result in actual military operations. The plan is based on circumstances existing at the time planning occurs. In either situation, deliberate or crisis planning, prescribed procedures are followed to formulate and implement a response. Deployment planning is a key element of both deliberate and crisis action planning and aims at delivering the right force, at the right place, and at the right time.

Planning for deployment is based primarily on mission requirements and the time available to accomplish the mission. During deployment operations, supported combatant commanders are responsible for building and validating movement requirements; determining predeployment standards; and balancing, regulating, and effectively managing the transportation flow. Supporting combatant commands and agencies source requirements not available to the geographic combatant commander and are responsible for verifying supporting unit movement data; regulating the support deployment flow; and coordinating effectively during deployment operations. The JOPES directives, guidance, and decision support tools serve as a common framework within the Joint Planning and Execution Community (JPEC) to facilitate the deployment planning processes.

B. Movement Planning

To meet contingency support requirements, units develop deployment movement plans and standard operating procedures (SOP). An effective movement plan contains sufficient detail to prepare units to execute strategic deployments while the SOP outlines functions that should initiate actions upon notification of unit mvmt.

1. Deployment Movement Plans

Deployment movement plans define responsibilities, functions, and details for each part of a unit deployment from origin to reception in theater. There may be more than one deployment plan required depending on the number of contingencies/operations plans (OPLAN) the unit must prepare to support. Movement plans are written in a five-paragraph OPLAN format.

Deployment related information is contained in the GCCS database and is accessible through GCCS ad hoc queries or via JOPES, a GCCS application. Units deploying under a JOPES OPLAN must increment their movements consistent with OPLAN TPFDD requirements, as delineated by ULNs. At the predeployment conference the unit requests a ULN for each element. The Army Command or ASCC assigns ULNs to units.

The transportation component commands (TCC) schedule lift against the ULN to meet the EAD-LAD window. AMC publishes airflow schedules to call forward personnel and equipment to the APOE. These schedules are in GCCS. The call forward schedules are movement directives that specify when units must have their equipment at the POE to meet available-to-load dates. Based on these schedules, deploying units and intermediate command levels backward plan movements to the POE to meet the ALD. Movement directives (if published) provide windows by mode for cargo arrival at the POE.

See following page (6-14 to 6-15) for discussion on developing a deployment movement plan. Refer to FM 4-01.011 for a sample plan.

2. Installation Readiness SOP

Each installation normally has a readiness SOP that outlines the local procedures in place to be used by the staff, deploying unit, and support elements in the different stages of the deployment process.

3. Unit Movement SOP

The unit movement SOP should be in enough detail to define the day-to-day as well as alert functions. The SOP defines the duties of subordinate units/sections that will bring the unit to a higher state of readiness. Functions addressed in the SOP could include unit property disposition, supply draw, equipment maintenance, vehicle and container loading, security, marshaling procedures, purchasing authorities, unit briefings, in-transit visibility (ITV), and other applicable deployment activities.

4. Deployment Binders

In addition to the unit movement SOP, units will maintain deployment binders containing the unit deployment plan, appointment orders, training certificates, recall rosters, a current OEL, and copies of load cards and container packing lists. Units will also maintain prepared copies of transportation requests, convoy movement requests and special handling permits, and BBPCT requirements.

5. Battle Books

The battle book documents how the commander accomplishes the mission in the specified area of operation. It should include the organization and responsibilities for the unit's RSOI in the theater. The following items, if applicable, should be included in the battle book: pertinent information from the OPLAN and the TPFDD; information and photographs/schematics concerning the layout and facilities of railheads, staging areas, APODs and SPODs; maps of convoy routes within the area of operation to its employment location; and plans and locations for drawing pre-positioned equipment.

Unit Movement Dates

Ref: FM 4-01.011, Unit Movement Operations (Oct '02), chap 2.

To meet their responsibilities to support operational, exercise, and contingency plans, units develop movement plans. Normally divisions, brigades, and battalions create movement plans and companies use extracts from battalion movement plans in company operation orders. Unit movement plans are tailored to the requirements for mobilization, deployments, and exercises, which have specific goals and missions. The plans are written in operation order format and are usually an annex to an operation order. The unit plans the move using the movement plan and executes the move under an operation order.

Unit line numbers available on JOPES OPLAN reports divide the unit by transportation mode, ports of embarkation or debarkation, and dates. Dates correspond to the established C-day for the designated plan TPFDD.

The unit movement is phased by the following dates relative to C-day:

1. Ready-to-Load Date (RLD)

The RLD is the TPFDD date when the unit must be prepared to depart its origin. For AC (Active Component) units, origin is the installation and for RC units origin is the mobilization station or site.

2. Available-to-Load Date (ALD)

The ALD is the TPFDD date when the unit must be ready to load on an aircraft or ship at the POE.

3. Earliest Arrival Date (EAD)

The EAD is the earliest date that a unit, a resupply shipment, or replacement personnel can be accepted at a POD during a deployment. It is used with the latest arrival date to describe a delivery window for transportation planning. The supported combatant commander specifies the EAD.

4. Latest Arrival Date (LAD)

The LAD is the latest date when a unit, a resupply shipment, or replacement personnel can be accepted at a POD to support the concept of operations. It is used with the EAD to describe a delivery window for transportation planning. The supported combatant commander specifies the LAD.

5. Required Delivery Date (RDD)

The RDD is the date when a unit, a resupply shipment, or replacement personnel must arrive at a POD and complete off-loading to support the concept of operations. The supported combatant commander specifies the RDD.

6. Schedules

Air Mobility Command publishes airflow schedules to call forward personnel and equipment from the APOE. The call-forward schedules are movement directives that specify when units must have their equipment at the POE to meet ALDs. Based on these schedules, deploying units and intermediate command levels backward-plan movements to the POE. Movement directives (if published) provide windows by mode for cargo arrival at the POE. MTMC performs the same functions for sealift.

Developing a Deployment Movement Plan

Ref: FMI 3-35, Army Deployment and Redeployment (June '07), pp. 2-4 to 2-6.

The following paragraphs describe a recommended step-by-step process for developing deployment movement plans:

1. Identify what needs to be moved

Based upon mission, enemy, terrain and weather, troops and support available, time available, and civil considerations (METT-TC) and command guidance, deployment planning must reflect personnel, equipment, supplies, and how the unit will accomplish the move. For planning purposes, units plan to deploy with assigned personnel and on-hand equipment. Upon execution, the plan may need to be modified if additional personnel are assigned or equipment cross-leveled to bring the unit to the required readiness level. Units should plan to move their basic load of supplies to sustain their operations upon arrival in the theater. The quantities to be deployed are normally defined in OPLANs, unit or Army Command SOPs, or ASCC instructions. The UMO must have a detailed listing of each piece of equipment to be deployed based on the automated unit equipment list (AUEL). All outsize, oversize, overweight, or hazardous equipment/cargo must be identified, as it will require special considerations.

2. Identify equipment to accompany troops (TAT)

- **Yellow to accompany troops (TAT)** equipment must accompany troops and be accessible enroute. Examples include Class I basic load items, individual carry-on baggage, and weapons. For personnel traveling via commercial air, this is generally only the baggage that will fit under the seat.
- **Red TAT** items must be available at the destination before or upon unit arrival. This equipment may be sensitive cargo that requires special security or handling at the POE or POD. Red TAT must be unitized/palletized and reported on the AUEL/ deployment equipment list (DEL). Not to accompany troops (NTAT) equipment is normally shipped by surface and does not accompany the troops. It consists of all other equipment required by the unit to perform its mission.

3. Identify what needs to move by air

This could include personnel, advance parties, baggage, and some equipment. The balance of equipment normally moves by sea. For deployments supporting combatant commander OPLANs/operation orders (OPORD), the TPFDD will designate the strategic movement mode.

4. Identify hazardous (also sensitive and classified) cargo for packaging

Identify hazardous (also sensitive and classified) cargo for packaging, labeling, segregation, and placarding for movement. Codes of Federal Regulation (CFR) 49 Transportation provides guidance on the packaging, labeling, placarding, and movement of hazardous materials.

5. Develop vehicle load plans for unit equipment

Equipment that cannot be loaded on organic vehicles should be planned for movement by other means (container, commercial rail or highway, or military assets). Vehicle load plans are recorded on DD Form 1750 (Packing List) for organic vehicles and trailers carrying secondary loads. FORSCOM units may use FORSCOM Form 285-R (Vehicle Load Card) for preparing vehicle load plans. The installation transportation officer (ITO) is the installation point of contact (POC) for obtaining commercial transportation to move equipment to POE that is beyond the unit's organic capability. Unit cargo (vehicles and equipment) is prepared for shipment according to the mode of transportation. Depending on the strategic lift full reduction may or may not be required. Reduction details are normally in the SDDC port call message or the operations order for sealift. For deployment by air, reduction is determined by type of aircraft. Transportation Engineering Agency (TEA) Pamphlet 55-24 provides guidance for preparing vehicles for airlift. Vehicle modifications (that is, shelters, bumper modifications, and so forth) made by the unit which change the vehicle configuration/dimensions/weight normally must be approved by the unit's Army Command/ASCC and ultimately by TEA. Vehicle modifications must be reflected on the AUEL and DEL. Information on dimensions, weights, and cubes for all Army equipment is in CD-ROM and World Wide Web versions of TB 55-46-1. The hard copy version only contains major end items.

6. Identify Blocking, Bracing, Packing, Crating & Tie-Down (BBPCT) requirements

All crates, containers, boxes, barrels, and loose equipment on a vehicle must be blocked, braced, and tied-down to prevent shifting during transit. The POC for blocking and bracing requirements is normally the unit movement coordinator (UMC). Chapter 6 of FORSCOM/ARNG Regulation 55-1 describes policy for obtaining and stocking blocking, bracing, packing, crating, and tie-down (BBPCT) materials and related railcar loading equipment for deploying units. FM 4-01.011 provides guidance for securing loads moving by air, and FM 55-17 provides guidance for securing loads by other modes.

Additional tie down guidance is in TEA Pamphlet 55-19 and TEA Pamphlet 55-20.

7. Translate what needs to be moved into transportation terms

Personnel and equipment data are translated into transportation terminology as unit movement data (UMD) and recorded on the OEL. Upon deployment execution, units use TC-AIMS II to update the OEL and create the DEL. The UMC provides assistance to deploying units for OEL updates and DEL development.

8. Determine how the personnel and equipment will move to the POEs

Determine how the personnel and equipment will move to the POEs. In continental United States (CONUS) wheeled vehicles and tracked vehicles move via commercial rail, truck, or barge. Unit personnel usually move to the POE by military or commercial buses. Army rotary wing aircraft normally self-deploy to the POE, where they will be disassembled for shipment.

9. Prepare the unit deployment plan

Prepare the unit deployment plan. The administrative, logistical and coordination requirements for the plan must be determined. Items such as enroute medical, messing, and maintenance for movement to POEs must be coordinated and documented.

10. Maintain the movement plan

Maintain the movement plan. Update the OEL as changes occur in the OPLAN, equipment, commander's intent, or upon mission execution. The OEL is used to produce the unit's equipment manifest and military shipping labels (MSL), and errors can result in the unit's equipment being lost in the transportation system.

II. Preparation for Movement

Ref: FM 4-01.011, Unit Movement Operations (Oct '02), chap 2.

Preparation for movement is an ongoing unit activity in peacetime that continues after the unit receives a warning or alert for movement. Units normally identify deployment as a mission essential task and annotate it on their mission essential task list (METL). Predeployment activities are those tasks accomplished by Army units and installations prior to movement to POEs. During normal peacetime operations, predeployment activities involve preparation for force projection, crisis response missions, and field exercises. Units conduct routine movement training to ensure they can meet the Joint Force Commander's mission requirements.

1. Unit Alert Procedures

Division and higher level headquarters are normally alerted for missions through the JOPES procedures. Procedures for alerting subordinate units for movement are contained in higher headquarters SOPs, deployment regulations, and unit movement or deployment SOPs. These SOPs normally contain unit alert reporting requirements. Units maintain alert rosters for contacting unit personnel. Alert procedures are validated and tested according to unit SOP or other direction.

2. Identifying Support Requirements

Units generally require extensive support to prepare for movements. This support can include assistance related to equipment inspection, maintenance, property transfer and loading. It also can include assistance in the marshaling and staging areas, and help with predeployment and life support activities. These support requirements are usually identified in division and installation SOPs. Installation and non-deploying units are normally tasked to provide this support to deploying units. Additional support is available from MTMC which dispatches deployment support teams where needed.

3. Soldier Readiness Processing (SRP)

The goal of the SRP program is that all soldiers are maintained administratively ready for deployment at all times. Soldier readiness is a continuous process that involves both the unit commander and the installation staff. Headquarters, Department of the Army requires that specific administrative deployment processing requirements be checked and updated prior to individual soldier or unit movement [SRP requirements are categorized by levels ranging from Level 1 (basic movement SRP requirements) to level 4 (deployment area and mission unique SRP)]. Prior to soldier or unit movement in support of combat or contingency operations, commanders with the assistance of a soldier readiness processing team (SRP Team) physically review on-site processing requirements in levels 1 through 4 within the 30 days prior to departure. AR 600-8-101, Personnel Processing, establishes readiness requirements for each of the levels, and MACOMs and installations ensure they are met.

4. Movement Training

Units are required to have an appropriate number of personnel trained to perform special movement duties previously discussed in chapter one. These special duties include the UMO, the unit loading teams, the hazardous cargo certifying officials, and the air load planners. Each MACOM has specific requirements and policies for appointing and training personnel in these positions. Many commands and installations maintain a local capability to provide deployment training because all deployable units require personnel trained to perform these duties.

Predeployment Activities

Predeployment activities are those that units accomplish based on initial notification, warning orders, and alert orders for operations. These activities may overlap in the deployment process or occur in a different order than presented here, depending on time available between initial notification and actual deployment execution.

1. Initial Notification Activities

Following warning order receipt, the deploying unit headquarters evaluates the ability of its subordinate units to meet mission requirements. If a unit needs reorganization or augmentation, a plan is developed to meet established requirements through cross-leveling or outside augmentation. Using TC-AIMS II, personnel adjustments are made to the OEL, then the UDL and equipment and supplies adjustments are made directly to the UDL.

The deploying unit creates a UDL by identifying items from the OEL for deployment. It verifies the shipping information (size, weight, line identification number [LIN], model, and configuration) of the equipment selected for the UDL. The deploying unit also begins preparing other required deployment documentation such as HAZMAT certification.

Upon notification of a potential deployment, the unit reviews its deployment readiness status. The deploying unit's higher headquarters confirms readiness status of all its units and identifies actions needed to raise deficient units to standard. The deploying unit also begins gathering information to identify any special needs (e.g., clothing, equipment) based upon climate, location, or current unit configuration.

2. Movement Order Activities

Receipt of the movement order causes the unit to refine its movement plan based on information provided in the alert order and verifies or updates the following:

- Maintenance lead times and maintenance priorities for deploying equipment
- Requisition and personnel fill times
- Train-up completion time (if required) for unit movement personnel
- Container availability (pack, load, certify, and transport to POE) time

If the deploying unit is drawing APS, the unit deploys or prepares to deploy the APS advance party and unit representatives to the survey, liaison, reconnaissance party, and the off-load preparation party. These soldiers coordinate with the gaining command and act as liaison in preparing for reception and staging.

During warning order activities, the deploying unit continues cross-leveling equipment and submits requisitions for needed supplies that were not identified earlier. As in the initial notification, supply levels may be directed in the alert order or by the deploying unit's higher headquarters. Requisitions may be filled at point of origin and incorporated into the UDL, received at the POE and added to the UDL, or shipped separately to arrive at the POD.

Refinement of the UDL is a continuing process with the deploying unit based on unit status and changes imposed as a result of force tailoring or higher headquarters guidance. The unit verifies equipment status compared to the UDL and updates load plans, equipment dimensions and weight, and HAZMAT shipping declarations. Once corrections are made, the unit prints and applies military shipping labels (MSLs; DD Form 1387) to supplies and equipment. Additionally, the red and yellow TAT, and NTAT equipment are identified.

Unit equipment must be safeguarded IAW governing regulations and SOPs, while it is being transported to and staged at installations, marshalling areas, and POEs. Beyond usual unit safeguarding provisions, certain cargo categories require care while in transit and some special cargo categories require extraordinary protection and monitoring while in transit.

III. Training

Unit deployment training is an essential element in developing the mental agility and knowledge required for strategic responsiveness. Training is a proven way for units and individuals to acquire and improve the skills required to increase the speed of projecting combat power.

A. Collective Training

Units train in peacetime to meet unit and individual training requirements for deployment operations. Deployments can occur rapidly leaving the deploying unit with little or no time to correct training deficiencies. The objective of collective deployment training is to implant the knowledge, skills, attitudes, and abilities so it becomes a reflex activity executed with precision. Units must identify deployment as a mission essential task, annotate it on their mission-essential task list (METL), and gain proficiency. Many Army training programs offer the opportunity to include deployment training in major training events.

Emergency Deployment Readiness Exercise (EDRE)/ Sealift Emergency Deployment Readiness Exercises (SEDRE)

EDRE/SEDRE events are designed to exercise unit or command movement plans for deployment. EDREs/SEDREs may involve the unit moving to POEs and loading unit equipment on strategic sealift/airlift assets. Army Command, ASCC, installation, and brigade level commands normally have SOPs and/or deployment regulations and policies establishing subordinate unit required activities in notification hour (N-Hour) sequence for deployments.

B. Deployment Training

Units with deployment missions are required to have an appropriate number of personnel trained to perform special deployment duties. These duties include unit movement officer, unit loading teams, hazardous cargo certifying officials, and air load planners. Army Commands and ASCCs have specific policies for appointing and training personnel to assume these positions, and this is a synopsis of the more important ones.

See facing page (p. 6-17) for further discussion.

C. Route and Location Reconnaissance and Rehearsal

Reconnaissance of the route of the POEs and to predesignated POEs should be done periodically. It may be accomplished through passive means such as map surveillance or, optimally, through site visits. Walking the terrain at the power projection platform and designated port facilities allows commanders to understand space limitation, see choke points, survey facilities, understand the simultaneous nature of the operation, and visualize the deployment operation. Terrain walks can be useful as a unit level activity but are more beneficial when they involve all participating and supporting units.

IV. Installation Support

Installations are an integral part of the deployed force from home station to the foxhole. Operational deployments and rotational assignments across the globe mean installation capabilities will transcend more traditional expeditionary support requirements associated with mobilizing, deploying, and sustaining the force. More than a jump point for projecting forces, installations serve a fundamental role in minimizing their footprint through robust connectivity and capacity to fully support reach back operations.

Deployment Training

Ref: FMI 3-35, Army Deployment and Redeployment (June '07), pp. 2-7 to 2-8 and FM 4-01.011, Unit Movement Operations (Oct '02), app. J.

Units with deployment missions are required to have an appropriate number of personnel trained to perform special deployment duties. These duties include unit movement officer, unit loading teams, hazardous cargo certifying officials, and air load planners. Army Commands and ASCCs have specific policies for appointing and training personnel to assume these positions, and this is a synopsis of the more important ones.

1. Unit Movement Officer

The commander is responsible for all aspects of deployment preparation, training, and execution, and the UMO is the commander's designated representative. The UMO should be a graduate of a resident Unit Movement Officer Planning Course, appointed on unit orders, and have at least twelve months retention in the unit. In most cases, a unit movement noncommissioned officer (NCO) may augment the UMO. Commanders should consider retention in selecting UMOs and unit movement NCOs.

2. Loading Teams

Units must have personnel trained in vehicle preparation and aircraft and rail loading/unloading techniques. The type and quantity of equipment to be loaded and the time available for loading determines the composition of the team. Training is arranged through the installation UMC, and once training is completed, the load teams are put on unit orders. Load team composition is tailored to the type and quantity of equipment being loaded and time available for loading:

- **For rail movements**, a well trained team of five operators, using prefabricated tiedown devices, can complete loading and lashing of equipment on a chain equipped flatcar. Units are normally provided 72 hours for loading once the cars are spotted.
- **For air movement**, a six person team can provide efficient loading and tie down of equipment. United States Air Force Mobility Command offers the Equipment Preparation Course to train unit load teams to prepare, load and tie down unit equipment on military aircraft.

3. Hazardous Cargo Certification

Each unit is required to have at two individuals to certify hazardous cargo. The hazardous cargo certifying official is responsible for ensuring the shipment is properly prepared, packaged, and marked. The certifying official is also responsible for personally inspecting the item being certified and signing the HAZMAT documentation. Hazardous cargo certifiers must be trained at a Department of Defense (DOD) approved school on applicable regulations for all modes within the past 24 months. Upon training completion, they are authorized to certify documentation for commercial and military truck, rail, sea, and air. A common mistake occurs when the HAZMAT certifier is sent with the advance party leaving no one to accomplish the HAZMAT inspections during departure operations.

4. Air Load Planning

Air load planners are trained at Air Mobility Command (AMC) sanctioned courses to prepare, check, and sign unit aircraft load plans. Upon course completion, the individual is authorized to sign load plans; however, they must be approved by an Air Force loadmaster. The resident course is taught at the U.S. Army Transportation School but similar courses are conducted at CONUS and outside the continental United States (OCONUS) locations.

Installation facilities must readily adapt to changing mission support needs, spiraling technology, and rapid equipment fielding. Installation connectivity must also support en route mission planning and situational awareness. Education and family support will use the same installation support connectivity to sustain the morale and emotional needs of our Soldiers and their families.

U.S. Army Installation Management Command (IMCOM)

The U.S. Army Installation Management Command (IMCOM) manages Army installations worldwide. IMCOM and its garrison commanders play a critical role in ensuring successful mobilization, demobilization, and force projection operations in CONUS and OCONUS. Garrisons participate in deployment training programs, deployment operations, and are instrumental in supporting reach-back operations during unit deployments.

Installation Transportation Officer (ITO)

In CONUS, the installation transportation officer (ITO) is a pivotal participant in the force projection process providing links to FORSCOM and USTRANSCOM services, commercial integration, power projection platform operational expertise, and planning support. The UMC is appointed by the ITO to:

- Assist and monitor unit movement plans, data, and documentation
- Coordinate transportation requirements with the supporting commands for movement from home station/mobilization station to POEs
- Request strategic lift from USTRANSCOM
- Source unit deployment information and data using TC-ACCIS or TC-AIMS II
- Provide technical assistance and monitor deployment training
- Coordinate requests for convoy clearances and special hauling permits with state highway authority and the defense movement coordinator
- Arrange for and monitors the status of commercial and military shipping containers and 463L pallets
- Report movement of units as directed
- Advise unit commanders/organizations on identifying and obtaining equipment in support of unit deployments, to include BBPCT materials

At OCONUS locations, a combination of units (IMCOM, the Army Command, ASCCs, support brigades, and movement control battalions) will come together to perform the deployment support functions of the CONUS ITO.

Unit Movement Officer (UMO)

The commander appoints a UMO to prepare the unit for deployment. The UMO must know the unit's mission and the commander's intent to prepare the unit for deployment, so appropriate coordination, planning, and execution can take place. The UMO will:

- Assemble and maintain unit movement plans and documentation
- Prepare the unit for movement
- Create the unit equipment list
- Supervise the load out of the unit

Installation Deployment Processing Site

Ref: FMI 3-35, Army Deployment and Redeployment (June '07), pp. 2-10 to 2-11.

The installation deployment-processing site is a centralized location where deploying units process and assemble their equipment for movement to the POE. Deploying units begin preparing their equipment for deployment in unit marshaling areas and motor pools. Activities include preparation of required documentation (for example, transportation control and movement documents (TCMD), MSLs, hazardous cargo, vehicle preparation, preparation of sensitive cargo, building 463L pallets, and containerization of equipment.

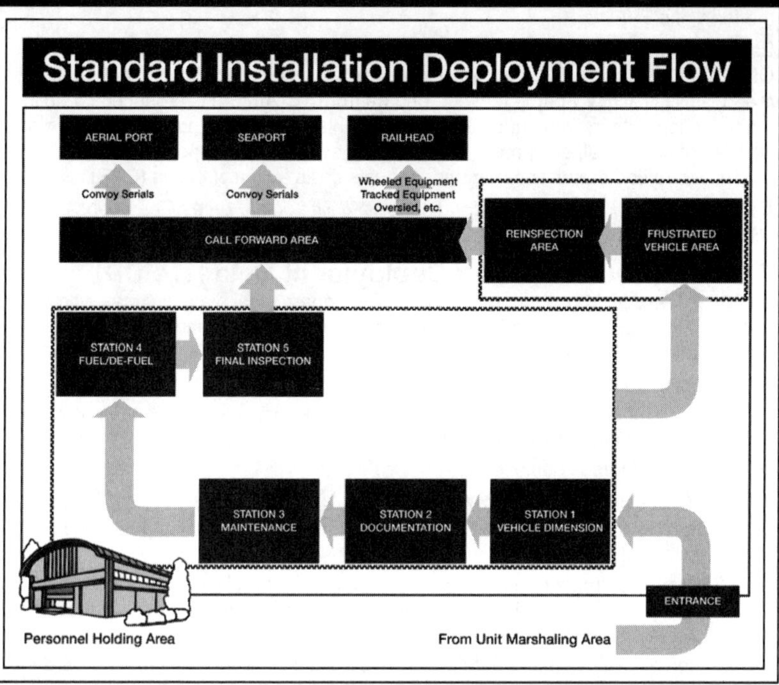

Ref: FMI 3-35, Army Deployment and Reployment, fig. 2-4, p. 2-11.

Units are sequenced by chalks or equipment type upon arrival at the deployment-processing site. Stations are setup by equipment type and in the notional site depicted above would use the following sequence to inspect the vehicles:

- **Station 1**. Conduct initial inspection for general condition of vehicle; perform a preliminary check of documentation; verify dimensions.
- **Station 2**. Inspect documentation for accuracy; ensure military shipping labels are accurate; ensure MSLs and radio frequency (RF) tags are applied properly; check documentation and location of HAZMAT, if appropriate; inspect secondary loads for documentation and ensure the loads are adequately secured.
- **Station 3**. Conduct thorough maintenance inspection of vehicles.
- **Station 4**. Fuel/refuel vehicles.
- **Station 5**. Verify completion of stations 1-4; direct vehicles to proper staging area.

Deployment/Movement Plans (JOPES/TPFDD/TC-AIMS II)

Ref: FMI 3-35, Army Deployment and Redeployment (June '07), pp. 2-2 to 2-3 and FM 4-01.011, Unit Movement Operations (Oct '02), chap. 2.

Joint Operational Planning & Execution System (JOPES)

JOPES is the integrated, joint command and control system used to support military operational planning, execution, and monitoring activities. JOPES incorporates policies, procedures, personnel and systems, and underlying Global Command and Control System (GCCS) information technology support to provide senior-level decision-makers and their staffs with enhanced capability to plan and conduct joint military operations. JOPES provides the mechanism to submit movement requirements to lift providers in the form of a TPFDD.

Each combatant commander conducts deliberate planning to produce a series of OPLANS that provide detail on how to execute potential operations in their area of responsibility (AOR). Source movement data from TC-AIMS II is produced by units to maintain accurate movement data in JOPES. This source data provides the warfighters visibility of personnel, equipment, and supplies available to support his OPLAN. During a crisis, each combatant commander conducts crisis action planning (CAP).

Detailed information on JOPES can be found in The Joint Forces Operations & Doctrine SMARTbook.

Time-Phased Force and Deployment Data (TPFDD)

TPFDD is the JOPES data-based portion of the operational plan; it contains time-phased force data, nonunit related cargo and personnel data, and movement data for the operational plan including:

- In-place units
- Units to be deployed to operational plan with a priority indicating the desired sequence for their arrival at the POD
- Routing of forces deployed
- Movement data associated with deploying forces
- Estimates of nonunit related cargo and personnel movements to be conducted concurrently with the deployment of forces
- Estimates of transportation requirements, which are fulfilled by common user lift resources, as well as those requirements that can be fulfilled by assigned or attached transportation resources

Flowing forces in accordance with a TPFDD results in the delivery of sequenced force packages to the combatant commander and makes the best use of the apportioned strategic lift, while at the same time providing situational awareness of arriving forces to the COCOM. During recent deployments supporting Operations Enduring Freedom and Iraqi Freedom, operational requirements and force flow decisions resulted in TPFDD sequence adjustments via requests for forces (RFF). In addition, revised or updated deployment orders (DEPORD) were used to alert and move affected units.

TC-AIMS II Movement Plan

The TC-AIMS II System creates a product called a movement plan. In it is a wealth of information and data that is both useful and crucial to the formulation of unit movement plans to support a JOPES OPORD or OPLAN and exercises. (Much of the information in the TC-AIMS II movement plan can be used as attachments in the unit movement plan.) However, a TC-AIMS II movement plan, while defined as a movement plan within the system, is not a unit movement plan *(see pp. 6-12 to 6-15)*. TC-AIMS II contributes much to the creation of unit movement plans, but does not produce them.

Chap 6

III. Fort-to-Port

Ref: FMI 3-35, Army Deployment and Redeployment, chap. 3 and FM 4-01.011, Unit Movement Operations (Oct '02).

The predeployment activities discussed in the previous section form the underpinning for successful deployment. Once the deployment notification is issued, the luxury of time is lost. The planning must be completed, the support network must be trained and in place, and the deploying unit must be prepared to execute its deployment plan.

I. Notification

A. Initial Notification and Warning Order Activities

The CJCS publishes a formal warning order to prepare for possible military response to a crisis situation. The force provider/Army Command then alerts the deploying units and installations with a warning order. Following receipt of the order, the unit determines its deployment readiness status and begins updating personnel and equipment status and deployment documentation, plans, and procedures. The installation adjusts their plans and procedures to support the anticipated deployment and notify the elements (personnel, supply, maintenance, transportation, and training) that are required to support the deploying unit. Based upon SDDC port call message or an AMC airflow message, the installation publishes schedules for movement to POE.

B. Alert Order Activities

The force provider/Army Command passes a detailed alert order to its subordinate elements. If not already provided, the Army Command receives the project code that allows units to commit resources toward the deployment. To improve its readiness posture, the deploying unit cross-levels equipment and submits requisitions for unit basic load and other needed supply classes. The filling and receipt of supplies is dependent upon the deployment timeline and availability of stocks. The unit also requests supplies to support movement operations (BBPCT, dunnage, containers, and 463L pallets). The unit verifies that assigned ULNs are consistent with the unit's movement increments (for example, main body, advance party) for deployment. The unit also verifies equipment status compared to the UDL and updates load plans, equipment dimensions and weight, and shipper's declaration of dangerous goods. Once corrections are made, the unit prints and applies MSLs (DD Form 1387, Military Shipping Label) and attaches RF tags. Additionally, identify equipment as yellow TAT, red TAT, or equipment that does not have to accompany troops (NTAT). Yellow TAT accompanies the troops, must be accessible enroute, and includes class I and individual carry on baggage and weapons. Red TAT must be available upon arrival and includes chemical, biological, and radiological equipment. The unit finalizes the UDL as early as possible.

The unit commander briefs key personnel concerning the mission, threat, task organization, movement plans, and the Status of Forces Agreement, force protection, and geopolitical situations. Additionally, the commander or a representative should brief family members on the availability of family support activities.

If not previously provided by the Army Command, the deploying unit activates derivative Department of Defense Activity Address Code (DODAAC) and derivative unit identification code (UIC). The UMO finalizes lift and load plans, shipping documentation, and convoy clearances as secondary loads and pallets are built and containers are stuffed.

The installation initiates the support for the deploying unit based on established policies and procedures. The installation publishes schedules for movement to the POE based on the SDDC port call message or an AMC airflow message. Designated activities prepare to provide support based on the installation SOP.

II. Movement to the Port of Embarkation (POE)

Movement to POE activities normally begin with Army Command receipt of an execute order. The order, along with any additional guidance is forwarded to the appropriate subordinate commands, deploying units, and installations. SDDC issues a port call message that identifies the date the unit must have their equipment at the POE to meet the available-to-load date. The port call message or the operations order for sealift will also normally include details for vehicle preparations and reducing their cubic dimensions. AMC enters the APOE and airflow scheduling information into Global Transportation Network (GTN). Scheduling information is also available in the JOPES. Based on port call messages and air schedules, the organization backward plans movements to the POEs. Deploying unit equipment normally moves from unit marshaling areas to a central staging area on the installation for further processing. The name, organization and responsibilities for these installation level staging or marshaling areas may differ; however, the functions performed to prepare units for movement are essentially the same.

Each power projection platform (PPP) has an associated strategic aerial port and/or seaport. The proximity of the port facilities to the PPP determines the type of movement and the numbers and types of assets required to complete the movement to the port. In some cases, the distance to APOEs and SPOE is short, allowing units to maximize the use of organic equipment and convoys. In other cases, the distance to the APOE or SPOE is longer; in which case, units may have to rely heavily on commercial road and/or rail transport to complete the move to the port.

In-Transit Visibility (ITV)

In-transit visibility preparation begins during predeployment and continues through the load out of vehicles and equipment. Ensuring the automated information technology (AIT) storage devices are accurate, properly attached, and readable facilitates in-transit visibility (ITV) throughout the transportation pipeline. AIT readers and interrogators report the movement to automated information systems, allowing deployment managers to track and control the flow of equipment.

Prior to departing their deployment stations, units must write radio frequency identification (RFID) devices and attach them to vehicles and equipment.

Refer to FMI 3-33, app. D for detailed guidance on writing and attaching RFID devices.

Convoy Operations

In some cases units convoy their vehicles and equipment to the POE. A convoy is a group of vehicles organized for the purpose of control and orderly movement under the control of a single commander. In the absence of policies to the contrary, convoys are considered six or more vehicles. Vehicles in a convoy are organized into groups to facilitate command, control, and security and normally move at the same rate.

Mobilization Movement Control Automation System (MOBCON)

To assist in the centralized convoy management, FORSCOM has implemented a Mobilization Movement Control Automation System (MOBCON) in each state. The unit will submit a convoy request (DD Form 1265, Request for Convoy Clearance) through the installation UMC. MOBCON uses the National Highway Network database to schedule and deconflict convoys within CONUS. The deconfliction process allows only one convoy to operate over a segment of highway at any given time. The program links the Defense Movement Centers and provides visibility of all convoys.

Units deploying from OCONUS locations must contact the supporting movement control team (MCT) for convoy march credits. The minimum number of vehicles in a convoy is dictated by theater policy, standardization agreement, or the host nation (HN), as are the procedures for processing convoy clearances.

Refer to FM 4-01.011 and FM 55-30 for specific guidance on planning and conducting convoy operations; and FORSCOM/ARNG Regulation 55-1 for specific information regarding MOBCON.

Rail Operations

Rail is also used to ship vehicles and equipment from power projection and power support platforms. Responsibility for planning and executing rail movements is split between the deploying units and the ITO. The deploying unit determines its movement requirements and submits them to the ITO. The deploying unit is also responsible for preparing their equipment for rail loading. Units load railcars and chock, block and tie down equipment under the technical supervision of the ITO, who is responsible for approving all rail loads.

The ITO is responsible for obtaining rail cars based on deploying unit requirements; validating railcar requirements based on unit rail load plans; and maximizing the use of the available rail assets. In addition, the ITO is the official liaison with SDDC and the railway agent and inspects all railcars for serviceability before units begin loading.

The MCT performs the ITO functions in OCONUS locations and obtains the rail cars, validates railcar requirements, serves as the liaison with the railway agent, and inspects the railcars before the units begin loading.

Refer to FM 4-01.011 for specific guidance on planning and moving by rail.

III. Movement to the Port of Debarkation (POD)

The combination of strategic airlift and sealift provides the capability to respond to contingencies. Each element of strategic lift has its own unique advantages and disadvantages. In general, airlift transports light, high priority forces required to rapidly form units with pre-positioned equipment and supplies. Strategic airlift support is the critical deployment enabler during the early stages of deployment and remains so until the sea line of communications is attained. In most cases, sealift accounts for the majority of the total cargo delivered to a theater of operations. As the principal means for delivering equipment and logistic support, sealift impacts the ability to conduct sustained operations and influence the outcome of the operation being conducted.

Refer to JP 3-17 and JP 4-01.2 for additional information regarding strategic lift.

IV. Seaport of Embarkation (SPOE)

Ref: FMI 3-35, Army Deployment and Redeployment, (June '07), pp. 3-4 to 3-6. *See p. 6-28 for discussion of SPOE responsibilities.*

A number of essential activities occur at the SPOE during deployment operations as units prepare for shipment by strategic sealift. The tasks are performed by a number of DOD and Army units and ad hoc organizations.

Notional Seaport of Embarkation

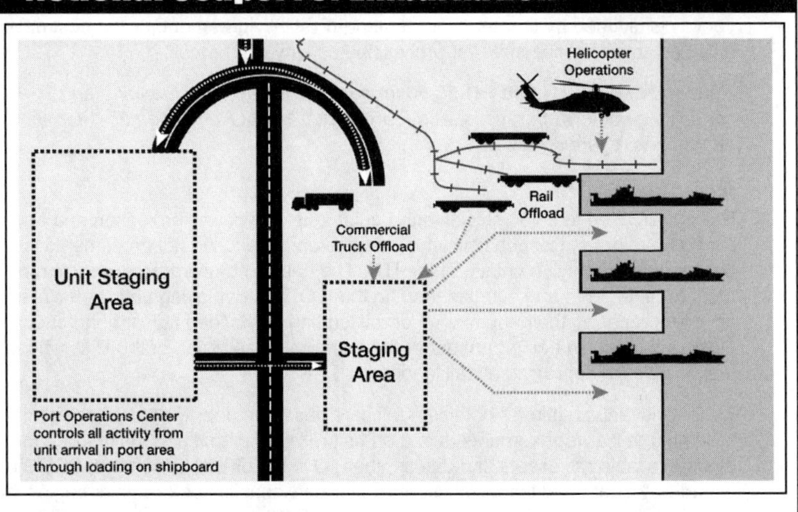

Ref: FMI 3-35, Army Deployment and Redeployment, fig. 3-3, p. 3-5.

1. Unit Staging Area

For movement to SPOEs, deploying units and equipment may use an enroute unit staging area, established and operated by the supporting installation. These areas are ideally located near the port staging area and in the immediate vicinity of rail and truck discharge sites. The SPOE marshaling area is the final enroute location for preparation of unit equipment for overseas movement prior to the equipment entering the port staging area.

Marshaling Area

Establishment of a marshaling area reduces congestion within the terminal area and provides space for sorting vehicles for vessel loading. The layout of a marshaling yard is not fixed but is contingent on available space and needs of the unit. Equipment arriving in the marshaling area is normally segregated in accordance with the vessel stow plan. When port call instructions are received from MTMC Operations Center, units are notified when and where to move their personnel, supplies, and equipment. This destination may be a port marshaling area or a port staging area. Support installations (SI), area support groups (ASG), or other organizations may be tasked to operate the marshaling area.

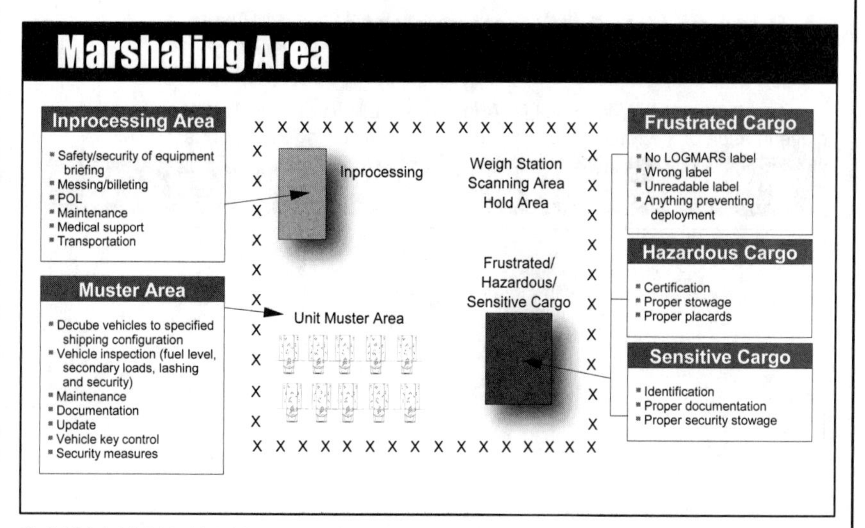

Ref: FM 4-01.011, Unit Movement Operations, fig. 4-1, p. 4-4.

2. Staging Area

The equipment is moved from the marshaling area or installation to the staging area based on the call forward plan and as directed by the port commander. The SDDC port commander assumes custody of the cargo at this point. Activities within the area include inspecting equipment for serviceability, packing lists/load card, determining the accuracy of dimensions and weights, inspecting properly secured secondary loads, and documenting any cargo requiring special handling. Military shipment labels affixed to equipment will be scanned using bar code readers. The data will then be loaded into the Worldwide Port System (WPS) to produce the ship manifests and serve as the basis for status reports.

In CONUS, a supporting installation is assigned responsibility for the logistical support of deploying units that transit the marshaling area. The installation element providing this support is not part of the port support activity (PSA). The PSA operates in the SPOE staging area. If marshaling areas are not used, units move directly to the port staging area.

In a mature OCONUS theater, an organization will be designated to provide PSA and associated logistic support for deploying units. Otherwise, the support group provides logistic support for deploying units, and the deploying force provides PSA support to augment the port commander's terminal operations force.

When processed, equipment may be segregated into different lots within the staging area by type, size, and any other special considerations such as hazardous materials, sensitive and classified items, and containerized equipment. From the staging area, vehicles are called forward to load the ship based on the stow plan and call forward schedules.

3. Supercargo

Supercargoes are unit personnel designated on orders to accompany, secure, and maintain unit cargo onboard ships. They perform liaison during cargo reception at the SPOE, vessel loading and discharge operations, and SPOD port clearance operations. The supercargoes are attached to the port operator and remain with the port operator at the SPOD until the offload is complete and they are released back to their units. Unit commanders recommend the composition of supercargoes based on several factors, including the amount and types of equipment loaded aboard the ship and the number of units with equipment on the ship; however, MSC determines the actual number of supercargo personnel permitted onboard, based on the berthing capacity on the ship.

V. Aerial Port of Embarkation (APOE)

Ref: FMI 3-35, Army Deployment and Redeployment, (June '07), pp. 3-6 to 3-10.

The APOE is the transition point for Army units deploying by air. The tasks that take place in each of the four areas are outlined in the following paragraphs.

Notional Aerial Port of Embarkation

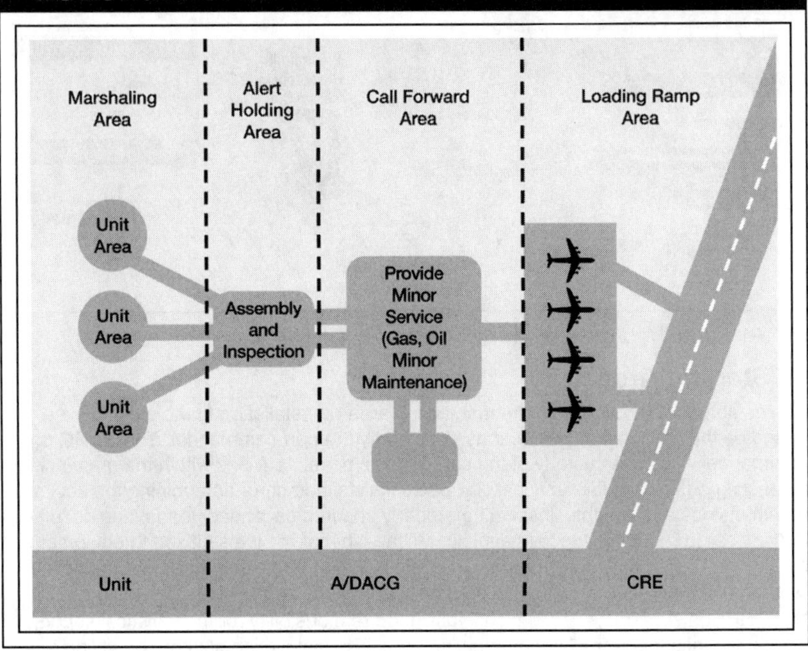

Ref: FMI 3-35, Army Deployment and Reployment, fig. 3-6, p. 3-8.

1. Marshaling Area

The primary purpose of the marshaling area is to provide a location near the APOE to assemble personnel, supplies, and equipment and make final preparations for air shipment. Unit marshaling areas are used to receiving convoys and processing vehicles before they are staged for loading. Marshaling areas are the responsibility of the deploying commander who is normally assisted by the ITO, supporting units, or other designated organizations, based on local policy.

The deploying unit:

- Establishes liaison with the DACG
- Coordinates a joint planning conference with the DACG and CRG to discuss aircraft allowable cabin load, pallet restrictions, aircraft configuration, equipment preparation requirements, airflow schedule, and any other issues impacting deploying unit preparation and processing
- Prepares vehicles and equipment
- Ensures adequate shoring material is available
- Prepares personnel and cargo manifests
- Assembles personnel, supplies, and equipment into aircraft loads
- Ensures planeload commanders are appointed and briefed
- Provides escorts for sensitive items

The DACG:

- Maintains liaison with the deploying unit and the CRG
- Coordinates with the CRG for technical assistance
- Calls aircraft loads fwd from the marshaling area and assumes control in the alert holding area

2. Alert Holding Area

The alert holding area is the equipment, vehicle, and passenger control area. It is normally located in the vicinity of the departure airfield and is used to assemble, inspect, hold, and service aircraft loads. Control of loads is transferred from the individual unit to the DACG at this point.

The deploying unit:

- Ensures the aircraft loads arrive on time
- Provides manifests to the DACG
- Corrects load discrepancies identified during pre-inspection
- Ensures vehicle drivers remain with the vehicles until released
- Passes control of aircraft loads to DACG

The DACG:

- Ensures loads arrive at the alert holding area at scheduled times
- Receives, inventories, and controls aircraft loads as they arrive.
- Inspects aircraft loads
- Ensures required shoring, floor protection materials, 463L pallets, and dunnage
- Verifies weight and balance markings
- Establishes a discrepancy correction area
- Inspects documentation
- Inspects HAZMAT for proper documentation
- Provides emergency maintenance, defueling, and related services
- Coordinates MHE support
- Directs aircraft loads to the call forward area

3. Call Forward Area

The call forward area is the joint responsibility of the CRG and the DACG and is the location for the joint inspection of deploying unit equipment and cargo. The DACG, the deploying unit, and the CRG conduct the inspection (DD Form 2133).

The DACG:

- Establishes communications with the CRG and deploying units
- Reviews HAZMAT documents & load plans
- Ensures loads and manifests are correct
- Ensures the deploying unit adheres to the established movement schedule
- Ensures that discrepancies found during the JI are corrected
- Provides loading team personnel/support
- Escorts chalks to the ready line and briefs personnel on flight safety
- Retains copies of manifests and inspection records

The CRG:

- Coordinates with the DACG on all changes required by aircraft configuration
- Conducts the JI together with the DACG and unit representatives
- Provides guidance to DACG and deploying Soldiers concerning flight line procedures
- Provides a team chief for each loading team
- Notifies the DACG to dispatch chalks to the loading ramp/ready line
- Provides airflow status to the DACG

4. Loading Ramp Area

The loading ramp area, including ready line area, is controlled by the CRG. At this point, control of units for movement purposes passes to AMC.

The chalk commander:

- Follows directions of load team chief
- Monitors and controls aircraft passengers; retains copy of the final manifests
- Provides assistance in loading and securing the load as required
- Ensures vehicle/equipment operators follow instructions

The DACG:

- Transfers control of the loads to the CRG
- Maintains coordination with the deploying unit and CRG
- Obtains chalk completion times (CRG) from the CRG

The CRG:

- Accepts chalks from the DACG at the ready line and loads them aboard aircraft
- Ensures each aircraft load is positioned
- Escorts passengers to the aircraft; briefs all personnel on flight line procedures
- Ensures each chalk is positioned
- Maintains liaison with the aircraft crew and the DACG
- Coordinates with the loadmaster to ensure the aircraft is loaded in time
- Provides MHE & special loading equipment
- Provides loadmaster with manifests and retains copies for file

The load team:

- Receives loads at the ready line
- Loads and secures vehicles and equipment under supervision of the loadmaster
- Provides the loadmaster with manifests
- Informs CRG of load completion time

SPOE Responsibilities

Ref: FM 4-01.011, Unit Movement Operations (Oct '02), chap 4. See also p. 6-24.

In organizing for reception of personnel, equipment, and supplies at its SPOEs, MTMC may be assigned any of the following to assist in the deployment mission: UMT, a TTB, a PSA, cargo transfer companies, FCDTs, and cargo documentation teams.

Transportation Terminal Brigade (TTB)

TTBs are Reserve Component (RC) units that allow the MTMC to expand the number and capability of seaports. TTBs conduct ocean terminal operations at established ports where existing manpower, equipment, and infrastructure are available. They may be deployed Outside Continental United States (OCONUS) to expand the number and capability of ports for sustainment or redeployment purposes.

A typical TTB operates two or three berths simultaneously (four or five berths for limited surge periods), provides traffic management, and supervises contracts. It employs Army information systems such as Integrated Computerized Deployment System (ICODES) and Worldwide Port System (WPS), and uses automated identification technology (AIT) to maintain in-transit visibility. (See new FM 3-35.4.)

Port Support Activity (PSA)

The PSA is a temporary military augmentation organization (or contracted organization) comprised of personnel with specific skills that aid the port commander in receiving, processing, and clearing cargo at the SPOE. It is under the operational control of the port commander. CONUS installations are tasked by FORSCOM to provide PSAs to specific ports. This includes the PSA and associated logistic support for deploying units. The PSA establishes the necessary communications to ensure the proper flow of cargo. It provides daily operational reports of cargo received, maintenance performed, and operational problems to the port commander. In an OCONUS area of operation (AO), the ASG provides the PSA and associated logistic support for deploying units. (See FM 3-35.4.)

Cargo Transfer Company (CTC)

A CTC is organized with four cargo transfer platoons and a documentation section. The four platoons have material handling equipment (MHE) to support transshipping cargo, containers, and unit equipment to ships and aircraft. Each platoon can operate independently at a remote site to support transshipment operations. The company assists in loading ships and operating a staging area. The small CTC Documentation Section, equipped with TC-AIMS II, cannot support each of the four Platoons simultaneously when they operate at remote terminals. When operating remote terminals, the CTC is augmented with one or more cargo documentation teams. (See FM 55-1; new FM 4-01.)

Cargo Documentation Team (CDT)

A cargo documentation team is staffed with 88N Documentation Specialists. The cargo documentation team has no MHE. The team is normally assigned to augment a cargo transfer company to prepare documentation for cargo and equipment being loaded on vessels. (See FM 55-1; new FM 4-01.)

Freight Consolidation and Distribution Team (FCDT)

The FCDT is staffed to operate its forklifts, loading ramps, and a TC-AIMS II computer with AIT devices and printers. The FCDT can be located at small terminals to provide independent loading and documentation services or at larger port complexes as a tailored augmentation to the TTB. The FCDT prepares documentation for cargo and equipment being loaded on vessels. (See FM 55-1; new FM 4-01.)

IV. Reception, Staging, Onward Mvmt, Integration

Ref: FMI 3-35, Army Deployment and Redeployment, chap. 4 and FM 4-01.011, Unit Movement Operations (Oct '02).

During force projection operations the process used to generate combat power is known as reception, staging, onward movement, and integration. It is the means by which commanders shape and expedite force closure in the theater of operations. While logistics is critical to the mission, the steps a unit must accomplish to complete the process should be discussed in the deploying unit's operations order and treated as an operational requirement. It should not be relegated to the administrative and logistics or service support annexes. Reception, Staging, Onward Movement and Integration (RSOI) is necessary to return units to combat-ready status in a controlled, orderly, and systematic process after being split to facilitate movement. Given proper operational priority, units will experience a much more successful execution. Effective, well conceived RSOI operations speed force closure; conversely, ineffective RSOI delays force closure and compromises the combatant commander's ability to implement a concept of operations. In a force projection environment, the ability to execute a mission quickly depends on the speed with which combat power can be assembled at required locations.

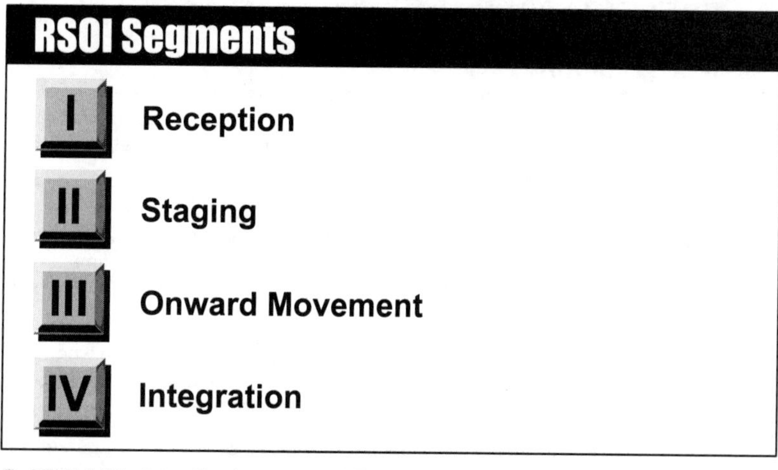

Ref:FMI 3-35, Army Deployment and Redeployment, chap. IV.

Reception is the unloading of personnel and equipment from strategic transport, marshaling them, transporting them to staging areas, and if required, providing life support services. *See pp. 6-31 to 6-33.*

Staging is the assembling, holding, and organizing arriving of personnel, equipment, and basic loads into units; preparing the units for onward movement; and providing life support until the unit becomes self-sustaining. *See pp. 6-33 to 6-38.*

Onward Movement is moving units from reception facilities and staging areas to TAAs or other theater destinations; placing arriving nonunit personnel to gaining commands; and providing sustainment to distribution sites. *See pp. 6-38 to 6-39.*

Integration is the synchronized transfer of authority of units to a designated component or functional commander for employment in the theater of operations. *See pp. 6-39 to 6-40.*

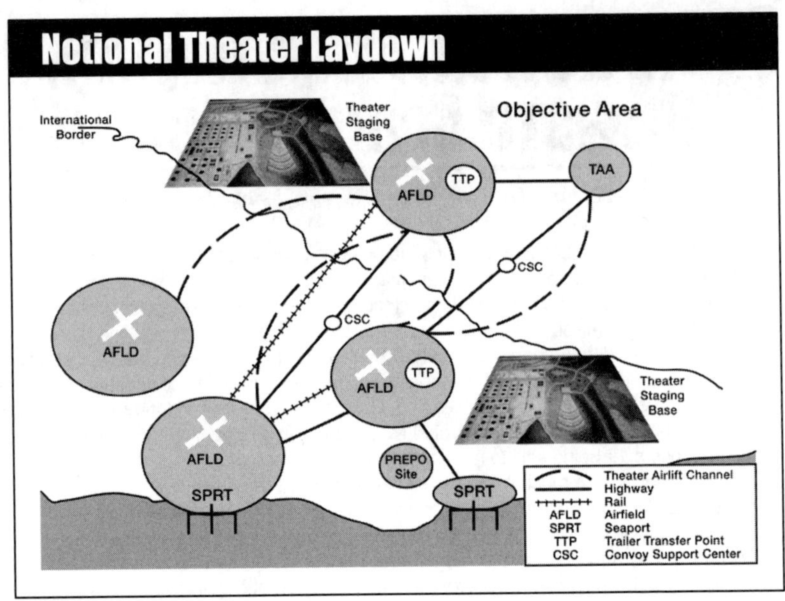

Ref: FM 3-35, Army Deployment and Reployment, fig. 4-5, p. 4-12.

RSOI Planning

Planners must consider the impact of the threats and operational environment on the conduct of RSOI operations. The assessment must address whether and to what degree a potential threat can disrupt or block operations as well as determine what infrastructure and support assets are available.

Military operations begin with an event that requires movement of forces somewhere in the world, either planned or no-notice. Analyzing the mission leads to the development of courses of action. For each COA, the staff identifies the estimated throughput, in terms of personnel and equipment. The throughput estimates provide a basis for developing the requirements necessary to support the flow.

Understanding capabilities of the theater infrastructure and the time when enabling assets become available are also essential elements to developing a successful RSOI operation. Theater RSOI infrastructure is divided into two categories:

- RSOI Organizational Infrastructure. The organizational infrastructure is normally composed of a mix of forward-deployed forces, Army pre-positioned stocks, deploying RSOI units, host nation/allied support, and contractor support
- RSOI Physical Infrastructure. The RSOI physical infrastructure consists of the theater's nodes and available modes of transportation. Nodes are a location in a mobility system where a movement requirement is originated, processed for onward movement, or terminated. During RSOI operations, nodes form wherever transportation modes are changed; for example, at airports, seaports, and at staging areas. The two modes of transportation are surface and air. Surface is further subdivided into sea, inland waterways, coastal waterways, highway, and rail.

Once the COA has been selected, the planners develop the plans identifying the location of the PODs, corresponding support requirements, resource allocations, lines of communication, and infrastructure to achieve the desired force closure.

Principles of RSOI

Ref: FMI 3-35 Army Deployment and Redeployment (June '07), pp. 4-2 to 4-3.

The following principles guide the planning and execution of RSOI operations:

1.Unity of command

One commander should control and operate the RSOI process—adjusting resources based upon deployment flows, controlling movements in the area of operations, and providing life support to arriving personnel.

2. Synchronization

Synchronization is managing the flow to ensure the arrival of personnel and their equipment coincide. This enhances command and control and helps maintain unit integrity. Force planners and transportation agencies must ensure that unit integrity is a fundamental consideration when planning unit deployments.

3. Unit integrity

Moving personnel and their equipment on the same strategic transportation asset from origin to destination simplifies force tracking and reduces RSOI requirements. An objective of Army movement planning is to maintain the maximum unit integrity possible within the DTS framework.

4. Balance

Proper sizing of the logistic structure required to support the deploying force is essential to balance. The goal is to deploy minimum assets necessary to optimize throughput of units and materiel. The balance is achieved by managing the TPFDD, specifically including theater opening resources together with the combat forces and incorporating sustainment in the flow with unit supplies and equipment.

Building Combat Power

The purpose of RSOI is to build the combat power necessary to support the COCOM's concept of operation. Force closure is that point which the combatant commander determines that an adequate combat-ready force is available. Force closure requires well-defined criteria by which unit commanders can measure their readiness.

Reporting the build of combat power begins with established standards for readiness. Assessments of combat power are based on unit capability, rather than simple tallies of numbers of vehicles and weapon systems on hand. Readiness and reporting are inherently operational matters, normally handled through operations channels.

Thus, RSOI operations must also be particularly flexible regarding force closure. Commanders may accelerate rates of force integration or change the sequence of unit integration at the risk of disrupting the flow. Nonetheless, due to both limitations of strategic lift and time delays inherent in intercontinental deployments, many decisions made at the beginning of the deployment process are practically irrevocable.

RSOI Port Selection

Ref: FMI 3-35 Army Deployment and Redeployment (June '07), pp. 4-9 to 4-11.

Seaport and airfield throughput capacities significantly influence the speed, order, and, to a large extent, the types of units that can deploy through them. Consequently, before thought is given to actual deployment of forces, planners must evaluate available airfield and port facilities within the area of operations, as well as the transportation networks linking them with each other and to the interior. As was the case during Operation Desert Storm, it may be better to use a world class port hundreds of miles away from TAAs rather than conduct an in-stream discharge operation or use a smaller, degraded port facility with limited capacity and throughput. Diplomatic and military contacts should be made at the earliest possible opportunity with the host nations controlling key facilities and rights-of-way.

The combatant commander in conjunction with USTRANSCOM selects the PODs that will be used for deployment. METT-TC considerations and the theater transportation infrastructure will drive the sequence, type, size of forces, and materiel arriving at ports of debarkation. These decisions impact the speed of combat power buildup and continued development of the theater. For example, when opposed entry is likely, commanders may have to seize and secure airfields and seaports to permit insertion of follow-on forces. Afterwards, it may be necessary to repair damaged facilities in order to process arriving units at the required rate. Even in the event of unopposed entry, ports of debarkation will still require improvement and repair to accommodate high throughput rates required for rapid force closure. Thus, the early entry of units such as transportation battalions, cargo transfer companies, Army watercraft, causeway detachments, and engineer assets can be critical to off-loading materiel, clearing ports and consequently speeding deployment.

Aerial Port of Debarkation (APOD)

The airlift challenge is often due to a lack of airports and associated capacity and not always the lack of aircraft. Consequently, maximum throughput at limited airports is paramount. The APOD is by its very nature a joint facility and will likely be a multinational facility. It is a POD for deploying forces, and a port of embarkation for forces moving to other theaters and noncombatant evacuation. The host nation may limit the APOD to coalition military use, or the military may be sharing the facility with commercial activities. Governmental, nongovernmental, and private organizations will likely be competing for use of the APOD along with military forces.

The APOD serves as the primary port of entry for all deploying personnel and for early entry forces normally airlifted into theater together with their equipment. Responsibility for APOD operations is divided between the Air Force and the Army, with the Air Force responsible for airfield operations, including air terminal control, loading, unloading, and servicing of aircraft. The Army is responsible for clearing personnel and cargo off the tarmac and for required life support. Air Force/Army interface occurs between the Air Force CRG and the Army A/DACG and movement control teams.

Necessary communication, personnel, and cargo handling equipment must be in place to facilitate rapid movement out of the airport. The CRG and the A/DACG must be included in the lead elements of the deploying force. The CRG controls all activities at the off-load ramp area and supervises aircraft off-loading. The A/DACG escorts loads and personnel to holding areas, thus clearing the airfield and ensures airfield operations and strategic airflow are not obstructed and limited due to the accumulation of cargo.

Two primary physical constraints for airfields are parking MOG and working MOG. Parking MOG is the number of aircraft that can actually fit on the ground without impeding other activities. Working MOG is the number of aircraft that can be worked simultaneously based on the available material handling equipment and maintenance and fueling capabilities. Optimally, working MOG should equal parking MOG.

Seaport of Debarkation (SPOD)

Activities at seaports are normally joint, multinational, and intermixed with commercial operations. Seaports can serve as ports of debarkation for arriving forces and simultaneously as ports of embarkation for forces deploying to other theaters of operations. The COCOM has several options for management of seaport operations in theater. These options include the use of deployable Active Army and U.S. Army Reserves transportation terminal units or SDDC to operate some or all of the theater water terminals. USTRANSCOM through SDDC is the DOD-designated single port manager (SPM) for all common user ports worldwide. The SPM performs those functions necessary to support the strategic flow of the deploying forces' equipment and sustainment supply in the SPOE and hand-off to the theater COCOM in the SPOD. The SPM is responsible for providing strategic deployment status information to the COCOM and to workload the SPOD port operator based on the COCOM's priorities and guidance. The SPM is responsible through all phases of theater port operational continuum from a logistics over-the-shore operation to a totally commercial contract-supported deployment.

The PSA is an important aspect of port operations and the TSC planners coordinate the requirements with SDDC. The PSA is an ad hoc organization comprised of personnel with the skills necessary to receive, process, and clear cargo from the SPOE or SPOD. The composition of the PSA varies with the mission to be supported and can be military, civilian, or host nation support or a combination of these. The PSA works directly for the port operator who reports to the port manager.

The volume of cargo arriving in the theater in a small window of time can drive the need for multiple seaports to meet deployment timelines. The physical size of the large medium-speed roll-on/roll-off (LMSR) ship and the draft requirement to bring vessels of this class pier side may also present a challenge. If world-class port facilities are available, off-loading can be rapidly accomplished. If facilities are less than world class or austere, then multiple ports and slower in-stream operations may be required.

The ability to project forces into an operational area, despite ports rendered unusable or inaccessible to deep draft vessels, is essential to the Army's force projection strategy. Army watercraft provides this capability through in-stream discharge. They allow the ship awaiting berthing space because of congestion or port denial to be off-loaded in-stream. In situations where world class ports are not available, Army watercraft can discharge the LMSRs in-stream and transship the cargo on smaller Army lighterage establishing an intra-theater sea line of communication to smaller coastal ports or directly over the shore.

The throughput capacity of a port (the ability to receive, process, and clear personnel and equipment) is a critical planning factor. The planner must check that the port is capable of receiving the planned strategic flow, considering not only the port's capability, state of repair, and congestion, but also its throughput capability. The ability to conduct in-stream (or offshore) discharge operations can expand a port's reception capability. A smaller port lacking the capability to receive large vessels can receive cargo discharged in-stream to increase the overall theater throughput. However, ability to perform in-stream off-loading is largely contingent on availability of Army watercraft and other assets moving cargo from ship to shore. In-stream off-load operations are sensitive to weather and sea conditions, and generally require a protected anchorage.

Seaport operations are similar to airport operations: vessels are off-loaded, cargo moved to a holding area, and then the port is cleared. Unit cargo clearing the port moves to a theater staging base (TSB) or directly to the TAA. Movement out of the port is controlled by movement control elements and must be integrated into the theater movement plan. Port clearance operations can involve one or more of the following transportation modes: highway, rail, and coastal/inland waterways.

I. Reception

As the initial step in introducing combat power, reception can determine success or failure of the RSOI operation. Reception from strategic lift is implemented at or near designated air and seaports of debarkation, normally under control of the combatant commander. It must be thoroughly planned and carefully executed. While the reception plan for each theater may vary, reception capacity should at least equal planned strategic lift delivery capability.

The intelligence preparation of the battlefield (IPB) and analysis of theater reception capability provide an understanding of how competition for reception at airfields and seaports could affect the force flow. Units may compete for off-loading berths with other services or multinational forces that are part of the supported COCOM plan.

For the initial period of deployment, the aerial port is the lifeline to the front-line. All that is not pre-positioned or available from the host nation comes through the aerial port. Then, the first surge sealift ships begin to arrive, dramatically increasing forces. Airlift remains a critical element regarding delivery of personnel, but most unit equipment to build the combat power arrives through seaports.

A. Functions

Synchronizing transportation reception activities are critical to facilitating throughput at the ports of debarkation. They include command and control, movement control, and port operations.

1. Command and Control

The combatant commander typically assigns the senior Army logistics commander with the responsibility to support reception of forces, there are other service organizations that provide theater opening capabilities, depending on the time and size of the deployment. The joint task force—port opening is composed of AF contingency support personnel and Army movement control and cargo handling capabilities to establish a POD and clear the cargo from the port to a nearby distribution hub. In addition, joint task force—port opening will conduct assessments of the existing distribution network to determine the resources required to support the continued flow of deploying forces and begin the distribution of sustainment. The Sustainment Brigade with the Transportation TO element are designed to provide the deploying force the capability to open multiple air and sea ports and establish RSOI capability in theater.

The arrival of strategic air and sealift will be controlled by the combatant commander through the USTRANSCOM element attached to the combatant commander's staff. The APOD and SPOD will normally be managed by AMC and SDDC respectively and operated by the designated logistics command under command and control of the lead service. It should be noted that reception activities continue after force closure is achieved to facilitate arrival and processing of sustainment and unit replacements.

2. Movement Control

Efficient movement control allows commanders to redirect forces and rapidly compensate for disruptions in the LOC. A movement control element must be positioned at each reception node and remain in constant communication with USTRANSCOM elements on-site and with other movement control elements in-theater. A well-disciplined and centralized system must be implemented to control movements along all LOCs. The movement control system is responsible for establishing protocols with host/allied nations concerning use of available transportation nodes and infrastructure.

Two factors determine reception throughput: reception capacity and clearance capability. All ports have finite processing and storage space. Unless personnel and equipment are cleared quickly, the port will become congested, cargo will be frustrated, and the infrastructure will be unable to receive forces at the required rate of delivery. Factors contributing to efficient port clearance are proper documentation, professional movement control expertise, adequate materiel/container handling equipment, and trained personnel. Port operators need timely and accurate documentation, including advance information on forces and equipment arriving in-theater. Efficient movement control assures best use of available infrastructure and proper metered flow of forces and equipment according to operational priorities.

3. Port Operations

Experience in contingency operations has shown the need to rapidly expand and improve port reception capability, regardless of the nature of ports being used. As the buildup of combat forces begins, capability for rapid expansion depends on well-synchronized arrival of personnel and equipment. The JFC must, therefore, control the deployment flow so that reception capabilities are not overwhelmed. APODs and SPODs should be considered integral parts of a single reception complex, unless the distance separating them precludes mutual support. Reception capacity depends on—

- Port and airfield infrastructure, condition, and characteristics
- Availability of host nation labor and port services
- Off-loading and holding space
- Condition and capacity of exit routes
- Efficiency of movement control systems
- Holding and support facilities at reception nodes
- Weather
- Enemy situation

II. Staging

Staging is that part of the RSOI operation that reunites unit personnel with their equipment and schedules unit movement to the TAA, secures or uploads unit basic loads, and provides life support to personnel.

Theater Staging Base (TSB)

These staging activities occur at multiple sites in controlled areas called TSBs that are required because space limitations normally preclude reassembly of combat units at seaports of debarkation. In general, there will be at least one TSB for each SPOD/APOD pairing.

See following pages (pp. 6-36 to 6-37) for a description of TSB functions.

A. Force Closure

In order to meet the force closure requirements, the time that units spend staging through the TSB must be minimized. Staging should not be a lengthy process, but inefficiencies can cause delays (for example, personnel arriving before their equipment, equipment arriving before its personnel, and gaps in matching troops with proper equipment). In fact, a battalion-sized unit should strive to spend no more than two days staging in the TSB.

TSBs should be located in areas convenient to both the SPOD and APOD, with good lines of communication back to ports of debarkation and forward to designated TAAs. In addition, the TSB should have sufficient space to accommodate the largest force scheduled to stage through it, together with facilities for vehicle marshaling, materiel handling, equipment maintenance and calibration, and possibly bore sighting and test firing of weapons. All of these are needed if the TSB is to fulfill its function of converting personnel and equipment into mission-ready combat units.

Theater Staging Base (TSB) Functions

Ref: FMI 3-35 Army Deployment and Redeployment (June '07), pp. 4-12 to 4-14.

Staging is that part of the RSOI operation that reunites unit personnel with their equipment and schedules unit movement to the TAA, secures or uploads unit basic loads, and provides life support to personnel.

These activities occur at multiple sites in controlled areas called TSBs that are required because space limitations normally preclude reassembly of combat units at seaports of debarkation. In general, there will be at least one TSB for each SPOD/APOD pairing.

Theater Staging Base (TSB) Layout

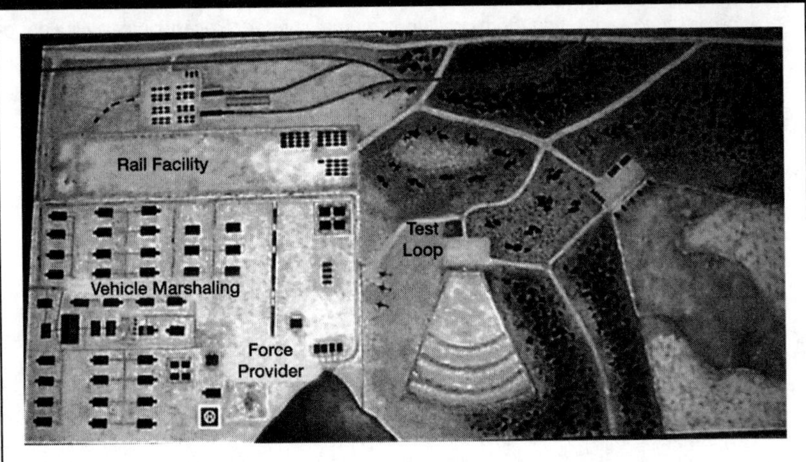

- Communication
- Life Support
- Arming, Fueling, Fixing
- Preparation for Onward Movement
- Security

Ref: FMI 3-35, Army Deployment and Reployment, fig. 4-7, p. 4-14.

1. Communications

Reliable, secure, and compatible communications are essential to operations in the theater staging base. The COCOM must know when forces are combat-capable and prepared for onward movement to give the COCOM the capability to control and employ these forces at the decisive point and time.

Force tracking provides situational awareness of combat-ready units within the operational area. While in transit, visibility begins at home station; the process of force tracking begins in the staging area, where equipment and personnel are reassembled into combat-ready units. Staging operations must have the communications, data processing equipment, and personnel assets to provide and manage force tracking data. Efficient movement control can provide force tracking information, but it must be able to communicate directly with operational commanders.

ITV acts as a staging enabler by providing commanders with clear pictures of locations of units and materiel in RSOI and deployment. For the TSB commander, ITV provides an awareness of the scheduled arrival of personnel and equipment; so the resources required to support them and time required to assemble the unit in a mission-ready configuration are available.

At present, there are a number of joint systems in various stages of development that provide visibility of force deployment and sustainment. Unfortunately, present systems do not completely satisfy the requirements of force tracking and much of the process must be accomplished manually.

2. Life Support

Regardless of time actually spent in the TSB, troops staging through it will require support, including housing, sustenance, sanitation, and health care. RSOI planners must ensure that the force provider units are sequenced early in the TPFDD to be in place and functioning by the time the first units arrive. Even if this requires displacement of some combat capability, it pays dividends later in the form of higher throughput, faster buildup of combat power, and earlier force closure in the operation. The Army's Force Provider modules (each designed to provide base camp support to 550 people) and the Air Forces Prime Beef and Prime Rib programs are viable options for providing field services to transient and permanent parties.

3. Arming, Fueling, and Fixing

Equipment arriving at the TSB may require maintenance before it becomes combat ready. This includes calibration of equipment; bore sighting of weapons; replacement of parts damaged in transit; and painting, fueling, and loading. The TSB should provide adequate facilities to support these activities, including marshaling areas, maintenance shelters, fuel and ammunitions storage, a test- driving loop, and range areas.

4. Preparation of Units for Onward Movement

In addition to preparing equipment, units at the TSB undergo training and reorganization. Communications networks are established, and tracking systems allow senior commanders to monitor the buildup of combat power. Commanders must participate in planning the onward movement, including route planning, unit tracking, and movement control.

5. Security

Theater staging bases are high-value targets. Their destruction or damage results in serious delays in force closure and disruption of the COCOM's concept of operations. Units in the TSB are vulnerable to attack by enemy air, missile, and ground forces. Being immobile and only partially combat ready, they possess limited capability for self-defense. Moreover, with many troops and their equipment concentrated into a relatively compact area, there is great potential for massive casualties, which could result in serious strategic consequences.

Other factors affecting selection of a TSB include geography and terrain (for example, water supply may be a factor in desert operations or land space in urban settings), and availability of organic and host nation assets. These factors, together with the size of the deploying force, may often necessitate multiple TSBs. The requirement for multiple staging bases is most evident in the urban sprawl of Europe and Korea, particularly around seaport facilities. In many cases, it is tremendously difficult to find even one square mile of open terrain much less the total space requirement for a TSB. Under normal circumstances, troops deploy by air while equipment deploys by sea and/or rail. The speed differential between air and sea surface transportation is the fundamental cause of complexity and potential difficulties in the staging process. Troops and equipment must be sequenced in the TPFDD so that both arrive (nearly) simultaneously, expeditiously unite, and ready themselves for onward movement. Troops arriving too early must be provided with meals and quarters while waiting for their equipment to arrive. The command responsible for RSOI would be expected to accommodate these needs. Moreover, the mass of immobile, unprotected troops presents an inviting and vulnerable target. On the other hand, if equipment arrives much earlier than the troops, ports of debarkation can become congested, and space management becomes critical.

Early deployment of essential theater opening support units at the expense of combat units pays dividends later by speeding the flow of the entire force, enhancing the COCOM's ability to build combat power and increase operational flexibility. Conversely, front loading the TPFDD with combat forces may jeopardize the COCOM's ability to build up forces as rapidly as required and thus reduce COCOM flexibility.

III. Onward Movement

Personnel and equipment reassembled as combat-ready units must be moved to the TAA based on the COCOM's priorities. Onward movement is a joint/multinational effort using capabilities and organizational structures of other services, allies, host nation and other governmental entities. It is an iterative activity in which units advance from one line of communications (LOC) node to another. Onward movement occurs when units move from ports to theater staging bases or forward to the TAA. Three primary factors affecting onward movement are movement control, transportation infrastructure, and force protection.

A. Movement Control

Movement control is the planning, routing, scheduling, and controlling forces and sustainment over lines of communication while maintaining in-transit visibility and force tracking. This is not a passive activity. Successful movement control requires continual analysis of requirements, capabilities, shortfalls, alternatives, and enhancements. Bottlenecks within the theater must be identified, and potential interruptions to the flow minimized. One of the biggest challenges of movement control is rapidly adjusting to changes in the operational environment and the commander's priorities. The challenge of a theater movements program is to merge the COCOM's concept of operations and priorities in a movement plan and execute them. This challenge can be met by employing an adequate number of movement control resources, appropriately enabled by communications, to anticipate and improvise. Efficient movement control enables the commander to redirect forces and rapidly overcome disruptions in the LOC.

B. Transportation Infrastructure

The total transportation infrastructure-modes, routes, control factors, host nation support, and specialized handling requirements must be coordinated to maximize speed of movement. Army Forces, other Services, and allied forces will be competing for the same networks, and congestion will result if proper coordination is not accomplished. Capabilities of the transportation network must be balanced against movement requirements, so that modes and routes are neither saturated nor under used. Planners should anticipate simultaneous demands on limited infrastructure, difficulties with communications, and differences in transportation capabilities.

During onward movement, mode selection is an operational issue, as it determines whether the commander of the unit in transit maintains control or whether control is lost and further staging required. Ideally, rail and heavy equipment transporters should transport tracked vehicles, and wheeled vehicles should convoy. In-land and coastal waterways should be used when available as they afford militarily useful solutions.

Establishment of convoy support centers and trailer transfer points along main supply routes and other support centers at temporary airfields, rail sites and waterway drop off points, further aids onward movement. These allow units and line haul drivers to rest, eat, perform vehicle maintenance, and contact unit/movement control personnel to receive operational updates, to revise priorities, and when necessary, to execute diversions.

C. Force Protection

The onward movement phase can provide the enemy with numerous opportunities to inflict serious losses and delay the build-up of combat power by exploiting vulnerability of units in transit. Force protection consists of those actions taken to prevent or mitigate hostile activities necessary to conserve the capability of the deploying force.

Adversaries seek to frustrate Army deployment operations by resorting to asymmetric means rather than challenge the Army in conventional combat. Insurgents may use improvised explosive devices, car bombs, and snipers to disrupt the flow of forces and sustainment moving over the lines of communication.

Enemy interdiction of onward movement presents special challenges that can be partially overcome by using alternative routing and mode substitution when feasible, but all units must be prepared to defend themselves. OIF convoys are organized and tightly controlled to afford a higher degree of security. Moreover, hardened gun trucks escort the convoys, and additional armed personnel ride in the vehicles to immediately engage insurgents as required.

IV. Integration

During integration, combat-ready units are transferred to the operational commander and merged into the tactical plan. The transfer may require interaction and familiarization among units and that arriving units meet certain standards before being completely integrated into the combat plan. Consequently, requirements for integration planning and coordination must occur early in the force projection process and modified according to METT-TC until force closure is achieved. Integration is complete when the COCOM establishes positive command and control over the arriving unit, usually in the TAA, and the unit is capable of performing its assigned mission.

Unit operational and mission-readiness are prerequisites before integration can occur. The unit must be able to move, fight, and communicate at acceptable levels of capability. Internal command and control must be re-established, and the unit must meet the readiness standard established by the tactical commander. The unit must be totally absorbed into the joint force and be able to communicate and receive command and control from its higher headquarters.

The time required for integration may vary, depending upon the size of the total force, contingency conditions, and amount of predeployment and ongoing planning and coordination. As the Forces are introduced into the theater, their movement will be in competition with resources required to support and sustain the existing and growing number of Forces deployed into the theater. Rapid integration, however, is critical to the success of combat operations, and adequate planning and coordination can reduce integration time.

The ability to accurately predict when integration will occur is critical to the commander's ability to conduct operations and maintain momentum. In order to accomplish this, the COCOM and component staffs must be able to build a TPFDD that meets the commander's intent. The COCOM can communicate intent through well-planned required delivery dates. Transportation feasibility will be conducted throughout the military decision making process as a means of checking acceptability and suitability of the courses of action. Once deployment begins, the COCOM monitors the TPFDD execution and any changes made must be analyzed for their impact on integration of mission essential capabilities. If changes are made, the TPFDD is revalidated by the COCOM to adjust for these changes.

Predeployment planning establishes force structure for the contingency and identifies units that must integrate. Once identified, units establish predeployment liaison and plan for theater integration. Coordination measures, ITV, and force tracking are used to predict the start of force integration and the time required for its completion. Criteria for unit mission readiness are an essential element of integration and must be included in the integration plan. Integration requirements are best defined using end-state analyses based on the COCOM's force requirements. The analysis identifies milestones for deploying units.

Plans must be open and flexible enough to adapt to reality on the ground. Technical problems, natural conditions, land space constraints, and enemy action all conspire to alter the commander's initial plan. The concept of operations should be broad enough to accommodate changes in strategic, operational, and tactical situations as they occur.

Deployment operations are time sensitive; compressed planning timelines and furious activity are the norm. Commanders need timely, accurate information to execute or modify initial plans in response to rapidly changing operational and tactical conditions. Confusion inherent to deployment often results in conflicting guidance, frequent planning changes, and inefficient task execution—all of which delay the build-up of combat power and the force closure.

Control measures, such as liaison officers or movement control teams can reduce confusion between integrating units, RSOI forces, and receiving headquarters. These measures act as guardians of the commander's intent and focus effort on force integration. These measures should be established immediately as part of the planning process and be maintained throughout the RSOI process.

The success of RSOI depends on:

- Responsive leadership
- Effective combat-power tracking system
- Reliable communications
- Force protection
- Transportation support
- Host-nation resources
- Soldier support service

V. Redeployment

Ref: FMI 3-35, Army Deployment and Redeployment, chap. 5.

Redeployment is the return of forces to home station or demobilization station. Commanders plan for redeployment within the context of the overall situation in the theater. The four phases of redeployment are redeployment preparation activities, movement to and activities at POE, movement to POD, and movement to home or demobilization station.

I. Redeployment Preparation Activities

Following completion of military operations, redeploying Forces move to designated assembly areas. Redeployment operations at the assembly areas are under the control and supervision of the TSC commander and include actions necessary to prepare the unit for movement. The COCOM issues a redeployment operations order releasing units from their missions and authorizing movement. Redeployment planning by the ASCC/Army Forces (ARFOR) normally precedes the actual issuance of an order and tentatively outlines information about the support network, follow-on operations, security requirements, and movement limitations imposed by infrastructure and resources. Redeployment operations must be conducted at a pace that does not disrupt the ability of the COCOM to execute continuing missions.

The redeployment plan conveys the commander's intent and includes responsibilities, priorities, and guidance for movement of forces, individuals, and materiel. Issues that must be addressed in the plan are:

- ARFORGEN
- Scheduling of redeployment activities
- Transfer of equipment (stay behind equipment or equipment designated for depot rebuild)
- Basic load turn-in
- Army pre-positioned stocks (APS) procedures
- Security of the force
- Availability of theater transportation assets
- Availability of strategic lift

A movement order may be issued sequentially for each movement or may be contained in one movement order designating the timing and means of transport to the POE. The theater movement control element issues movement tables that give detailed movement instructions to redeploying units. The TSC usually manages the redeployment support that can be performed by a subordinate organization.

The unit begins the redeployment process by identifying requirements and determining current unit status. Other actions include:

- Submitting personnel and pay actions
- Conducting medical screening
- Performing equipment checks and services
- Conducting equipment inventory
- Refining the UDL

II. Movement to and Activities at the Point of Embarkation (POE)

Ref: FMI 3-35 Army Deployment and Redeployment (June '07), pp. 5-2 to 5-4.

Redeployment planning results in a network of transit areas designed to efficiently move forces from their area of operations to their final destinations. Use of these areas may vary with the situation.

1. Assembly Area

Units move to an assembly area to prepare for redeployment after being relieved from their operational mission. The assembly area should be away from the immediate employment area. Movement to and within the area is under control of the Distribution Management Center (DMC). Units in the assembly area inventory, inspect, and process equipment for turn-in or transfer; load containers; prepare documentation; conduct U.S. Customs inspections; finalize unit movement data; and plan for movement to a POE. Units update UDLs and generate documentation and MSLs using TC-AIMS II. Equipment moving from the assembly area to the POE must have MSLs applied prior to loading. Documentation includes hazardous shipping declarations, labels, placards, secondary cargo load plans/cards, packing lists, and MSLs.

Units wash major end items to satisfy Department of Agriculture standards. Customs and agricultural inspection standards are based on the destination and types of equipment being redeployed. Units should make plans to perform the activities necessary to meet these standards. The time required to wash vehicles can be considerable and likely will be the overriding factor in redeployment scheduling. For example, a high mobility multipurpose wheeled vehicle can take approximately 12 hours to wash to meet the agricultural standards, and larger equipment can take a day or more to wash. Considerations in computing the estimated time to wash unit equipment should include the equipment density, estimated time for each piece of equipment, the number of wash points, and the staffing at each location. Once the equipment is cleared by customs inspectors, it will be held in a secure sterile area until it is moved to the POE.

2. Activities at the SPOE

Units normally move to the SPOE staging area from assembly areas. Some SPOEs may not have total use of the port area. Port managers and operators must closely coordinate their activities with host nation authorities as well as joint and multinational elements. Joint-use facilities and limited real estate availability may require port authorities and redeploying forces to modify processes to accommodate port capabilities.

SDDC as the single port manager directs water terminal operations to include supervising contracts, cargo documentation, security operations, and the overall flow of information. SDDC is responsible for providing strategic redeployment information to the COCOM and to workload the port operator based on the COCOM's priorities and intent.

3. Activities at the APOE

The agencies and processes involved in moving Army units through an APOE during a deployment are similar to those at an APOE during redeployment. Customs and agricultural inspections are the principal difference, and the inspection standards are based on the country the unit is departing from, the unit's destination, and the type of equipment.

Customs Procedures

All DOD-sponsored cargo is inspected at the overseas point of origin by military customs inspectors. Military equipment is inspected at the time it is placed in boxes, crates, containers, sea vans, or similar receptacles for movement and secured until departure from the overseas area. Vehicles and similar items to be shipped are inspected and secured immediately prior to loading on the departing aircraft or vessel. The packing list and TCMD for each container replace the DD Form 1253 Military Customs Inspection (Label) and DD Form 1253-1 Military Customs Inspection (Tag).

Inspectors normally check a minimum of 10 percent of all baggage 24 hours before the departure time. Once inspected, baggage is stored in a sterile area until transported and loaded at the APOE. Approximately 4-6 hours prior to the scheduled departure, Soldiers process through customs with their carry-on bags. Once cleared through customs, Soldiers remain in the sterile area until they board the aircraft.

Unit representatives coordinate inspection times with customs officials not later than five days prior to departure date.

4. POE Staging Area

Intra-theater transportation assets may move units directly to a POE staging area or to an intermediate staging area. These movements are largely determined by the distance to be traveled, size of the redeploying force, and theater capabilities. Units that were issued APS equipment usually turn it in at a separate location prior to moving to the POEs. Procedures for return of APS to storage locations are established during redeployment planning.

Refer to FM 100-17-1 and FM 100-17-2 for specific information regarding APS.

SPOE staging operations prevent congestion within the terminal area and provide space for segregating vehicles for vessel loading. This is the final en route location for preparation of unit equipment for strategic movement prior to the equipment entering the port holding area. The redeployment coordination cell monitors the flow of vehicles and equipment into the port and notifies the theater movement element when there is a backlog. The ESC/TSC establishes and operates the POE staging area and assists with opening the staging area at the POE.

Movements into the POE staging area must be carefully managed to preclude congestion and to avoid exceeding the capacity of the facility. Early planning in the assembly area ensures that units arrive at the POE on time and fill scheduled modes of transportation. Instructions directing movement to the port will come in the form of a call forward message from SDDC and is based on the availability of space in the port and the TPFDD timelines.

The theater personnel manager is responsible for establishing processing centers. The centers are responsible for strength reporting and providing personnel services and casualty operations support for all personnel departing the theater. Moreover, they verify unit manifests, coordinate manifest changes with the Air Force, and transmit final flight manifests to the appropriate major commands, personnel agencies, and destination installation commanders.

III. Mvmt to Home or Demobilization Station

Ref: FMI 3-35 Army Deployment and Redeployment (June '07), pp. 5-2 to 5-4.

Movement from Point of Debarkation (POD)

The combination of strategic airlift and sealift provides the capability to redeploy forces, albeit in different timeframes and along separate routes. Personnel are transported by strategic airlift to the destination APOD and bussed to the destination installation. Vehicles, unit equipment, and containers are moved by strategic sealift to the designated SPOD, unloaded, and transported by commercial truck or rail to the destination installation.

Movement to Home or Demobilization Station

Once cargo arrives at the SPOD, the destination installation has the primary role of coordinating with SDDC for reception and onward movement of the cargo. The supporting installation has the following responsibilities at the POD:

- Provide download teams and drivers to support POD operations
- Stage equipment for movement to the final destination. The equipment may be configured into unit sets, organized by type of equipment, or configured for movement by a certain type of transport.
- Provide rail loading teams
- Coordinate for customs clearance inspections
- Complete equipment inspections and process movement documentation

The supporting installation is responsible for supporting arriving forces until they reach their destination and developing a reception plan. In preparation for redeployment, the installation coordinates the actions and location of required support for the arrival ports and airfields and establishes en route support sites as required by the redeployment plan. The destination for Active Army units is normally their home station; whereas, U.S. Army Reserves and Army National Guard units return through a demobilization station.

The installation coordinates with SDDC and other affected agencies to provide commercial transportation and MHE as needed and monitors operations, resolves problems, and complete reports as required to higher headquarters and other coordinating organizations. Specific functions of the destination installation include:

- Activate emergency operations center as required
- Notify supporting units and key agencies, including Public Affairs offices and family readiness groups
- Activate Soldier readiness point
- Open billets, dining halls, and morale, welfare, and recreation (MWR) facilities as required
- Conduct reception for returning units
- Process personnel
- Provide maintenance, transportation, and MHE support
- Establish turn-in of weapons and special equipment

The unit performs the following tasks upon arrival at the destination:

- Conduct personnel processing (including legal, financial, and medical processing), review personnel records, and conduct personal affairs briefings
- Download and receive unit equipment
- Develop and implement a maintenance plan to return equipment to predeployment condition, including technical inspections, preventive maintenance, oil analysis, and calibration

Index